mODa 8 - Advances in Model-Oriented Design and Analysis

Contributions to Statistics

V. Fedorov/W.G. Müller/
I.N. Vuchkov (Eds.)
Model-Oriented Data Analysis,
XII/248 pages, 1992

J. Antoch (Ed.)
Computational Aspects of Model Choice,
VII/285 pages, 1993

W.G. Müller/H.P. Wynn/
A.A. Zhigljavsky (Eds.)
Model-Oriented Data Analysis,
XIII/287 pages, 1993

P. Mandl/M. Hušková (Eds.)
Asymptotic Statistics,
X/474 pages, 1994

P. Dirschedl/R. Ostermann (Eds.)
Computational Statistics,
VII/553 pages, 1994

C.P. Kitsos/W.G. Müller (Eds.)
MODA 4 - Advances in Model-Oriented
Data Analysis,
XIV/297 pages, 1995

H. Schmidli
Reduced Rank Regression,
X/179 pages, 1995

W. Härdle/M.G. Schimek (Eds.)
Statistical Theory and Computational
Aspects of Smoothing,
VIII/265 pages, 1996

S. Klinke
Data Structures for Computational Statistics, VIII/284 pages, 1997

C.P. Kitsos/L. Edler (Eds.)
Industrial Statistics,
XVIII/302 pages, 1997

A. C. Atkinson/L. Pronzato/
H. P. Wynn (Eds.)
MODA 5 - Advances in Model-Oriented
Data Analysis and Experimental Design,
XIV/300 pages, 1998

M. Moryson
Testing for Random Walk Coefficients in
Regression and State Space Models,
XV/317 pages, 1998

S. Biffignandi (Ed.)
Micro- and Macrodata of Firms,
XII/776 pages, 1999

W. Härdle/Hua Liang/J. Gao
Partially Linear Models,
X/203 pages, 2000

A.C. Atkinson/P. Hackl/W. Müller (Eds.)
MODA 6 - Advances in Model-Oriented
Design and Analysis,
XVI/283 pages, 2001

W.G. Müller
Collecting Spatial Data, 2nd edition,
XII/195 pages, 2001

C. Lauro/J. Antoch/V. Esposito Vinzi/
G. Saporta (Eds.)
Multivariate Total Quality Control,
XIII/236 pages, 2002

P.-A. Monney
A Mathematical Theory of Arguments
for Statistical Evidence
XIII/154 pages, 2003

Y. Haitoysky/H.R. Lerche/Y. Ritov (Eds.)
Foundations of Statistical Inference,
XI/230 pages, 2003

A. Di Bucchiancio/H. Läuter/H.P. Wynn
(Eds.)
MODA 7 - Advances in Model-Oriented
Design and Analysis,
XIII/240 pages, 2004

S. Sperlich/W. Härdle/G. Aydınlı (Eds.)
The Art of Semiparametrics,
VIII/178 pages, 2006

Jesús López-Fidalgo
Juan Manuel Rodríguez-Díaz
Ben Torsney

(Editors)

mODa 8 - Advances in Model-Oriented Design and Analysis

Proceedings of the 8th International Workshop in Model-Oriented Design and Analysis held in Almagro, Spain, June 4-8, 2007

With 27 Figures and 37 Tables

Physica-Verlag
A Springer Company

Series Editors

Werner A. Müller
Martina Bihn

Editors

Professor Jesús López-Fidalgo
E.T.S.I. Industriales, Universidad de Castilla-La Mancha
Departamento de Matemáticas
Avda. Camilo José Cela, 3
13071 Ciudad Real
Spain
jesus.lopezfidalgo@uclm.es

Dr. Juan Manuel Rodríguez-Díaz
Universidad de Salamanca
Departamento de Estadística, Facultad de Ciencias
Plaza de los Caídos s/n
37008 Salamanca
Spain
juanmrod@usal.es

Dr. Ben Torsney
University of Glasgow
Department of Statistics
15 University Gardens
Glasgow G12 8QW
United Kingdom
b.torsney@stats.gla.ac.uk

Library of Congress Control Number: 2007925054

ISSN 1431-1968

ISBN 978-3-7908-1951-9 Physica-Verlag Heidelberg New York

Production: LE-TEX Jelonek, Schmidt & Vöckler GbR, Leipzig
Cover-design: WMX Design GmbH, Heidelberg

SPIN 12030019   154/3100YL - 5 4 3 2 1 0   Printed on acid-free paper

The volume is dedicated to Anthony Atkinson
on the occasion of his 70th birthday

Preface

This volume contains a substantial number of the papers presented at the mODa 8 conference in Almagro, Castilla–La Mancha, Spain in June 2007. mODa stands for Model Oriented Data Analysis. Previous conferences have been held in Wartburg, then in the German Democratic Republic (1987), St Kirik Monastery, Bulgaria (1990), Petrodvorets, St Petersburg, Russia (1992), the Island of Spetses, Greece (1995), the Centre International de Rencontres Mathematiques, Marseille, France (1998), Puchberg/Schneeberg, Austria (2001) and Heeze, Netherlands (2004).

The purpose of these workshops has traditionally been to bring together two pairs of groups: firstly scientists from the East and West with an interest in optimal design of experiments and related topics; and secondly younger and senior researchers in the field. These traditions remain vital to the health of the series. In recent years Europe has seen increasing unity. Indeed since the last mODa, and even this year, the EU has expanded to admit several of the countries which have long been strong participants in mODa, with one, Bulgaria, hosting the second conference. One might argue that this expansion has seen the fruition on a wider scale of our mODa ideals, and maybe has been fostered, in a small way, by them.

One might also argue that this expansion has seen mODa come of age. Several recent conferences were financially supported by the EU. Now competition is so wide that such sources were not available. In some sense we have had to become 'self sufficient' and look elsewhere, with the exception that GlaxoSmithKline has again very generously continued its support. All of the various new sources have been Spanish based. These are mainly the University of Castilla-La Mancha, the Spanish Ministerio de Educación y Ciencia (contract MTM2006-27463-E), Caja Castilla–La Mancha, Junta de Comunidades de Castilla–La Mancha, Diputación provincial de Ciudad Real, the Department of Mathematics and the Institute of Mathematics Applied to Science and Engineering of the University of Castilla-La Mancha.

We are very grateful for these substantial contributions.

The major optimal design topics featuring in these proceedings include models with covariance structures, generalised linear models, applications in clinical trials, and designs for discrimination; also new models and criteria appear, and classical design topics feature too. A breakdown is as follows:

1. The most common theme is that of covariance structures with the papers by Fathy and Müller; by Fedorov and Müller; by Fedorov and Wu; by Pepelyshev; by Rodrigues–Pinto and Ponce de Leon; and by Stehlík focusing or touching on this.
2. Generalised Linear Models feature in the contributions of Dorta-Guerra et al; of Fedorov and Wu; of Haines et al; and of Nguyen and Torsney.
3. The topic of clinical trials arises either in the form of adaptive patient allocation rules as in Biswas and Mandal and in Moler and Flournoy; or in the timing of observation taking as in each of the two contributions co-authored by Anisimov and Fedorov plus a differing third author: Downing in one, Leonov in the other.
4. New models appear in Laycock and López–Fidalgo, in Melas and in Ortiz et al; while new design concepts or criteria are seen in Pázman and Pronzato; and in Patan and Bogacka.
5. Bayesian approaches feature in two of the papers so far cited: in Patan and Bogacka; and in Fathy and Müller.
6. The topic of designing for non-parametric testing continues from the last conference in the papers of Basso et al; of Arborett–Giancristofaro et al; and of Rasch and Simeckova.
7. The classical design topics of random or mixed effects features in Schmelter et al; and in Patan and Bogacka; while that of factorial designs is seen in Dorta–Guerra et al; in Stanzel and Hilgers; and in Ye et al.
8. Topics covered by one paper and the authors thereof are: design for discrimination between models by Tommasi; micro-array experiments by Stanzel and Hilgers; model selection in mixture experiments by Maruri–Aguilar and Riccomagno; paired comparisons experiments by Grossman et al; and a unique contribution on optimal design of Bell (or quantum nonlocality) experiments by Gill and Pluch, a longstanding topic in the theory of quantum mechanics.
9. Finally one possibly emerging theme is an emphasis on distance between design points in a one dimensional context, a natural consideration, when at most one observation can be taken at any design point, as happens under some covariance structures. The paper by Stehlík is a case in point. Interestingly there is a similar feature in the paper by Nguyen and Torsney on cut-point determination.

Jesús López–Fidalgo, Ciudad Real
Juan Manuel Rodríguez–Díaz, Salamanca
Ben Torsney, Glasgow

January 2007

Contents

Efficient Sampling Windows for Parameter Estimation in Mixed Effects Models

Quantile and Probability-level Criteria for Nonlinear Experimental Design

Optimal Designs for the Exponential Model with Correlated Observations

Determining the Size of Experiments for the One-way ANOVA Model I for Ordered Categorical Data

Bayesian D_s-Optimal Designs for Generalized Linear Models with Varying Dispersion Parameter

Recruitment in Multicentre Trials: Prediction and Adjustment

Vladimir V. Anisimov[1], Darryl Downing[2], and Valerii V. Fedorov[2]

[1] GlaxoSmithKline, New Frontiers Science Park (South),Third Avenue, Harlow, Essex, CM19 5AW, U.K.   Vladimir.V.Anisimov@gsk.com
[2] GlaxoSmithKline, 1250 So Collegeville Rd, PO Box 5089, Collegeville, PA 19426-0989, U.S.A.
darryl.j.downing@gsk.com   Valeri.V.Fedorov@gsk.com

Summary. There are a few sources of uncertainty/variability associated with patient recruitment in multicentre clinical trials: uncertainties in prior information, stochasticity in patient arrival and centre initiation processes. Methods of statistical modeling, prediction and adaptive adjustment of recruitment are proposed to address these issues. The procedures for constructing an optimal recruitment design accounting for time and cost constraints are briefly discussed.

Key words: patient recruitment, optimal design, multicentre trial, adaptive adjustment

1 Introduction

The recruitment time (time required to complete patient recruitment) is one of the key decision variables at the design stage of clinical trials. Existing techniques of recruitment planning are mainly deterministic and do not account for various uncertainties in input information and stochastic fluctuations of the recruitment process.

We consider multicentre trials and propose a recruitment model, where the patients enter different centres according to Poisson processes with time-constant rates. This assumption seems to be well accepted; cf. Senn (1997, 1998); Anisimov et al (2003). We suggest viewing these rates as a sample from a gamma distributed population. Other mixing distributions can be used in a similar setting, but a gamma distribution is conjugate to a Poisson distribution and is a natural candidate for describing prior uncertainties when the Bayesian approach is used. The analysis of a number of completed trials has shown Anisimov and Fedorov (2005) that a Poisson-gamma model is in good agreement with existing data.

The model allows the development of methods for predicting the number of recruited patients and the recruitment time together with confidence/credibilty intervals, accounting for the randomness of the recruitment process, the variability in recruitment rates, the random delays in centre initiation and uncertainties in prior information.

The paper is organized as follows. Section 2 deals with patient recruitment modeling. Recruitment prediction is considered in Section 3. Methods of adaptive adjustment are introduced in Section 4. Study design is discussed in Section 5.

2 Recruitment modeling

Let n be the total number of patients needed to be recruited by N clinical centres and let $T(n, N)$ be the recruitment time. We assume that patients arrive at centre i according to a Poisson process with rate λ_i and the recruitment rates $\{\lambda_i\}_1^N$ are viewed as a sample of size N drawn from a gamma distributed population. One may also think about the Bayesian setting with a gamma prior for recruitment rates. We consider only a homogeneous case, i.e. $\{\lambda_i\}_i^N$ do not vary in time. We acknowledge that reality could be more complex but generalization to the non-homogeneous case is beyond of the scope of this paper; cf. Thall (1988), where the estimation problem for count data with time-dependent rates and a random-effect parameter described by a gamma distribution is studied.

Note that in our notation $Ga(\alpha, \beta)$ stands for a gamma distributed random variable and its probability density function (p.d.f.) is $p(x, \alpha, \beta) = e^{-\beta x}\beta^\alpha x^{\alpha-1}/\Gamma(\alpha)$ with mean α/β and variance α/β^2, and $\Pi_\lambda(t)$ stands for the Poisson process with rate λ.

Assume first that all centres are initiated at time $t_0 = 0$. Let $n_i(t)$ be the number of patients recruited by centre i up to time t. Then $n_i(t) \sim \Pi_{\lambda_i}(t)$ and the total number of recruited patients $n(t, N) = \sum_{i=1}^N n_i(t)$, conditioned on $\{\lambda_i\}_1^N$, is a Poisson process with rate $\Lambda = \sum_{i=1}^N \lambda_i$. Observing that $\Lambda \sim Ga(\alpha N, \beta)$, one may conclude that the corresponding marginal distribution of $\Pi_\Lambda(t) \bigwedge Ga(\alpha N, \beta)$, (compare with Johnson et al (1993), pp. 204, 308) is:

$$\mathbf{P}(n(t, N) = k) = \frac{1}{k\mathbf{B}(k, \alpha N)} \frac{t^k \beta^{\alpha N}}{(t + \beta)^{k+\alpha N}}, \tag{1}$$

where $\mathbf{B}(a, b)$ is the beta function, and the distribution of $T(n, N)$ is a Pearson type VI distribution with p.d.f.

$$p(t, n, N, \alpha, \beta) = \frac{1}{\mathbf{B}(n, \alpha N)} \frac{t^{n-1}\beta^{\alpha N}}{(t + \beta)^{n+\alpha N}}, \; t \geq 0; \tag{2}$$

see Anisimov and Fedorov (2005, 2006), Johnson et al (1994), ch. 8.2. If the parameters (α, β) were known, then (1) and (2) allow solution of the following problems:

1. Given n and N find the least t^* such that

$$\mathbf{P}(n(t, N) \geq n) \geq p . \tag{3}$$

2. Given T and n find the least N^* such that

$$\mathbf{P}(T(n, N) \leq T) \geq p , \tag{4}$$

where p is some prescribed probability. Solutions can be found numerically. To speed up computing one may use normal approximations. For example, in the later case for sufficiently large N and n the random variable $(T(n, N) - M_1(n, N))/S_1(n, N)$ is approximately normally distributed, where $M_1(n, N) = \mathbf{E}[T(n, N)] = \frac{\beta n}{\alpha N - 1}$, $S_1^2(n, N) = \mathbf{Var}[T(n, N)] = \frac{\beta^2 n(n + \alpha N - 1)}{(\alpha N - 1)^2(\alpha N - 2)}$, $\alpha N > 2$. Consequently (4) can be replaced by the (computationally) much simpler equation

$$M_1(n, N) + z_p S_1(n, N) = T, \tag{5}$$

where z_p is the p-quantile of the standard normal distribution. As the functions $M_1(n, N)$ and $S_1(n, N)$ with respect to N are monotonically decreasing, a unique solution of (5) exists and can easily be found numerically.

If centres are initiated at different times $\{u_i\}_1^N$, then the process $n(t, N)$ is a non-homogeneous Poisson process with cumulative rate on the interval $[0, t]$

$$\Sigma(t) = \sum_{i=1}^{N} \lambda_i \cdot [t - u_i]_+, \tag{6}$$

where $[t - u_i]_+ = t - u_i$, if $t > u_i$ and $= 0$ otherwise. Using the properties of a Poisson process and a gamma distribution, we can establish the relations:

$$\mathbf{P}(T(n, N) \leq T) = \mathbf{P}(\Pi_{\Sigma(T)} \geq n) = \mathbf{P}(Ga(n, 1) \leq \Sigma(T)). \tag{7}$$

Computationally (7) is not much more difficult than (4). Note that u_i can also be viewed as a sample from some random population (i.e. we have two mixing levels in each centre, randomness in rate and in centre initiation date).

3 Prediction of recruitment

There are two basic stages of recruitment prediction: before the study is started (the initial stage) and during the study, when some interim information becomes available (the intermediate stage).

3.1 Initial stage

At this stage the recruitment rates for centres that potentially may be involved in a trial are solicited from site managers. The rates can be also evaluated using historical data. This information is used to estimate the parameters (the mean rate $m = \alpha/\beta$ and the variance $s^2 = \alpha/\beta^2$) of a Poisson-gamma model. The estimation can be done separately for different clusters of centres; for example, for different regions. To make the model more flexible we admit a possibility that centres may be initiated at different times. Let $\{u_i\}_1^N$ be the centre initiation times where u_i is generated in a specified time interval $[a_i, b_i]$ using a uniform distribution. In practice intervals $[a_i, b_i]$ may be the same for different centres. Thus, patients enter centre i according to a delayed Poisson process $n_i(t) = \Pi_{\lambda_i}(t - u_i)$, where the rates λ_i are sampled from a gamma population with the mean m and the variance s^2.

Assume for simplicity that $\max b_i < T$, i.e. all centres are initiated before deadline T. Denote $M(t, N) = \mathbf{E}[\Sigma(t)]$, $S^2(t, N) = \mathbf{Var}[\Sigma(t)]$. Then

$$M(t, N) = \sum_{i=1}^{N} M(t, a_i, b_i, m), \ \ S^2(t, N) = \sum_{i=1}^{N} S^2(t, a_i, b_i, m, s^2), \tag{8}$$

where $\Sigma(t)$ is defined in (6) and

$$M(t, a, b, m) = mt - m(a + b)/2, \tag{9}$$
$$S^2(t, a, b, m, s^2) = (m^2 + s^2)(b - a)^2/12 + s^2(t - (a + b)/2)^2,$$

for $t > b$. Similar formulae are derived for $t < b$.

As $n(t, N)$ is a doubly stochastic Poisson process with cumulative rate $\Sigma(t)$, then $\mathbf{E}[n(t, N)] = M(t, N)$, $\mathbf{Var}[n(t, N)] = M(t, N) + S^2(t, N)$. Thus, formulae (8),(9) provide us with the mean and the variance of the predicted $n(t, N)$ and at large enough N one can also calculate the approximate prediction confidence boundaries over time using the normal approximation.

To find the approximation for t^* in Problem 1 (i.e. the solution of (3) where n and N are given) one can build approximate quantiles for $n(t, N)$ or solve the equation

$$\frac{M(t, N) - n}{\sqrt{M(t, N) + S^2(t, N)}} = z_p. \tag{10}$$

To solve Problem 2 we can use the fact that $\Sigma(T)$ is a sum of independent random variables and $\Sigma(T) \approx M(T, N) + S(T, N)\mathcal{N}(0, 1)$ for sufficiently large N. Observing that $Ga(n, 1) \approx n + \sqrt{n}\mathcal{N}(0, 1)$ for large n and applying (7) one may approximate N^* by solving the following equation

$$\frac{M(T, N) - n}{\sqrt{n + S^2(T, N)}} = z_p \tag{11}$$

with respect to N. Solutions of (10) and (11) can easily be found numerically.

3.2 Updating prediction at interim looks

Let us assume that at some interim time point t_1 the information about ongoing recruitment in N centres is available; i.e. we know how many patients k_i were recruited and initiation times, u_i, in the i-th center. Let τ_i be the actual duration of recruitment in centre i, $(\tau_i = t_1 - u_i)$, $K_1 = \sum_{i=1}^{N} k_i$ be the total number of patients recruited up to time t_1 and $K_2 = n - K_1$. Using interim data we want to update the distribution of $T(K_2, N)$ - the remaining time until the recruitment target n will be met.

Assume all centres are of the same type or belong to the same group (country, region, healthcare system, etc.). This means that the parameters (α, β) of a Poisson-gamma model are the same for all of them. In this case k_i, as a doubly stochastic Poisson variable with gamma distributed rate, has a negative binomial distribution with parameters $(\alpha, \tau_i/\beta)$ (Johnson et al, 1993, p. 199). Thus, given data $\{k_i, \tau_i\}_1^N$, the log-likelihood function up to a constant has the form

$$\mathcal{L}(\alpha, \beta) = \sum_{i=1}^{N} \ln \Gamma(k_i + \alpha) - N \ln \Gamma(\alpha) - K_1 \ln \beta - \sum_{i=1}^{N} (k_i + \alpha) \ln(1 + \tau_i/\beta),$$

and the ML estimator $(\widehat{\alpha}, \widehat{\beta})$ is the point maximising $\mathcal{L}(\alpha, \beta)$.

Suppose now that (α, β) are known. As λ_i has a prior gamma distribution with parameters (α, β), then the posterior estimator of λ_i is $\widehat{\lambda}_i \sim Ga(\alpha + k_i, \beta + \tau_i)$ with mean $m_i = (\alpha + k_i)/(\beta + \tau_i)$ and variance $s_i^2 = (\alpha + k_i)/(\beta + \tau_i)^2$. Therefore, confidence boundaries of the updated prediction of the number of recruited patients can be built using the re-estimated m_i and s_i^2 in each centre and formulae similar to (8) and (9).

Let us now consider an updated distribution of the predicted remaining recruitment time. Assume for simplicity that all N centres after time t_1 continue to recruit without interruption. In this case the re-estimated overall recruitment rate is $\widehat{\Lambda} = \sum_{i=1}^{N} \widehat{\lambda}_i$, and as $T(K_2, N)$ is the time of the K_2-th event of a Poisson process with rate $\widehat{\Lambda}$, its updated distribution coincides with the distribution of $Ga(K_2, 1)/\widehat{\Lambda}$.

If $\tau_i = \tau$ (the same duration of recruitment in all centres) then $\widehat{\Lambda} \sim Ga(\alpha N + K_1, \beta + \tau)$ and $T(K_2, N)$ is distributed as

$$\frac{Ga(K_2, 1)}{Ga(\alpha N + K_1, 1)}(\beta + \tau).$$

Thus, $T(K_2, N)$ has a Pearson type VI distribution. If the τ_i are different, then for large N, $\widehat{\Lambda}$ can be well approximated by a gamma distributed random variable with the same mean and variance as $\widetilde{\Lambda}$. Thus, the distribution of $T(K_2, N)$ can be approximated by a Pearson type VI distribution with corresponding parameters and the inequality:

$$\mathbf{P}(T(K_2, N) \leq T - t_1) \geq p \tag{12}$$

corresponding to Problem 2 (recruitment will be completed before deadline with probability p) can easily be verified.

The impact of additional errors in estimating the parameters (α, β) on the precision of predictions cannot be precisely evaluated as the estimators cannot be written in closed form. However, simulation results show that if $t_1/T > 2/3$, the difference in predicting using either known or estimated parameters is practically negligible Anisimov and Fedorov (2007).

Example. Consider a real multicentre trial. Initially it was planned to recruit 366 patients by 75 centres. In the event 372 patients were recruited by 107 centres. To provide a retrospective analysis, the total duration of recruitment was divided into 4 equal periods (64 days). For each period the following data were given: the date of initiation of each centre and the number of patients recruited in the period by this centre. During the 1st period 15 patients were recruited by 60 centres; so the estimates are: $m = 0.02$, $\alpha = 0.5$. After the 2nd period 118 patients were recruited by 84 centres; so $m = 0.021$, $\alpha = 0.73$. After the 3rd period 254 patients were recruited by 109 centres; so $m = 0.026$, $\alpha = 0.9$.

Figures below show the initial 95% prediction area (upper band) and 95% confidence boundaries for the predicted number of patients after each of the first three periods, constructed using real data and shown by the dotted line. One can see that predictions become narrower as more data are available. As more additional centres are added over time recruitment is accelerated.

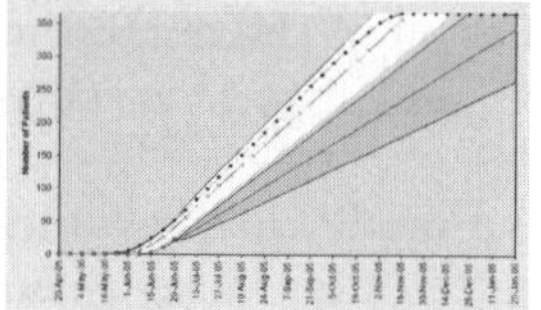 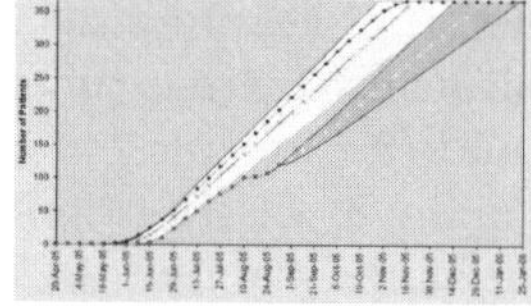 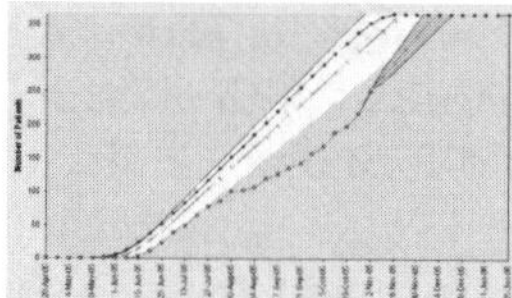

4 Adaptive adjustment

If (12) is not satisfied, then we need to solve the updated Problem 2 to find how many centres should be added. To attain a closed-form (but approximate) solution we assume that $\tau_i = \tau$ and that all new centres will be initiated with the same delay $d < T - t_1$. Assume also that the parameters (α, β) are known.

Then to complete recruitment before time T with probability p we need to add M new centres, where

$$M \geq \frac{A + Bz_p^2/2 + z_p\sqrt{AB + Q + B^2z_p^2/4}}{\alpha B}, \tag{13}$$

$A = K_2 - (\alpha N + K_1)(T - t_1)/(\beta + \tau)$, $B = (T - t_1 - d)/\beta$,
$Q = K_2 + (\alpha N + K_1)(T - t_1)^2/(\beta + \tau)^2$.

Indeed, at $t \geq t_1$ the existing N centres continue to recruit with the overall rate Λ. If we add M new centres with the same delay d, then after time $t_1 + d$ these centres add the total rate $\Lambda_+ \sim Ga(\alpha M, \beta)$ and the overall rate becomes $\widehat{\Lambda} + \Lambda_+$, where we use the estimated rate $\widehat{\Lambda} \sim Ga(\alpha N + K_1, \beta + \tau)$ instead of Λ. Thus, the number of patients recruited in the interval $[t_1, T]$ can be represented as $X \sim \Pi_{\widehat{\Lambda}}(T - t_1) + \Pi_{\Lambda_+}(T - t_1 - d)$. The recruitment will be completed before the deadline if $X \geq K_2$. Hence we need to find M such that $\mathbf{P}(X \geq K_2) \geq p$. Assuming that the values N and K_2 are large enough and using the normal approximations of the Poisson and gamma distributions and relation (7) we can prove (13).

If the parameters (α, β) are not known, we can replace them by their estimators and formula (13) provides the first approximation. Simulations show that at large enough N ($N \geq 20$) and $K_1 \geq 50$, this value is very close to the actual number of centres required.

If M new centres are added with delay d, then the remaining time $\widehat{T}(M)$ can be represented as

$$\widehat{T}(M) = d + \frac{Ga(K_3, 1)}{\widehat{\Lambda} + Ga(\alpha M, \beta)}, \tag{14}$$

where $K_3 = K_2 - \Pi_{\widehat{\Lambda}}(d)$ is the number of patients left to recruit after time $t_1 + d$. As $d < T - t_1$, then at large N, $K_3 > 0$ with probability close to one.

Different characteristics of $\widehat{T}(M)$ can be calculated very quickly numerically using simulation of the gamma random variables involved in (14). In this case for a large number of simulation runs we recommend using the expression $K_3 = [K_2 - \Pi_{\widehat{\Lambda}}(d)]_+$ to avoid possible negative values.

5 Study design

Optimal patient recruitment design should account for such major factors as the total recruitment time, the total number of centres and costs per centre and per study delay. It is clear that adding more centres will decrease the recruitment time, however, it will increase study costs.

Consider the class of admissible designs satisfying (4). As the left-hand side in (4) is monotonically increasing in N, then for any number of patients n, there exists a minimal number of centres N_n^* satisfying (4) which can be found using a quick numerical procedure.

Suppose now that the purpose is to find an optimal design minimizing some risk function $R(n, N)$ given restriction (4). Then a solution of this problem can be found numerically using a two-step optimization procedure:
1) for any n find the optimal N: $N_{opt}(n) = \arg\min_{N \geq N_n^*} R(n, N)$;
2) find the optimal n: $n_{opt} = \arg\min_n R(n, N_{opt}(n))$.

Finally $N_{opt} = N_{opt}(n_{opt})$ and the pair (n_{opt}, N_{opt}) provide the optimal solution for this study design problem given restriction (4).

Consider, as an example, the risk function $R(n, N) = C_1 n + C_2 N + C_3 \mathbf{E}[(T(n, N) - T)]_+$, where C_1 and C_2 are patient and centre costs, respectively, and C_3 is the cost per unit excess of the recruitment deadline. $R(n, N)$ cannot be evaluated analytically. However, as the predicted $T(n, N)$ has either a Pearson type VI distribution or can be represented in the form (7) or (14), then for any values n and N, $R(n, N)$ can be calculated very quickly either numerically or using Monte Carlo simulation.

Additional realistic constraints might be $N \leq N_{Max}$, or $C_1 n + C_2 N \leq C_{Max}$. In this case, a search at step 1) above should be performed in the admissible region $\{N \geq N_n^*, \ N \leq N_{Max}\}$, or in the region $\{N \geq N_n^*, \ C_1 n + C_2 N \leq C_{Max}\}$, and empty regions should be excluded from consideration.

Note that in real trials the number of patients n cannot exceed several hundred (or possibly a couple of thousands for very large trials). Thus, each step can be performed in a very short time.

References

Anisimov V, Fedorov V (2005) Modeling of enrolment and estimation of parameters in multicentre trials. GSK BDS Technical Report 2005-01

Anisimov V, Fedorov V (2006) Design of multicentre clinical trials with random enrolment, Advances in Statistical Methods for the Health Sciences, Birkhäuser, Boston, chap 25

Anisimov V, Fedorov V (2007) Modeling, prediction and adaptive adjustment of recruitment in multicentre trials. Statistics in Medicine (to appear)

Anisimov V, Fedorov V, Jones B (2003) Modeling, prediction and adaptive adjustment of recruitment in multicentre trials. GSK BDS Technical Report 2003-03

Johnson N, Kotz S, Adrienne W (1993) Univariate Discrete Distributions (2nd edn). Wiley, New York

Johnson N, Kotz S, Balakrishnan N (1994) Continuous Univariate Distributions (2nd edn), v. 1. Wiley, New York

Senn S (1997) Statistical Issues in Drug Development. Wiley, Chichester

Senn S (1998) Some controversies in planning and analysis multi-centre trials. Statistics in Medicine 17:1753–1756

Thall P (1988) Mixed poisson likelihood regression models for longitudinal interval count data. Biometrics 44:197–209

Optimal Design of Pharmacokinetic Studies Described by Stochastic Differential Equations

Vladimir V. Anisimov[1], Valerii V. Fedorov[2], and Sergei L. Leonov[2]

[1] GlaxoSmithKline, New Frontiers Science Park (South),Third Avenue, Harlow, Essex, CM19 5AW, United Kingdom. `Vladimir.V.Anisimov@gsk.com`
[2] GlaxoSmithKline, 1250 So Collegeville Rd, PO Box 5089, Collegeville, PA 19426-0989, U.S.A.
`Valeri.V.Fedorov@gsk.com`   `Sergei.2.Leonov@gsk.com`

Summary. Pharmacokinetic (PK) studies with serial sampling which are described by compartmental models are discussed. We focus on intrinsic variability induced by the noise terms in stochastic differential equations (SDE). For several models of intrinsic randomness, we find explicit expressions for mean and covariance functions of the solution of the system of SDE. This, in turn, allows us to construct optimal designs, i.e. find sequences of sampling times that guarantee the most precise estimation of unknown model parameters. The performance of optimal designs is illustrated with several examples, including cost-based designs.

Key words: pharmacokinetic models, stochastic differential equations, intrinsic randomness, optimal sampling times, cost-based designs

1 Introduction

Repeated measures models are often encountered in biopharmaceutical applications. Examples include pharmacokinetic (PK) studies with serial blood sampling which are traditionally described by ordinary differential equations (ODE). Deterministic PK models include two sources of variability: (1) measurement, or observational error, and (2) population, or between-patient variability. The focus of this paper is the third source, namely within-patient variability which comes into play from considering stochastic differential equations (SDE) instead of ODE. We address the optimal design problem, i.e. selection of sequences of sampling times that maximize the information in an experiment, and show that when costs are taken into account, designs with smaller numbers of samples may become better than more "dense" sampling schemes, with respect to selected optimality criterion. In this paper we focus on the design problem; for a discussion on parameter estimation in stochastic PK models, see Overgaard et al (2005).

The paper is organized as follows. In Section 2 we introduce models described by ODE (deterministic) and SDE (stochastic) and derive mean and covariance functions for the stochastic systems. In Section 3 basic concepts of optimal design theory are mentioned and examples of optimal sampling schemes are presented. Section 4 outlines potential extensions and future work.

2 Models

2.1 Deterministic open one-compartment model

For illustration purposes, we select an open one-compartment model:

$$d\eta_1(t) = -\theta_1\eta_1(t)dt, \quad d\eta_2(t) = \theta_1\eta_1(t)dt - \theta_2\eta_2(t)dt, \tag{1}$$

$\eta_1(0) = D$, $\eta_2(0) = 0$, where D is the dose of the drug, $\eta_1(t)$ is the amount of drug at the site of administration, $\eta_2(t)$ is the amount of drug in the central compartment; and θ_1 and θ_2 are absorption and elimination rate constants, respectively. Functions η_1 and η_2 depend on parameters θ_1 and θ_2, but we drop these arguments to simplify notation. The solution of system (1) is given by

$$\eta_1(t) = De^{-\theta_1 t}, \quad \eta_2(t) = \frac{\theta_1 D}{\theta_1 - \theta_2}\left(e^{-\theta_2 t} - e^{-\theta_1 t}\right), \tag{2}$$

see Gibaldi and Perrier (1982). Measurements $\{z_{ij}\}$ of drug concentration are taken from the central compartment at times t_{ij},

$$z_{ij} = \eta_2(t_{ij})/V_i + \varepsilon_{ij}, \quad j = 1,\ldots,k_i, \tag{3}$$

where V_i is the volume of distribution for patient i, k_i is the number of measurements for patient i, and ε_{ij} are measurements errors, the variance of which may depend on the mean response, as in power models. We use the notation $\gamma = (\theta_1, \theta_2, V)^T$ for the combined vector of PK parameters. In population modeling, it is assumed that the parameters γ_i are independently sampled from a given distribution.

2.2 Stochastic open one-compartment model

To account for intrinsic within-patient variability, we introduce a system of SDE with additional noise terms described by Wiener processes:

$$dy_1(t) = -\theta_1 y_1(t)dt + \sigma_1(t)dw_1(t), \; y_1(0) = D, \tag{4}$$

$$dy_2(t) = \theta_1 y_1(t)dt - \theta_2 y_2(t)dt + \sigma_2(t)dw_2(t), \; y_2(0) = 0,$$

where $\sigma_i(t) \geq 0, i = 1, 2$, are deterministic functions, and $w_i(t)$ are independent Wiener processes, for which $\mathbf{E}[w_i(t)] = 0$, $\mathrm{Cov}[w_i(t)w_i(s)] = \min(t, s)$.

We use Ito's concept, so that for any given deterministic function f, the process $\xi(t) = \int_0^t f(u)dw(u)$ is a Gaussian process satisfying conditions

$$\mathbf{E}[\xi(t)] = 0, \ \operatorname{Cov}[\xi(t), \xi(t+s)] = \int_0^t f^2(u)du, \ s > 0,$$

see, for example, Gardiner (1997). The solution of system (4) is given by

$$y_1(t) \ = \ \eta_1(t) \ + \ \frac{1}{D} \int_0^t \eta_1(t-u) \, \sigma_1(u)dw_1(u), \tag{5}$$

$$y_2(t) \ = \ \eta_2(t) \ + \ \frac{1}{D} \int_0^t \eta_2(t-u)\sigma_1(u)dw_1(u) \ + \ \int_0^t e^{-\theta_2(t-u)} \, \sigma_2(u)dw_2(u),$$

with the functions $\eta_i(t)$ defined in (2). Denote $\tilde{S}(t, t+s) = \mathbf{Cov}[y_2(t), y_2(t+s)]$. Using properties of Ito's integral and independence of $w_1(t)$ and $w_2(t)$, it is straightforward to show that $\mathbf{E}[y_1(t)] = \eta_1(t)$, $\mathbf{E}[y_2(t)] = \eta_2(t)$, and

$$\tilde{S}(t, t+s) = \frac{\theta_1^2}{(\theta_2 - \theta_1)^2} \int_0^t \sigma_1^2(u)du \left[e^{-\theta_1 s}e^{2\theta_1(u-t)} + e^{-\theta_2 s}e^{2\theta_2(u-t)} - \right.$$

$$\left. - \left(e^{-\theta_1 s} + e^{-\theta_2 s}\right) e^{(\theta_1+\theta_2)(u-t)} \right] + e^{-\theta_2(2t+s)} \int_0^t \sigma_2^2(s)e^{2\theta_2 u} \, du, \ s > 0. \tag{6}$$

If $\sigma_i(t) \ = \ \sigma_i e^{-v_i t}$, then the integrals in (6) can be evaluated explicitly. For example, if $v_1 \neq \theta_j, j = 1, 2; \ v_2 \neq \theta_2$, and $v_1 \neq (\theta_1 + \theta_2)/2$, then

$$\tilde{S}(t, t+s) \ = \ \frac{\theta_1^2 \sigma_1^2}{(\theta_1 - v_1)^2} \left[e^{-\theta_1 s} \frac{e^{-2v_1 t} - e^{-2\theta_1 t}}{2(\theta_1 - v_1)} + e^{-\theta_2 s} \frac{e^{-2v_1 t} - e^{-2\theta_2 t}}{2(\theta_2 - v_1)} \right.$$

$$\left. - \left(e^{-\theta_1 s} + e^{-\theta_2 s}\right) \frac{e^{-2v_1 t} - e^{-(\theta_1+\theta_2)t}}{\theta_1 + \theta_2 - 2v_1} \right] + \sigma_2^2 \, e^{-\theta_2 s} \frac{e^{-2v_2 t} - e^{-2\theta_2 t}}{2(\theta_2 - v_2)}. \tag{7}$$

When $v_i > 0$, then $\operatorname{Var}[y_2(t)] \to 0$ as $t \to \infty$, and $\tilde{S}(t, t+s) \to 0$ as $s \to \infty$ for any fixed $t > 0$. However, if at least one $v_i = 0$, then it follows from (6) that $\operatorname{Var}[y_2(t)] \to v^* > 0$, even though $\mathbf{E}[y_2(t)] \to 0$ as $t \to \infty$. This seems counterintuitive from physiological considerations, since in this case the trajectories of the process $y_2(t)$ become negative with positive probability.

2.3 Three sources of variability

Let $t_1, \ldots, t_k$ be a sequence of k sampling times and $\tilde{\mathbf{S}} = \{\tilde{S}_{j_1 j_2}\} = \left\{\tilde{S}(t_{j1}, t_{j2}), \ j_1, j_2 = 1, \ldots, k\right\}$. The traditional sources of variability include (a) additive measurement errors at times t_j, e.g. independent $\varepsilon_j \sim N(0, \sigma_{obs}^2)$, so that measurements $Z(t_j)$ satisfy $Z(t_j) = y_2(t_j)/V + \varepsilon_j$, and

12   Vladimir V. Anisimov, Valerii V. Fedorov, and Sergei L. Leonov

(b) population variability, e.g. $\gamma_i \sim N(\gamma^0, \Lambda)$, with γ defined in Section 2.1. If one takes into account these two sources in addition to the intrinsic variability and assumes that the variance $\sigma_i(t)$ is rather small, then the first-order approximation techniques lead to the following formula for the variance-covariance matrix of the process $\{Z(t), t = t_1, \ldots, t_k\}$, cf. Gagnon and Leonov (2005):

$$\mathbf{S} \approx \tilde{\mathbf{S}}/V^2 + \sigma_{obs}^2 \mathbf{I}_k + \mathbf{G}\,\Lambda\,\mathbf{G}^T, \tag{8}$$

where $\mathbf{I}_k$ is a $(k \times k)$ identity matrix; $\mathbf{G}$ is the $(k \times m_\gamma)$ matrix of the partial derivatives of the function $g(t, \gamma) = \eta_2(t)/V$ with respect to the parameters γ_l, where $m_\gamma = dim(\gamma)$ and $\mathbf{G} = \left\{ G_{jl} = \left[\partial g(t_j, \gamma^0)/\partial \gamma_l \right] \right\}$, $j = 1, \ldots, k$; $l = 1, \ldots, m_\gamma$.

3 Optimal designs

Optimal experimental design techniques were exploited in a number of recent publications to select optimal sampling schemes for population compartmental models described by ODE; see Fedorov et al (2002), Gagnon and Leonov (2005), Retout and Mentré (2003). The key for constructing optimal schemes is to derive the information matrix $\boldsymbol{\mu}(\mathbf{x}, \boldsymbol{\vartheta})$ of a single multidimensional "point" $\mathbf{x}$ for the observed $(k \times 1)$-vector of responses $\mathbf{Z}$:

$$\mathbf{E}[\mathbf{Z}|\mathbf{x}] = \mathbf{f}(\mathbf{x}, \boldsymbol{\vartheta}) = [f_1(\mathbf{x}, \boldsymbol{\vartheta}), \ldots, f_k(\mathbf{x}, \boldsymbol{\vartheta})]^T, \quad \text{Var}[\mathbf{Z}|\mathbf{x}] = \mathbf{S}(\mathbf{x}, \boldsymbol{\vartheta}),$$

where $\boldsymbol{\vartheta}$ includes all estimated parameters, $m = \dim(\boldsymbol{\vartheta}) \geq m_\gamma$, and $\mathbf{x} = (t_1, t_2, \ldots, t_k)$ is a $(k \times 1)$ vector of sampling times. In the context of this paper, in general, $\boldsymbol{\vartheta} = (\gamma;\ \sigma_1, \sigma_2, v_1, v_2;\ \sigma_{obs}, \Lambda)^T$.

When the vector $\mathbf{Z}$ is normally distributed, then $\boldsymbol{\mu}(\mathbf{x}, \boldsymbol{\vartheta})$ can be calculated explicitly; see Magnus and Neudecker (1988), Ch. 6, or Fedorov et al (2002):

$$\mu_{\alpha\beta}(\mathbf{x}, \boldsymbol{\vartheta})|_{\alpha,\beta=1}^{m} = \frac{\partial \mathbf{f}}{\partial \vartheta_\alpha}\,\mathbf{S}^{-1}\,\frac{\partial \mathbf{f}}{\partial \vartheta_\beta} + \frac{1}{2}\,\mathrm{tr}\left[\mathbf{S}^{-1}\,\frac{\partial \mathbf{S}}{\partial \vartheta_\alpha}\,\mathbf{S}^{-1}\,\frac{\partial \mathbf{S}}{\partial \vartheta_\beta}\right], \tag{9}$$

where $\mathbf{S} = \mathbf{S}(\mathbf{x}, \boldsymbol{\vartheta})$ is a $(k \times k)$ variance-covariance matrix introduced in (8). Then the construction of optimal designs becomes straightforward once the design region $\mathcal{X}$, or the set of admissible sampling sequences $\mathbf{x}$, is defined.

If n_i patients are assigned to sequence $\mathbf{x}_i$ and $\sum_i n_i = N$, one can define a design $\xi_N = \{(\mathbf{x}_i, n_i),\ \sum_i n_i = N,\ \mathbf{x}_i \in \mathcal{X}\}$ together with the information matrix $\mathbf{M}_N(\boldsymbol{\vartheta}) = \sum_i n_i \boldsymbol{\mu}(\mathbf{x}, \boldsymbol{\vartheta})$, and a normalized design ξ with the normalized information matrix $\mathbf{M}(\xi, \boldsymbol{\vartheta})$ (*information per observation*)

$$\xi = \{(\mathbf{x}_i, w_i),\ w_i = n_i/N,\ \mathbf{x}_i \in \mathcal{X}\}, \quad \mathbf{M}(\xi, \boldsymbol{\vartheta}) = \sum_i w_i \boldsymbol{\mu}(\mathbf{x}_i, \boldsymbol{\vartheta}). \tag{10}$$

In convex design theory weights w_i vary continuously, and various criteria depending on the normalized matrix $\mathbf{M}(\xi, \boldsymbol{\vartheta})$ are optimized; in particular the D-criterion $\Psi = \log|\mathbf{M}^{-1}(\xi, \boldsymbol{\vartheta})|$. For nonlinear models, the information matrix $\boldsymbol{\mu}(\mathbf{x}, \boldsymbol{\vartheta})$ depends on the values of the parameters $\boldsymbol{\vartheta}$, which leads to the concept of locally optimal designs; see Atkinson and Donev (1992), Fedorov and P. (1997).

An alternative normalization of the information matrix $\mathbf{M}_N(\boldsymbol{\vartheta})$ is possible if costs are taken into account. If $c(\mathbf{x}_i)$ is a cost of taking measurements at the sequence $\mathbf{x}_i$ and C is the total cost allowed, thereby imposing the constraint $\sum_i c(\mathbf{x}_i) \leq C$, then the information matrix may be normalized by the total cost (*information per unit cost*),

$$\mathbf{M}_C(\xi, \boldsymbol{\vartheta}) = \mathbf{M}_N(\boldsymbol{\vartheta})/C = \sum_i w_i \tilde{\boldsymbol{\mu}}(\mathbf{x}_i, \boldsymbol{\vartheta}), \tag{11}$$

with $w_i = n_i c(\mathbf{x}_i)/C$, $\tilde{\boldsymbol{\mu}}(\mathbf{x}, \boldsymbol{\vartheta}) = \boldsymbol{\mu}(\mathbf{x}, \boldsymbol{\vartheta})/c(\mathbf{x})$, and optimal cost-based designs may be constructed; see Mentré et al (1997), Fedorov and Leonov (2005).

Sampling times. Traditionally in PK studies more samples are taken immediately after administering a drug and then samples become more sparse after the anticipated maximum of the time-concentration curve. In this paper we consider a grid which possesses this property: take a uniform grid on the vertical axis with respect to values of the response function and project points on the response curve to the X-axis, to obtain sampling times. More precisely, let T be the right end of the time interval. Divide the interval $[0, \max_{t \in [0,T]} \eta_2(t)]$ into p equal parts which generates sampling sequences $\mathbf{x}_{2p}$ with $2p$ sampling times. For the final sampling time, take $t_{2p} = T$; see Fig. 2, top panel. The construction of such a grid requires preliminary estimates of parameters θ_1 and θ_2. However, this is also true when using traditional sampling schemes; for details, see Fedorov and Leonov (2006). For all examples in this section, $\theta_1 = 2$, $\theta_2 = 1$, $V = 1$, $D = 1$, and $v_1 = v_2 = 0.5$ in (7) for all SDE examples.

3.1 Saturation effect for SDE

When σ_{obs}^2 is small enough, then it may be expected that the correlation between measurements $Z(t)$ and $Z(t')$ is substantial for small $(t - t')$. In this subsection we compare information matrices generated by ODE (1) and SDE (4) with $\sigma_1 = \sigma_2 = 0.4$ in (7). For a meaningful comparison, we consider $\sigma_1^2, \sigma_2^2, v_1, v_2, \sigma_{obs}^2$ and Λ as given constants, so that, in calculating the information matrix in (9), derivatives are not taken with respect to those parameters. Thus, here the number of unknown parameters is $m = 3$ and $\boldsymbol{\vartheta} = (\theta_1, \theta_2, V)$.

The following information matrices were calculated for sampling sequences $\mathbf{x}_{2p}$: (1) $\boldsymbol{\mu}_1$ generated by ODE with $\sigma_{obs} = 0.1$ and no population variability, i.e. $\sigma_1 = \sigma_2 = 0$, $\Lambda = \mathbf{0}$; (2) $\boldsymbol{\mu}_2$ generated by SDE with no measurement errors and no population variability, i.e. $\sigma_{obs} = 0$, $\Lambda = \mathbf{0}$; (3) $\boldsymbol{\mu}_3$ generated by SDE with $\sigma_{obs} = 0.1$ and no population variability, $\Lambda = \mathbf{0}$; (4) $\boldsymbol{\mu}_4$ generated by SDE with $\sigma_{obs} = 0.1$ and $\Lambda = \text{diag}(0.25, 0.25, 0.25)$. Fig. 1 presents values

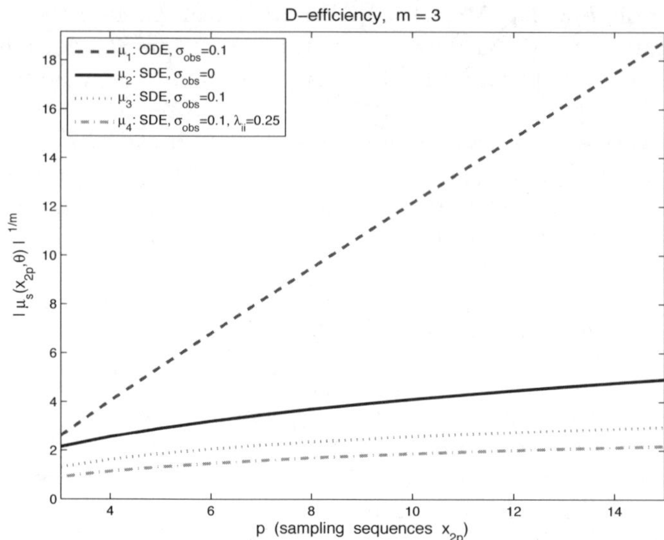

Fig. 1. Function $|\boldsymbol{\mu}_s(\mathbf{x}_{2p}, \boldsymbol{\vartheta})|^{1/m}$; $\sigma_{obs} = 0.1$ for $\boldsymbol{\mu}_{1,3,4}$ and $\sigma_{obs} = 0$ for $\boldsymbol{\mu}_2$

of $|\mu_s(\mathbf{x}_{2p}, \boldsymbol{\vartheta})|^{1/3}$, $s = 1, 2, 3, 4$, $p \in [3, 15]$. While the dashed line corresponding to $\boldsymbol{\mu}_1$ increases as a linear function of p, the solid line corresponding to $\boldsymbol{\mu}_2$ increases with p much more slowly. This effect may be explained by the strong correlation in the stochastic system which reduces the need for taking extra observations. When the observational error and population variability are added to SDE, then the information "amount" goes down; see dotted and dashed-dotted curves corresponding to $\boldsymbol{\mu}_3$ and $\boldsymbol{\mu}_4$, respectively.

3.2 Examples of optimal designs

To illustrate the performance of the first-order optimization algorithm for the standard approach as in (10), we selected $p = 8$, so that $[t_1, \ldots, t_{16}] = [0.033, 0.069, 0.111, 0.158, 0.215, \ldots 3.294, 5]$,; see Fig. 2, top panel, with a log-scale on the X-axis. For the design region, we take all combinations of $r \leq 7$ sampling times from the 16-time sequence $\{t_1, \ldots, t_{16}\}$; see Gagnon and Leonov (2005) for more details on such selections. In this subsection $\boldsymbol{\Lambda} = \text{diag}(0.25, 0.25, 0.25)$ and $v_1 = v_2 = 0.5$ are considered as given constants, so that $m = 6$ and $\boldsymbol{\vartheta} = (\theta_1, \theta_2, V; \sigma_1^2, \sigma_2^2; \sigma_{obs}^2)$.

When $\sigma_1 = \sigma_2 = 0.2$, $\sigma_{obs} = 0.3$, then the locally D-optimal design is

$$\xi_1 = \{\mathbf{x}_{11} = (t_{4-6,9-10,13-14}), w_{11} = 0.74, \mathbf{x}_{12} = (t_{5-6,9-10,13-15}), w_{12} = 0.26\}$$

(see Fig. 2, bottom panel, top design). Note that sampling times are clustered in optimal sequences which may be explained by a relatively small correlation of observations when the ratio σ_{obs}/σ_i is moderate. The three squares on the

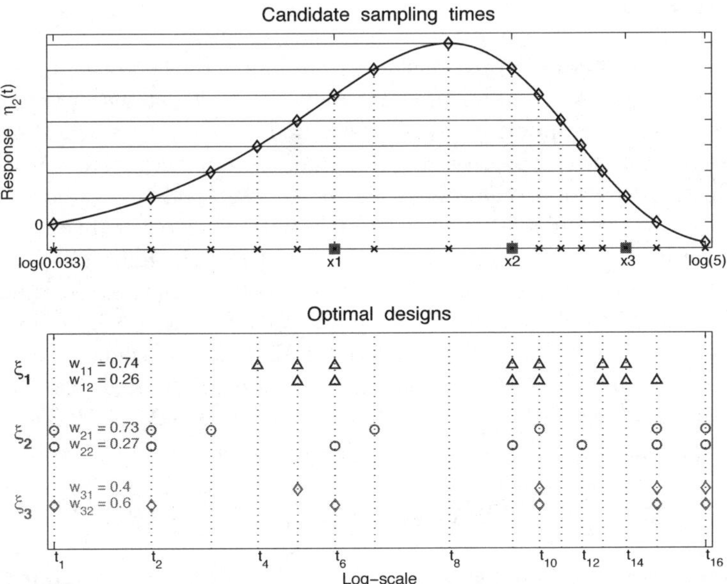

Fig. 2. Top panel: sampling times generated by uniform grid on the Y-axis. Bottom panel: optimal designs, up to 7-sample sequences allowed, $\sigma_1 = \sigma_2 = 0.2$. Design ξ_1: $\sigma_{obs}/\sigma_i = 1.5$. Design ξ_2: $\sigma_{obs}/\sigma_i = 0.05$. Design ξ_3: $\sigma_{obs}/\sigma_i = 0.05$, cost-based

X-axis in the top panel correspond to support points x_i of the D-optimal design for the 3-parameter fixed effects model $g(t,\gamma) = \eta_2(t)/V$, when the design region consists of 16 single points $\{t_1, t_2, \ldots, t_{16}\}$. Note that sampling times from sequences $\mathbf{x}_{11}$ and $\mathbf{x}_{12}$ "surround" points $x_i, i = 1, 2, 3$.

As the ratio σ_{obs}/σ_i decreases, so does the correlation between observations, and clusters of sampling times break down. When $\sigma_{obs} = 0.01$, then

$$\xi_2 = \{\mathbf{x}_{21} = (t_{1-3,7,10,15,16}), w_{21} = 0.73, \mathbf{x}_{22} = (t_{1-2,6,9,12,15,16}), w_{12} = 0.27\};$$

see Fig. 2, bottom panel, middle design.

Cost-driven designs. As demonstrated in our previous work for ODE, once costs are taken into account as in (11), then sampling schemes with smaller numbers of samples may outperform more "dense" schemes; for details, see Gagnon and Leonov (2005), Fedorov and Leonov (2005). In this subsection, the cost function is selected as $c(\mathbf{x}_k) = c_v + c_s k$, where c_v is the cost of a patient's enrolment and c_s is the cost of analyzing a sample. When $\sigma_1 = \sigma_2 = 0.2$, $\sigma_{obs} = 0.01$, $c_v = 1$, $c_s = 2$, the cost-based D-optimal design is supported on two sequences, one with 4 and another with 6 sampling times:

$$\xi_3 = \{\mathbf{x}_{21} = (t_{5,10,15,16}), w_{31} = 0.4, \mathbf{x}_{32} = (t_{1-2,6,10,15,16}), w_{32} = 0.6\};$$

see Fig. 2, bottom panel/design: though up to 7 samples were allowed, costs constraints led to the selection of sequences with smaller numbers of samples.

4 Discussion

The stochastic system (4) provides an example where the covariance function $\tilde{S}$ can be written in a closed form, which allows us to construct optimal designs using the resulting explicit expression for $\tilde{S}$. However, even when $\sigma_i(t) \to 0$ in (4) as $t \to \infty$, the trajectories $y_2(t)$ may become negative with positive probability which cannot happen with drug dose levels. Therefore it would be desirable to consider stochastic models with positive solutions $y_i(t)$, for example replacing the noise terms $\sigma_i dw_i(t)$ on the right-hand side of (4) with $\sigma_i y_i(t) dw_i(t)$, i.e. making them proportional to the signal. Unfortunately for this case we were unable to derive the covariance function $\tilde{S}$ in a closed form. Resorting to numerical methods is a potential solution for such cases.

We remark that for multicompartmental models one may use the matrix representation of vector SDE: $d\mathbf{Y}(t) = \boldsymbol{\Theta}\mathbf{Y}(t)dt + \mathbf{B}d\mathbf{W}(t)$, $\mathbf{Y}(t) = \mathbf{Y}_0$, see Gardiner (1997), for which $\mathbf{E}[\mathbf{Y}(t)] = e^{\boldsymbol{\Theta}t}\mathbf{Y}_0$, and

$$\mathrm{Cov}[\mathbf{Y}(t), \mathbf{Y}(t+s)] = e^{\boldsymbol{\Theta}t}\,\mathbf{Y}_0\mathbf{Y}_0^T\,e^{\boldsymbol{\Theta}t} + \int_0^t e^{\boldsymbol{\Theta}(t-u)}\mathbf{B}\mathbf{B}^T e^{\boldsymbol{\Theta}(t+s-u)}du \ .$$

In conclusion, we emphasize that all three sources of variability should be considered in stochastic PK models: within-patient, between-patient, and observational errors. We recommend cost-driven designs which allow for a meaningful comparison of sampling schemes with unequal numbers of samples.

References

Atkinson A, Donev A (1992) Optimum Experim. Design. Clarendon Press, Oxford

Fedorov V, Hackl P (1997) Model-Oriented Design of Experiments. Springer, N.Y.

Fedorov V, Leonov S (2005) Response driven designs in drug development. In: Wong W, Berger M (eds) Applied Optimal Designs, Wiley, Chichester, pp 103–136

Fedorov V, Leonov S (2006) Estimation of population PK measures: Selection of sampling grids. GlaxoSmithKline GSK BDS TR 2006-02

Fedorov V, Gagnon R, Leonov S (2002) Design of experiments with unknown parameters in variance. Appl Stoch Mod Bus Ind 18:207–218

Gagnon R, Leonov S (2005) Optimal population designs for PK models with serial sampling. J Biopharm Stat 15:143–163

Gardiner C (1997) Handbook of Stochastic Methods. Springer, Berlin

Gibaldi M, Perrier D (1982) Pharmacokinetics, 2nd Ed. Marcel Dekker, New York

Magnus J, Neudecker H (1988) Matrix Differential Calculus with Applications in Statistics and Econometrics. Wiley, New York

Mentré F, Mallet A, Baccar D (1997) Optimal design in random-effects regression models. Biometrika 84:429–442

Overgaard R, Jonsson N, Tornoe C, Madsen H (2005) Non-linear mixed-effects models with stochastic differential equations: implementation of an estimation algorithm. J Pharmacokinet Pharmacodyn 32:85–107

Retout S, Mentré F (2003) Further developments of the Fisher information matrix in nonlinear mixed effects models with evaluation in population pharmacokinetics. J Biopharm Stat 13:209–227

Comparisons of Heterogeneity: a Nonparametric Test for the Multisample Case

Rosa Arboretti Giancristofaro[1], Stefano Bonnini[2], and Fortunato Pesarin[3]

[1] Department of Mathematics, University of Ferrara, Via Macchiavelli, 35, 44100 Ferrara, Italy. `rosa.arboretti@unife.it`
[2] Center for Modelling, Computing and Statistics, University of Ferrara, Via Macchiavelli, 35, 44100 Ferrara, Italy. `bnnsfn@unife.it`
[3] Department of Statistical Sciences, University of Padova, Via C. Battisti, 241-243, 35121 Padova, Italy. `fortunato.pesarin@unipd.it`

Summary. In several scientific disciplines it is often of interest to compare the concentration of the distribution of a categorical variable between two or more populations. The aim is to establish if the heterogeneities of the distributions are equal or not. We propose a nonparametric solution based on a permutation test. The main properties of the test and a Monte Carlo simulation in order to evaluate its behaviour will be discussed.

Key words: permutation tests, heterogeneity, categorical variables

1 Introduction

The concept of heterogeneity is mostly used in descriptive statistics and is closely linked to the concept of homogeneity. Homogeneity normally means the disposition of a statistical variable X to occur in the same category A_i, $1 \leq i \leq k, 1 < k < \infty$. A set of statistical units is therefore homogeneous if all units that make it up are characterized by the same category. If this does not occur, that is if at least two categories in the set of statistical units are found, then we have heterogeneity. Thus heretogeneity is indicated by the absence of homogeneity. In other words heterogeneity is the dispersion of a categorical variable. Therefore the degree of heterogeneity measures the uniformity in distribution of cases among the available categories of a variable, i.e. the opposite tendency to that of concentration of frequencies among categories. The degree of heterogeneity depends on the number of categories observed as well as on their associated frequencies. In fact heterogeneity increases both with the number of categories and with the uniformity in the distribution of cases across categories. In particular heterogeneity is at a minimum if the distribution of the observed variable is degenerate, i.e. it assigns a single category with a relative frequency equal to 1 and all the others with a frequency equal to 0. On the other hand heterogeneity is at a maximum if the variable is equally distributed across all categories. Consequently an index of the degree of heterogeneity of the observed phenomenon must have the following characteristics: 1) To assume a minimum value when the phenomenon under study is manifested with a single category, i.e. in the presence of maximum homogeneity; 2) To assume increasingly greater values the more one approaches equidistribution; 3) To assume a maximum value in the presence of equidistribution. Heterogeneity can be associated not only with the concept of concentration but also with that of diversity, that is the tendency of a qualitative variable to assume different modalities. It is directly associated with the concept of uncertainty and that of information because in the case of minimum heterogeneity the uncertainty of a decision is also at a minimum and the information derivable from a single observation is at a maximum. In the opposite case of maximum heterogeneity one has maximum uncertainty and minimum information derivable from a single unit. Starting from this notion, various indicators were proposed of which only the most commonly used will be mentioned.

2 Permutation tests for heterogeneity: two-sample case

Consider the inferential problem which consists of testing the hypothesis that the heterogeneity of one population is greater than that of another. From a formal point of view, given two populations X_1 and X_2, if we denote by $Het(X_j)$ the degree of heterogeneity of the population $X_j, (j = 1, 2)$, the hypotheses can be expressed as follows

$$H_0 : Het(X_1) = Het(X_2) \qquad \text{against} \qquad H_1 : Het(X_1) > Het(X_2).$$

We denote by $p_i, i = 1, 2, ...k$, the parameters of the underlying distribution that is $p_i = Pr\{X \in A_i\}$. We take into consideration the indices of Gini, $G = \Sigma_i p_i(1 - p_i)$, of Shannon, $S = -\Sigma_i p_i log(p_i)$ and the index of Rényi, $R_\alpha = 1/(1 - \alpha)log(\Sigma_i p_i^\alpha)$ for $\alpha = 3$ and for $\alpha \to \infty$, i.e., $R_3 = -1/2log(\Sigma_i p_i^3)$ and $R_\infty = -log[sup_i(p_i)]$. The choice of R_3 instead of R_2 is dictated by the fact that R_2 is one-to-one related with G, and therefore the two indices imply the same inferential conclusions when using permutation tests. In other words, the two indices are permutationally equivalent (Pesarin (2001)). With $p_{(i)}, i = 1, 2, ..., k$, we denote the parameters p_i arranged in non-increasing order: $p_{(1)} \geq p_{(2)} \geq ... \geq p_{(k)}$. The four indices G, S, R_3 and R_∞ are order invariant, i.e. their value do not change if they are calculated with ordered parameters $p_{(i)}$ instead of with parameters p_i. If we denote by $p_{j(i)}, i = 1, 2, ..., k$, the ordered probabilities for population $j, j = 1, 2$, the fact that the indices of heterogeneity are order invariant allows us to express heterogeneity through ordered parameters: two populations such that $\{p_{1(i)} = p_{2(i)}, i = 1, 2, ..., k\}$, i.e. with the same ordered distribution, are equally heterogeneous. Moreover, if $\{p_{1(i)} = p_{2(i)}, i = 1, 2, ..., k\}$ the data of the two samples are exchangeable and so the permutation testing principle applies. In addition, if $P_{j(i)} = \Sigma_{s \leq i} p_{j(s)})$ are the cumulative probabilities for population j referred to the ordered parameters, the null hypothesis of our problem is equivalent to:

$$H_0 : P_{1(i)} = P_{2(i)}, i = 1, 2, ..., k.$$

Viceversa, considering the link between heterogeneity and the concentration of probabilities, we can use the dominance criterion proposed by Lorenz to define the case of greater heterogeneity of population X_1 with respect to population X_2 (Lorenz (1905)). Instead of using Lorenz's curve to measure concentration we can similarly use one of the indices described below. For example the index of Gini measures the deviation of the Lorenz curve from the diagonal representing the ideal case of homogeneity. The definition of the problem by means of cumulative probabilities makes it very similar to the problem of stochastic dominance for ordered categorical variables, with the peculiarity that the order is determined according to the values of the parameters p_i and not according to the categories the variable X can assume. Therefore, our problem can be referred to as one of *dominance in heterogeneity*. We may observe that X in problems of heterogeneity can be a nominal variable because heterogeneity is a property that concerns probabilities and does not involve the categories A_i of X, whereas the problems of stochastic dominance assume that classes $A_1, A_2, ..., A_k$ are ordered. For problems of stochastic dominance the literature offers quite a long list of exact and approximate solutions. For the univariate case most of the proposed solutions are based on the restricted maximum likelihood ratio test (Cohen et al (2000), Silvapulle and Sen (2005), Wang (1996)). In general these solutions are criticized because the distributions under the null and alternative hypotheses are asymptotically mixtures of chi-squared variables with weights essentially dependent on the unknown distribution P of the population. Nonparametric proposals are those of Troendle (2002), Brunner and Munzel (2000), Pesarin (2001, 1994), and Pesarin and Salmaso (2006). The latter, based on the nonparametric combination of dependent permutation tests (NPC), are exact, unbiased, and consistent tests. As far as we are concerned, when considering the difference in heterogeneity, it is reasonable to use as test statistics the difference of sampling indices. If we indicate a generic index of heterogeneity with I, the test statistic useful for the comparison of 2 samples is

$$T_I = I_1 - I_2,$$

where I_j indicates the sampling value of the index calculated for sample j, where $j = 1, 2$. Clearly the tests will be significant for large values, i.e. large values observed in the test statistic can lead to rejection of the null hypothesis in favour of the alternative. In order to apply the tests according to the usual approach, it is necessary to know their sampling distributions subject to a proper estimate under H_0 of the vector of the marginal ordered probabilities $(p_{j(1)}, p_{j(2)}, ..., p_{j(k)})'$, because the vectors of the probabilities $(p_{j1}, p_{j2}, ..., p_{jk})'$ as well as those of the ordered probabilities $((p_{j(1)}, p_{j(2)}, ..., p_{j(k)})', j = 1, 2,$

are unknown. In reality this question is not easy to solve exactly, with perhaps the exception where $k = 2$. For this purpose, instead of the true ordering of the unknown parameters $\{p_{j(1)}, p_{j(2)}, ..., p_{j(k)}; j = 1, 2\}$, we utilize their estimates. To this end we observe the contingency table where f_{ji} is the relative frequence for category A_i in the distribution of sample j so that $\sum_{i=1}^{k} f_{ji} = 1$ and we utilize an estimate based on ordered sampling frequencies $f_{ji} = \hat{p}_{ji}$ $i = 1, , k$, such that $f_{j(1)} \geq f_{j(2)} \geq ... \geq f_{j(k)}$, j=1,2 (data-driven ordering). Thus we obtain the ordered table:

Table 1. Estimation of the ordered probabilities by relative frequencies

	Classes				Sample sizes
Population	(1)	(2)	...	(k)	
X_1	$f_{1(1)}$	$f_{1(2)}$	...	$f_{1(k)}$	n_1
X_2	$f_{2(1)}$	$f_{2(2)}$	...	$f_{2(k)}$	n_2
	$f_{\cdot(1)}$	$f_{\cdot(2)}$	...	$f_{\cdot(k)}$	N

We note that the order is realized separately for each sample and as it is based on relative frequencies rather than on classes, it could be that the $i-th$ column of table 1 refers to two diverse classes for the two samples. In other words class (i) corresponds to the class whose observed relative frequency occupies the $i-th$ position in the ordered sequence and can be different for the two samples. Obviously the order imposed by the frequencies presents a random component and may vary depending upon sampling variations. Therefore under H_0 data are not exactly exchangeable as would be the case if the true order of population parameters were known and used. The exchangeability property can only be obtained asymptotically. Therefore, permutation solutions are approximate for finite sample sizes and exact only asymptotically. Using the data in table 1, the observed value T^o of the test statistic is calculated. For each permutation of the dataset one obtains a new permuted table (as in table 2), with different values from those of the observed table but with fixed marginal frequencies.

Table 2. Absolute frequencies after a permutation of data

	Classes				Sample sizes
Population	(1)	(2)	...	(k)	
X_1	$n^*_{1(1)}$	$n^*_{1(2)}$	...	$n^*_{1(k)}$	n_1
X_2	$n^*_{2(1)}$	$n^*_{2(2)}$	...	$n^*_{2(k)}$	n_2
	$n_{\cdot(1)}$	$n_{\cdot(2)}$	...	$n_{\cdot(k)}$	N

Using the data of the permutated table in the calculations of test statistics, one obtains the permutation values T^*. Calculating the values that can be obtained making all the possible permutations, one obtains the permutation distribution of each test statistic. Alternatively it is possible to extract from the set of all possible permutations a random sample, thus obtaining conditional Monte Carlo estimates. In this way, for each of the four tests, it is possible to calculate the p-value, that, if B is the number of considered permutations, is given by

$$\lambda = \sharp(T^* \geq T^o|X)/B,$$

where $\sharp(T^* \geq T^o|X)$ indicates the number of times permutation values are not lower than the observed ones, conditionally on the dataset $X = \{X_{ji}, i = 1, \ldots, n_j; j = 1, 2\}$. Therefore, according to the usual decision rule, if the p-value is less than or equal to a fixed significance level, the null hypothesis is rejected in favour of the alternative; otherwise the null hypothesis cannot be rejected.

3 The multisample case

In several scientific disciplines researchers might be interested in comparing more than two populations $X_1, X_2, \ldots, X_C$ with $C > 2$. For example, the hypothesis could be that some populations of individuals resident in a certain area present a diverse genetic heterogeneity or it could be of interest to test the hypothesis that the heterogeneity of the professional opportunities offered by C distinct doctorate courses is not the same. In these cases the hypotheses of the problem are:

$H_0 : Het(X_1) = Het(X_2) = \ldots = Het(X_C)$ and $H_1 : H_0$ is not true.

Also for this problem the proposed solution implies the construction of the ordered $C \times k$ table, with respect to the observed relative frequencies (C empirical orderings) and so also in this case the solution is approximate. If we indicate a generic index of heterogeneity with I, the test statistic useful for the comparison of C samples is

$$T_I^2 = \sum_{j=1}^{C} (I_j - \bar{I})^2 n_j,$$

where I_j denotes the sampling value of the index calculated for sample j, where $j = 1, 2, \ldots, C$; n_j denotes $j - th$ sample-size; $\bar{I}$ denotes the sampling value of the index calculated in the pooled sample sized $N = n_1 + n_2 + \ldots + n_C$. Large values of this test statistic are significant. Once the observed value of the test statistic is calculated , we perform B permutations of the dataset, exchanging the units among the C samples, obtaining for each permutation a table of relative frequencies from which the permutation values of the tests are calculated. These values allow us to calculate the p-value of the test as in section 2. In some cases it could be of interest to test for a given predefined ordering of the heterogeneity of the C populations. In this case the alternative hypothesis can be expressed as:

$$H_1 : Het(X_1) \geq Het(X_2) \geq \ldots \geq Het(X_C)$$

where at least one inequality is strict. The latter is configured as a problem of *monotonic stochastic ordering* whose theory and testing methods in the field of the permutation approach are extensively dealt with in Pesarin (2001).

4 Simulation study

In order to assess the properties of the test we considered a simulation study in which data are generated according to the following model:

$$X \sim 1 + Int[K \times U^\delta]$$

where $\delta \in \Re$, $U \sim U(0, 1)$, K is a positive integer, and $Int[\cdot]$ denotes the integer part of $[\cdot]$. The random variable X is therefore discrete and its domain consists of the first K positive integers. The situation of maximum heterogeneity can be simulated making $\delta = 1$, because in this case $X \sim U(1, 2, ..., K)$ and $f_i = \sharp(X = i)/n \cong 1/K$, where n is the sample size. Generating the data as described, for different sample sizes, different values of the parameters $(\delta_1, \delta_2, \delta_3)$ (to simulate different degrees of heterogeneity respectively of the populations X_1, X_2 and X_3) and different values of the nominal significance level, we calculated the rejection rates of the tests in order to evaluate their degree of approximation to the nominal significance levels, as well as the power. Table 3 reports the rejection rates of the tests under the null hypothesis, for a discrete variable with $K = 8$ categories, and for different degrees of heterogeneity represented by the values 1,2,3, respectively, of the parameters δ_1, δ_2 and δ_3 (1 = maximum heterogeneity). 1000 datasets were generated each by 1000 Monte Carlo Conditional simulations in order to approximate the related permutation distribution. In general the performances of the four tests are very similar and power increases with an increase in homogeneity of the distributions, i.e. for high values of the parameters δ_i. It is evident that the tests are substantially well approximated even if the test based on R_∞ violates the nominal alpha value more frequently than its competitors.

Table 4 reports the rejection rates under the alternative hypothesis. Obviously the power of the tests increases with increases in the difference of the values of the parameters δ_i, i.e. of the heterogeneities of the samples, and with increases in sample sizes (see table 4). Comparing the performances of the four tests it emerges that if the differences of the values of the parameters are large then test T_{R_∞} seems slightly better than the others, whereas when the differences are not large, test T_{R_3} seems better.

5 Conclusions

This work consists of an inferential procedure for a solution of the problem of hypothesis testing, in which the objective is that of comparing the heterogeneity of two or more populations on the basis of sampling data, i.e. to test the hypothesis that the heterogeneity of the populations are different. Such a

Table 3. Rejection rates of nonparametric tests under H_0 and $K = 8$

δ_1	δ_2	δ_3	n_1	n_2	n_3	α 0.01	0.05	0.10	Test
1	1	1	20	20	20	0.000	0.000	0.004	T_S
						0.000	0.000	0.002	T_G
						0.000	0.000	0.000	T_{R_3}
						0.000	0.002	0.006	T_{R_∞}
			30	20	10	0.000	0.000	0.006	T_S
						0.000	0.000	0.006	T_G
						0.000	0.000	0.002	T_{R_3}
						0.000	0.000	0.006	T_{R_∞}
2	2	2	20	20	20	0.010	0.018	0.044	T_S
						0.008	0.026	0.052	T_G
						0.014	0.042	0.086	T_{R_3}
						0.018	0.058	0.110	T_{R_∞}
			30	20	10	0.010	0.028	0.072	T_S
						0.008	0.030	0.070	T_G
						0.010	0.050	0.090	T_{R_3}
						0.010	0.056	0.114	T_{R_∞}
3	3	3	20	20	20	0.020	0.060	0.098	T_S
						0.020	0.058	0.096	T_G
						0.020	0.074	0.120	T_{R_3}
						0.008	0.064	0.108	T_{R_∞}
			30	20	10	0.018	0.054	0.090	T_S
						0.016	0.052	0.092	T_G
						0.020	0.060	0.116	T_{R_3}
						0.018	0.054	0.090	T_{R_∞}

proposal consists of finding appropriate test statistics and a general methodology of hypothesis testing based on the ordering of probabilites as well as on a methodological solution inspired by that proposed in Pesarin (2001) for problems of stochastic dominance. The test statistic consists of the comparison of the sampling indices of heterogeneity and it can vary according to the index of heterogeneity considered. The test statistics taken into consideration are those based on Shannon's index of entropy, on Gini's index of heterogeneity and on Rényi's indices of entropy both of order 3 and of infinite order. The simulation study allowed for the assessment of the degree of approximation to nominal significance levels and the power behaviour of the proposed non parametric tests of heterogeneity. The rejection rates increase with increases in the homogeneity of the distributions. When the differences in the values of the parameters are large, test T_{R_∞} seems better than the others even if it violates the nominal significance level more frequently than its competitors, whereas when the differences are not large test T_{R_3} seems better. The rejection rates under the alternative hypothesis are in any case satisfactory for all considered tests.

Table 4. Power of nonparametric tests (K = 8)

δ_1 δ_2 δ_3 n_1 n_2 n_3	0.01	α 0.05	0.10	Test
2 2 3 20 20 20	0.032	0.114	0.186	T_S
	0.032	0.116	0.184	T_G
	0.052	0.152	0.232	T_{R_3}
	0.050	0.132	0.216	T_{R_∞}
40 30 20	0.090	0.240	0.336	T_S
	0.090	0.240	0.334	T_G
	0.104	0.238	0.352	T_{R_3}
	0.074	0.174	0.304	T_{R_∞}
1 2 3 20 20 20	0.104	0.218	0.336	T_S
	0.104	0.222	0.334	T_G
	0.160	0.304	0.428	T_{R_3}
	0.204	0.372	0.472	T_{R_∞}
40 30 20	0.312	0.552	0.696	T_S
	0.316	0.550	0.694	T_G
	0.394	0.590	0.702	T_{R_3}
	0.458	0.638	0.712	T_{R_∞}
90 60 20	0.644	0.858	0.932	T_S
	0.642	0.856	0.930	T_G
	0.774	0.906	0.940	T_{R_3}
	0.798	0.900	0.928	T_{R_∞}

References

Brunner E, Munzel U (2000) The nonparametric behrens-fisher problem: asymptotic theory and small-sample approximation. Biometrical Journal 42:17–25

Cohen A, Kemperman J, Madigan D, Sakrowitz H (2000) Effective directed tests for models with ordered categorical data. Australian and New Zealand Journal of Statistics 45:285–300

Lorenz M (1905) Methods of measuring the concentration of wealth. Publications of the American Statistical Association 9 (70):209–219

Pesarin F (1994) Goodness-of-fit testing for ordered discrete distributions by resampling techniques. Metron LII:57–71

Pesarin F (2001) Multivariate Permutation Test With Application to Biostatistics. Wiley, Chichester

Pesarin F, Salmaso L (2006) Permutation tests for univariate and multivariate ordered categorical data. Austrian Journal of Statistics 35:315–324

Silvapulle M, Sen P (2005) Constrained Statistical Inference, Inequality, Order, and Shape Restrictions. Wiley, New York

Troendle J (2002) A likelihood ratio test for the nonparametric behrens-fisher problem. Biometrical Journal 44 (7):813–824

Wang Y (1996) A likelihood ratio test against stochastic ordering in several populations. Journal of the American Statistical Association 91:1676–1683

On Synchronized Permutation Tests in Two-Way ANOVA

Dario Basso[1], Luigi Salmaso[2], and Fortunato Pesarin[3]

[1] Department of Statistics, University of Padova, Via C. Battisti 241, 35121 Padova - Italy. `dario@stat.unipd.it`
[2] Department of Management and Engineering, University of Padova, Stradella S. Nicola 3, 36100 Vicenza - Italy. `salmaso@gest.unipd.it`
[3] Department of Statistics, University of Padova, Via C. Battisti 241, 35121 Padova - Italy. `fortunato.pesarin@unipd.it`

Summary. In $I \times J$ balanced factorial designs units are not exchangeable between blocks since their expected values depend on received treatments. It does not seem possible, therefore, to obtain exact and separate tests to respectively assess main factor and interaction effects. Parametric two-way ANOVA F tests are exact tests only under assumption of normal homoschedastic errors, but they are also positively correlated. Instead, it is possible to obtain exact, separate and uncorrelated permutation tests at least for main factors by introducing a restricted kind of permutations, named *synchronized* permutations. Since these tests are conditional on observed data, they are distribution-free and may be shown to be almost as powerful as their parametric counterpart under normal errors. We obtain the expression of the correlation between the main factor ANOVA tests as a function of the number of replicates in each block, the number of main factor levels and their noncentrality parameters.

Key words: synchronized permutations, noncentral F distribution

1 Introduction to synchronized permutations

It is well known from the literature that the parametric F test is not robust when applied to heavy-tailed or asymmetric distributions (see, for instance, Arnold (1948), Mardia (1970)). In these situations, nonparametric solutions are to be preferred. In such a context, however, the problem of testing for interaction (Mansouri and Chang (1995)) is considered too hard unless ranks are assigned to residuals with respect to suitable estimates of effects, leading to asymptotic solutions. As far as we know the only exact solution for finite sample sizes is that based on synchronized permutations (Pesarin (2001),

Salmaso (2003)). In a $I \times J$ replicated factorial design, data are assumed to follow the linear model:

$$y_{ijk} = \mu + \alpha_i + \beta_j + \gamma_{ij} + \varepsilon_{ijk} \qquad \begin{cases} i = 1, \ldots, I \\ j = 1, \ldots, J \\ k = 1, \ldots, n \end{cases} \qquad (1)$$

where y_{ijk} are experimental responses, μ is the overall mean, α_i, β_j are main factor effects, γ_{ij} are interaction effects, ε_{ijk} are exchangeable experimental errors, with zero mean, from an unknown distribution P_ε, and n is the number of replicates in each cell. We also assume the side-conditions $\sum_i \alpha_i = 0$, $\sum_j \beta_j = 0$, $\sum_i \gamma_{ij} = 0 \,\forall j$, $\sum_j \gamma_{ij} = 0 \,\forall i$. The hypotheses which should be separately tested are $H_{0A} : \alpha_i = 0 \ \forall i$ against $H_{1A} : \exists i : \alpha_i \neq 0$, $H_{0B} : \beta_j = 0 \ \forall j$ against $H_{1B} : \exists j : \beta_j \neq 0$ and $H_{0AB} : \gamma_{ij} = 0 \ \forall i, j$ against $H_{1AB} : \exists i, j : \gamma_{ij} \neq 0$. Since, due to other potentially active effects, the observations from different cells may have different means, they are not exchangeable in each null hypothesis. Then, to perform nonparametric testing on main factor effects, we need to introduce a restricted kind of randomization. Suppose we want to test for H_{0A}: the totals of blocks $A_i B_j$ and $A_s B_j$ ($j = 1, \ldots, J$), after exchanging $\nu^*_{is|j}$ units between this pair of blocks, can respectively be written as:

$$\sum_k^n y^*_{ijk} = \nu^*_{is|j} [\mu + \alpha_s + \beta_j + \gamma_{sj}] + (n - \nu^*_{is|j}) [\mu + \alpha_i + \beta_j + \gamma_{ij}] + \sum_k^n \varepsilon^*_{ijk},$$

$$\sum_k^n y^*_{sjk} = \nu^*_{is|j} [\mu + \alpha_i + \beta_j + \gamma_{ij}] + (n - \nu^*_{is|j}) [\mu + \alpha_s + \beta_j + \gamma_{sj}] + \sum_k^n \varepsilon^*_{sjk}.$$

Let us consider the structure of the difference of these totals, to compare α_i and α_s at level j of factor B:

$$^a T^*_{is|j} = \sum_k y^*_{ijk} - \sum_k y^*_{sjk} = (n - 2\nu^*_{is|j})[\alpha_i - \alpha_s + \gamma_{ij} - \gamma_{sj}] + \varepsilon^*_{is|j}$$

where "*" denotes that we have randomly exchanged $\nu^*_{is|j}$ units between two cells, and $\varepsilon^*_{is|j}$ is a linear combination of corresponding permuted errors. Thus, the sum over all columns:

$$\sum_j {}^a T^*_{is|j} = \sum_j (n - 2\nu^*_{is|j})[\alpha_i - \alpha_s + \gamma_{ij} - \gamma_{sj}] + \sum_j \varepsilon^*_{is|j}, \qquad (2)$$

only depends on effects α_i and α_s and on exchangeable errors if and only if $\nu^*_{is|j} = \nu^*_{is}, \forall j$ (synchronized permutations with respect to the levels of factor B), because of the side-conditions on interaction effects. Otherwise, if $\nu_{is|j_1} \neq \nu_{is|j_2}$ for some $j_1 \neq j_2$, the factor A effects would be confounded with the interaction effects in (2). Therefore, as H_{0A} depends on $\alpha_1, \alpha_2, \ldots, \alpha_I$, by synchronizing permutations also with respect to row indexes i and s, i.e. by setting $\nu^*_{is} = \nu^*$, the test statistic:

$${}^{a}T_{A}^{*} = \sum_{i<s}[\sum_{j} {}^{a}T_{is|j}^{*}]^2, \tag{3}$$

only depends on effects of factor A and on exchangeable errors, hence it is a separate exact test on factor A. In a similar way, define ${}^{b}T_{jh|i}^{*} = \sum_{k} y_{ijk}^{*} - \sum_{k} y_{ihk}^{*}$ the intermediate statistic for comparing factor B effects β_j and β_h at level i of factor A. Then the test statistic:

$${}^{b}T_{B}^{*} = \sum_{j<h}[\sum_{i} {}^{b}T_{jh|i}^{*}]^2, \tag{4}$$

only depends on effects of factor B and on exchangeable errors, hence it is a separate exact test on factor B. We can also define two test statistics for the interaction:

$${}^{a}T_{AB}^{*} = \sum_{i<s}\sum_{j<h}[{}^{a}T_{is|j}^{*} - {}^{a}T_{is|h}^{*}]^2 \quad \text{and} \quad {}^{b}T_{AB}^{*} = \sum_{j<h}\sum_{i<s}[{}^{b}T_{jh|i}^{*} - {}^{b}T_{jh|s}^{*}]^2. \tag{5}$$

Note that ${}^{a}T_{AB}^{*}$ is obtained from synchronized permutations involving the row factor A, whereas ${}^{b}T_{AB}^{*}$ is obtained from permutations involving the column factor B. Each statistic for interaction only depends on interaction effects and on exchangeable errors; moreover, and so they are jointly and equally informative. Thus, their linear combination $T_{AB}^{**} = {}^{a}T_{AB}^{*} + {}^{b}T_{AB}^{*}$ is a separate exact test for interaction.

2 Constrained and unconstrained synchronized permutations

The basic concept of synchronized permutations is exchanging the same number $0 \le \nu^{*} \le n$ of units between each pair of blocks that are considered. Note that ν^{*} is itself a random variable, which identifies distinct sets of synchronized permutations. Of course, different synchronized permutations can lead to the same ν^{*}. For a given ν^{*}, there are two ways to obtain a random synchronized permutation: exchanging units in the same original positions in each block (Constrained Synchronized Permutations, CSPs) or exchanging units without considering their original position (Unconstrained Synchronized Permutations, USPs). It is easy to see that the total number of CSPs only depends on which exchanges have been made in the first pair of blocks $A_i B_1$ and $A_s B_1$. There are $C_{CSP} = \binom{2n}{n}$ possible ways to exchange units in the first pair of blocks, therefore C_{CSP} is the cardinality of the CSPs. Another point to take into account is the cardinality of distinct values of the permutation test statistic (e.g., the number of distinct values of ${}^{a}T_{A}^{*}$s if we are testing for factor A): the squaring operation in formulas (3), (4) and (5) produces a symmetry, i.e., there are two distinct synchronized permutations generating the same value of the test statistic, therefore the cardinality of the support of the test statistic is $C_{CSP}/2$. Adopting CSPs, the exact null distribution of the test statistic can be obtained by considering all possible combinations of $2n$ units in groups of n (e.g. by using the R library *combinat*).

Adopting USPs, units being exchanged in each pair of blocks may differ from the original positions of units exchanged in the first pair of blocks. The only requirement is that the number of exchanges is the same. Hence, for any pair of levels of factor A and a given number of exchanges ν^*, there are $\binom{n}{\nu^*}^{2J}$ possible ways to choose the same number of units to be exchanged in the $2J$ cells. Since the number of possible pairs of levels of factor A is $I(I-1)/2$, the total number of USPs when testing for factor A is:

$$\sum_{x=0}^{n} \binom{n}{x}^{J \times I(I-1)}$$

Things get harder if we wish to calculate the total number of distinct values of the test statistic (e.g. ${}^{a}T_A^*$). The symmetry in the test statistic plays a different role when n is odd or even. Let us consider the case when n is odd first: it is easy to see that we can obtain the same values of ${}^{a}T_A^*$ when $\nu^* = x$ or $\nu^* = n - x, 0 \leq x \leq (n-1)/2$. Hence, the cardinality of distinct values of ${}^{a}T_A^*$ when n is odd is:

$$C_{USP}^o = \sum_{x=0}^{(n-1)/2} \binom{n}{x}^{J \times I(I-1)}.$$

By first choosing $0 \leq \nu^* \leq (n-1)/2$, randonly shuffling the units within each block and by exchanging the first ν^* units within each pair of blocks, we can obtain a random USP. The probability of making ν^* exchanges is:

$$P[N = \nu^*] = \frac{\binom{n}{\nu^*}^{J \times I(I-1)}}{\sum_{x=0}^{(n-1)/2} \binom{n}{x}^{J \times I(I-1)}}. \tag{6}$$

A two-step algorithm can guarantee the values of the test statistic to be equally likely by first choosing the number of exchanges to be made in accordance with (6), then shuffling the units within each cell and finally exchanging the first ν^* units between each pair of cells. The shuffling inside each block guarantees the values of ${}^{a}T_A^*$ given ν^* to be equally likely: $P[{}^{a}T_A^* = {}^{a}t_A^*|\nu^*] = \binom{n}{\nu^*}^{-J \times I(I-1)}$. When n is even, it can be demonstrated that the cardinality of the support of ${}^{a}T_A^*$ is:

$$C_{USP}^e = \sum_{\nu=0}^{n/2-1} \binom{n}{\nu}^{J \times I(I-1)} + \left[\binom{n}{n/2}^{2J}/2 \right]^{I(I-1)/2}, \tag{7}$$

and we can apply the same strategy as before by choosing the number of exchanges to be made in accordance with (7). The cardinality of distinct values of $^{a}T_A^*$ rapidly increases with n, I and J, so we recommend to use USPs when few replicates are available (say $n \leq 3$). If the number of replicates is greater than 3, one can easily apply the CSPs. A different choice between CSPs and USPs affects the minimum achievable significance level, which is equal to $1/C$, and C is the cardinality of the support of the test statistic.

3 Correlation between test statistics

It is known that the two-way ANOVA test statistics are ratios between independent quadratic forms. In particular, the main factor test statistics are defined as:

$$F_A = \frac{SS_A/(I-1)}{SS_\varepsilon/IJ(n-1)} \qquad \text{and} \qquad F_B = \frac{SS_B/(J-1)}{SS_\varepsilon/IJ(n-1)},$$

where $SS_A/(I-1)$ and $SS_B/(J-1)$ are unbiased estimates of σ^2 only if the related null hypothesis is true, whereas $SS_\varepsilon/IJ(n-1)$ is always an unbiased estimate of σ^2. Since both statistics have the same denominator, they are positively correlated. In this section we obtain the correlation between F_A and F_B as a function of I, J, n, and noncentrality parameters α and β. We want to evaluate:

$$\rho(F_A, F_B) = \frac{COV(F_A, F_B)}{\sqrt{Var(F_A)}\sqrt{Var(F_B)}},$$

where F_A and F_B are noncentral F if the corresponding alternative is true. Given that SS_A, SS_B and SS_ε are independent χ^2 variables, we have:

$$\begin{aligned}
COV[F_A, F_B] &= E[F_A \cdot F_B] - E[F_A]E[F_B] \\
&= E[SS_A/(I-1)]E[SS_B/(J-1)]E[I^2 J^2 (n-1)^2 SS_\varepsilon^{-2}] + \\
&\quad - E[SS_A/(I-1)]E[SS_B/(J-1)]E[IJ(n-1)SS_\varepsilon^{-1}]^2 \\
&= E[MS_A]E[MS_B]Var[IJ(n-1)SS_\varepsilon^{-1}] > 0,
\end{aligned}$$

even in the null hypothesis $H_{0A} \cap H_{0B}$. Being a χ^2 random variable, SS_ε belongs to the Gamma family, so it is easy to obtain $Var[SS_\varepsilon^{-1}]$. More precisely, $SS_\varepsilon \sim Ga(ab(n-1)/2, 1/2\sigma^2)$, hence:

$$Var[IJ(n-1)SS_\varepsilon^{-1}] = \frac{2I^2 J^2 (n-1)^2}{[IJ(n-1)-2]^2 [IJ(n-1)-4]\sigma^4}$$

and:

$$Var(F_A) = \frac{2I^2 J^2 (n-1)^2}{(I-1)^2} \times \frac{\alpha^2 + [(I-1)+2\alpha][I-1+IJ(n-1)-2]}{[IJ(n-1)-2]^2 [IJ(n-1)-4]}$$

$$Var(F_B) = \frac{2I^2 J^2 (n-1)^2}{(J-1)^2} \times \frac{\beta^2 + [(J-1)+2\beta][J-1+IJ(n-1)-2]}{[IJ(n-1)-2]^2 [IJ(n-1)-4]}.$$

Hence:

$$\rho(F_A, F_B) = \frac{E[MS_A]E[MS_B]Var[IJ(n-1)SS_\varepsilon^{-1}]}{[Var(F_A)Var(F_B)]^{1/2}}$$
$$= \frac{[(I-1)+\alpha][(J-1)+\beta]}{\sqrt{\tilde{\sigma}^2(F_A)\tilde{\sigma}^2(F_B)}}, \tag{8}$$

where:

$$\tilde{\sigma}^2(F_A) = \alpha^2 + [(I-1)+2\alpha][(I-1)+ab(n-1)-2], \qquad \alpha = Jn\sum_i \alpha_i^2/\sigma^2;$$

$$\tilde{\sigma}^2(F_B) = \beta^2 + [(J-1)+2\beta][(J-1)+ab(n-1)-2], \qquad \beta = In\sum_j \beta_j^2/\sigma^2.$$

Let us consider the synchronized permutation test statistics for main factors now. The tests on main factors are defined on different permutation spaces. In particular, $^aT_A^*$ is obtained by exchanging units between pairs of rows and within columns, $^bT_B^*$ is obtained by exchanging units between pairs of columns and within rows. Hence, without loss of generality, consider the case when we are testing for $H_{0A} : \alpha_i = 0 \forall i$. This hypothesis allows us to exchange units between rows and within columns. The test statistic $^aT_A^*$ is a random variable and its probability function $P(^aT_A^*)$ is defined on the related support. Units are exchanged within columns, so the total of each column is constant. Let $Y_{\cdot j}, j = 1, \ldots, J$ be the total of column j. Then:

$$\sum_i {}^bT_{jh|i}^* = \sum_i [\sum_k y_{ijk}^* - \sum_k y_{ihk}^*] = \sum_i \sum_k y_{ijk}^* - \sum_i \sum_k y_{ihk}^* = Y_{\cdot j} - Y_{\cdot h}$$

hence:

$$^bT_B^* = \sum_{j<h} [\sum_i T_{jh|i}^*]^2 \equiv \sum_{j<h} (Y_{\cdot j} - Y_{\cdot h})^2 = {}^bT_B$$

where ^bT_B is the observed value of the test statistic for factor B. Since $^bT_B^*$ is constant, $\rho(^aT_A^*, {}^bT_B^*) = 0$.

4 Power comparison

Table 1 reports a comprehensive power simulation comparison between CPS and ANOVA test statistics. Four types of error distribution have been considered: the normal, the Gamma (with one and two degrees of freedom) representing asymmetric error distributions, and the student t_3 representing heavy-tailed errors. Nominal significance levels have been chosen to be as close as possible to the usual ones (1%, 5% and 10%) from the achievable significance levels: when $n = 4$ the nominal levels are 0.029, **0.058** and *0.114*. When $n = 7, 10$, the nominal levels are 0.016, **0.048** and *0.104*. Recall that the achievable significance levels are multiple of $2/C_{CSP}$. Then, for each error distribution, the observed rejection rates are reported for main factors and interaction at the correspondent nominal level. The effects of main factors/interaction were set in accordance to a 3×2 full factorial design matrix. The true sizes of the effects (whose label is "T.S.") are displayed in third column, and were set equal to $\sigma/2, 3\sigma/2, \sigma$ for respectively factor A, factor B and interaction. The error variance σ^2 is held fixed to one. Note how the power of synchronized permutation tests is very close to that of the ANOVA test for any considered distribution. As regards the interaction, we have applied the linear combination of the test statistics in (5). See Pesarin (2001) for details on nonparametric combination. Figure 1 shows the relationship between main factor test statistics when synchronized permutation (top) and two-way ANOVA (bottom) tests are applied. Synchronized permutation tests statistics are denoted with "T.A" and "T.B", and their related p-values with "pa" and "pb". Two-way ANOVA test statistics are denoted with "F.A" and "F.B", and their related p-values with "pfa" and "pfb". The four scatterplots on the left-hand side are the results of 1000 independent sample generations with normal error under $H_{1A} \cap H_{0B} \cap H_{0AB}$ in a 2×2 design with $n = 5$. We have set $\alpha_1 = -\alpha_2 = 1.5\sigma$ and applied the CSPs. Note the discrete nature of CSPs: while the points representing factor B p-values assume any value that is a multiple of $1/126$ in the interval $[0,1]$, factor A p-values only assume the values $1/126, 2/126$ and $3/126$.

The four scatterplots on the right-hand side are the results of 1000 independent sample generations with normal error under $H_{1A} \cap H_{1B} \cap H_{1AB}$ in a 2×2 design with $n = 5$. We have set $\alpha_1 = -\alpha_2 = \beta_1 = -\beta_2 = \sigma$ and $|\gamma_{ij}| = 0.5\sigma, i, j = 1, 2$. Here the correlation between F.A and F.B is evident, while the pairs (T.A, T.B) are still randomly spread. Synchronized permutation tests are separate because they only depend on the effects of the factor/interaction under testing. Moreover, they allow for separate inferences because the tests are uncorrelated to each other.

References

Arnold H (1948) Permutation support for multivariate techniques. Biometrika 35:88–96

Mansouri H, Chang G (1995) A comparative study of some rank tests for interaction. Computational Statistics and Data Analysis 19:85–96

Mardia K (1970) Measures of multivariate skewness and kurtosis with applications. Biometrika 57:519–530

Pesarin F (2001) Multivariate permutation tests with applications in biostatistics, $1^s t$ edn. Wiley-Chichester

Salmaso L (2003) Synchronized permutation tests in factorial designs. Communication in Statistics - Theory and Methods 32:1419–1437

Table 1. Power simulation in a 3×2 design. Synchronized permutation and two-way ANOVA tests.

n	Factor	P_ε T.S.	Norm			Exp			t_3			Ga_2		
			Constrained Synchronized Permutation Tests											
4	A	.5	.226	**.353**	.507	.309	**.442**	.579	.360	**.516**	.664	.278	**.405**	.548
	B	1.5	.996	**.999**	1.00	.996	**.998**	1.00	.983	**.992**	.996	.992	**.998**	1.00
	AB	1	.814	**.906**	.958	.857	**.906**	.960	.867	**.907**	.953	.850	**.922**	.964
7	A	.5	.387	**.606**	.763	.439	**.615**	.734	.514	**.685**	.797	.426	**.620**	.741
	B	1.5	1.00	**1.00**	1.00	.615	**1.00**	1.00	.997	**.999**	1.00	1.00	**1.00**	1.00
	AB	1	.997	**.998**	1.00	.981	**.992**	.998	.974	**.981**	.994	.988	**.997**	1.00
10	A	.5	.521	**.691**	.796	.523	**.678**	.794	.592	**.746**	.841	.482	**.652**	.792
	B	1.5	1.00	**1.00**	1.00	1.00	**1.00**	1.00	.999	**1.00**	1.00	1.00	**1.00**	1.00
	AB	1	1.00	**1.00**	1.00	.997	**.998**	1.00	.992	**.993**	.994	.999	**1.00**	1.00
n	Factor	T.S.	**Two-way ANOVA Tests**											
4	A	.5	.244	**.373**	.516	.344	**.455**	.582	.401	**.524**	.650	.299	**.409**	.544
	B	1.5	1.00	**1.00**	1.00	1.00	**1.00**	1.00	.994	**.997**	.997	1.00	**1.00**	1.00
	AB	1	.871	**.928**	.961	.845	**.914**	.949	.882	**.920**	.951	.856	**.919**	.959
7	A	.5	.431	**.636**	.763	.427	**.613**	.736	.517	**.686**	.783	.433	**.614**	.755
	B	1.5	1.00	**1.00**	1.00	1.00	**1.00**	1.00	.958	**.999**	1.00	1.00	**1.00**	1.00
	AB	1	.955	**.997**	1.00	.965	**.992**	.995	.974	**.987**	.993	.981	**.994**	1.00
10	A	.5	.658	**.803**	.892	.663	**.795**	.877	.749	**.847**	.907	.637	**.796**	.880
	B	1.5	1.00	**1.00**	1.00	1.00	**1.00**	1.00	1.00	**1.00**	1.00	1.00	**1.00**	1.00
	AB	1	1.00	**1.00**	1.00	.995	**.997**	1.00	.987	**.992**	.993	.999	**1.00**	1.00

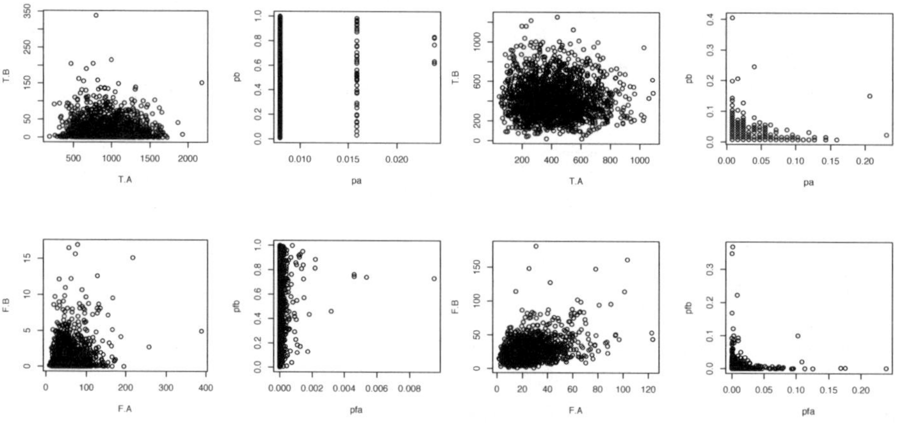

Fig. 1. Test statistic and p-value scatterplots of main factors in synchronized permutation and two-way ANOVA tests.

Optimal Three-Treatment Response-Adaptive Designs for Phase III Clinical Trials with Binary Responses

Atanu Biswas[1] and Saumen Mandal[2]

[1] Indian Statistical Institute, Applied Statistics Unit, 203 B.T. Road, Kolkata – 700 108, India. `atanu@isical.ac.in`

[2] Department of Statistics, University of Manitoba, Winnipeg, MB, R3T 2N2, Canada. `saumen_mandal@umanitoba.ca`

Summary. Response-adaptive designs may be used in phase III clinical trials to allocate a larger number of patients to the better treatment. Optimal response-adaptive designs are used for the same purpose, but the design is derived from some optimal points of view. The available optimal response-adaptive designs are only for two treatment trials. In the present paper, we extend this idea and derive some optimal response-adaptive designs for phase III clinical trials for more than two treatments. In particular, we work on three treatments. The extension is not trivial, as the designs for three treatments are often iterative, and they need specific algorithms for computation. The proposed approaches are numerically illustrated.

Key words: ethics, minimization, objective function, sequential estimation, urn models

1 Introduction

Response-adaptive designs are used in phase III clinical trials to achieve the *ethical* goal of treating a larger number of patients by the better treatment arm. Several such adaptive designs are available for binary treatment responses, e.g., the play-the-winner (PW) rule (Zelen (1969)), the randomied play-the-winner (RPW) rule (Wei and Durham (1978)), the generalized Pòlya urn design (GPU) (Wei (1979)), the success driven design (Durham et al (1998)), the birth and death design (Ivanova et al (2000)), the drop-the-loser (DL) (Ivanova (2003)). These are suggested primarily from intuition, and then some theoretical properties are illustrated. It is important to note that such designs allocate a larger proportion of patients to the better treatment. None of these designs are suggested from an optimal point of view, yet some of them are quite popular. In fact, almost all the real applications available in the literature are based on the PW (Rout et al (1993)) and the RPW (Bartlett et al (1985), Tamura et al (1994), Biswas and Dewanji (2004)). Some of the designs can also be easily extended to more than two treatments (e.g. GPU, RPW, success driven design, birth and death, DL).

Recently, there has been interest in deriving optimal response-adaptive designs for binary responses. Note that the Neyman allocation is an optimal allocation which allocates in proportion to the standard deviations of treatment responses. If n_A and n_B be the number of allocations to the two competing treatments A and B, with $n_A + n_B = n$, and $p_k (= 1 - q_k)$ is the probability of success under treatment k, $k = A, B$, the Neyman allocation proportion to treatment A is

$$\pi_A = \frac{\sqrt{p_A q_A}}{\sqrt{p_A q_A} + \sqrt{p_B q_B}}. \tag{1}$$

This allocation (1) may not be *ethical*, but maximizes power (see Rosenberger and Lachin (2002), p. 197). With two treatments at hand, Rosenberger et al (2001) extended the approach of Hayre (1979) and derived optimal response-adaptive designs for binary responses by minimizing the expected number of failures subject to a fixed variance of the estimated treatment difference. In their optimal design, Rosenberger et al (2001) minimized

$$q_A n_A + q_B n_B$$

subject to

$$Var(\widehat{p}_A - \widehat{p}_B) = \frac{p_A q_A}{n_A} + \frac{p_B q_B}{n_B} = K, \tag{2}$$

for a preassigned constant K. [Note that the Neyman allocation minimizes $n_A + n_B$ subject to (2).] Writing $R = n_A/n_B$, the optimal proportion for treatment A, $\pi_A = R/(R+1)$, turns out to be

$$\pi_A = \frac{\sqrt{p_A}}{\sqrt{p_A} + \sqrt{p_B}}. \tag{3}$$

The design suggests sequential estimates of p_A and p_B based on the available data, and a plug-in estimate of π_A by which to allocate any entering patient to treatment A with probability $\widehat{\pi}_A$. Note that the urn designs like the RPW or the DL has a limiting allocation of

$$\pi_A = \frac{\frac{1}{q_A}}{\frac{1}{q_A} + \frac{1}{q_B}}. \tag{4}$$

Note that any allocation design minimizes

$$n_A \Psi_A + n_B \Psi_B, \tag{5}$$

subject to

$$\frac{\sigma_A^2}{n_A} + \frac{\sigma_B^2}{n_B} = \frac{p_A q_A}{n_A} + \frac{p_B q_B}{n_B} = K, \tag{6}$$

where Ψ_k, $k = A, B$, is a function of p_A and p_B such that Ψ_A is decreasing in p_B (for fixed p_A), or decreasing in p_A (for fixed p_B), and Ψ_A is positive. A similar interpretation holds for Ψ_B (by interchanging the roles of A and B). The minimization problem (5) subject to (6) is quite similar to the formulation of Jennison and Turnbull (2000) (p. 328), which is for continuous treatment responses, and Ψ_A and Ψ_B were functions of treatment differences. However, we present this as a function of individual treatment parameters, which is easy to extend to more than two treatments.

Now, based on this Ψ, the optimal allocation is

$$\rho_\Psi = \frac{\sqrt{\Psi_B}\,\sigma_A}{\sqrt{\Psi_B}\,\sigma_A + \sqrt{\Psi_A}\,\sigma_B} = \frac{\sqrt{\Psi_B}\sqrt{p_A q_A}}{\sqrt{\Psi_B}\sqrt{p_A q_A} + \sqrt{\Psi_A}\sqrt{p_B q_B}} = \frac{\frac{\sqrt{p_A q_A}}{\sqrt{\Psi_A}}}{\frac{\sqrt{p_A q_A}}{\sqrt{\Psi_A}} + \frac{\sqrt{p_B q_B}}{\sqrt{\Psi_B}}}.$$

It is necessary to choose an appropriate Ψ_A and Ψ_B. This can be done in several ways. The RPW or DL rule (4) considers

$$\Psi_A = p_A q_A^3. \tag{7}$$

The optimal rule (3) of Rosenberger et al (2001) considers

$$\Psi_A = q_A. \tag{8}$$

Note that the popular urn designs like the RPW and DL are easy to extend to three or more treatments. Unfortunately the above optimal designs are not so easy to extend to more than two treatments, and no optimal design is available in the literature for more than two treatments and binary responses. The present paper attempts to fill this gap.

2 Optimal design for three treatments

Here we extend the optimal designs of Rosenberger et al (2001) for more than two treatments. But, as expected, the level of difficulty will increase remarkably. In fact, the optimal design is obtained in an iterative way. For simplicity, we illustrate our proposed design for three treatments, A, B and C. There is no optimal response-adaptive design available in the literature for more than two treatments satisfying any standard optimality criterion. One can easily extend (3) and (4) in intuitive ways to suggest, e.g., an allocation proportion of

$$\pi_j = \frac{\sqrt{p_j}}{\sqrt{p_A} + \sqrt{p_B} + \sqrt{p_C}}, \tag{9}$$

or

$$\pi_j = \frac{\frac{1}{q_j}}{\frac{1}{q_A} + \frac{1}{q_B} + \frac{1}{q_C}}, \tag{10}$$

for the jth treatment, $j = A, B, C$. In fact, (10) is the limiting proportion of the urn designs like the RPW or DL in a three-treatment scenario. But, we do not know whether the rules (10) and (9) are optimal or not, in some sense.

Suppose we want to minimize

$$n_A \Psi_A + n_B \Psi_B + n_C \Psi_C, \tag{11}$$

subject to

$$l_A \frac{\sigma_A^2}{n_A} + l_B \frac{\sigma_B^2}{n_B} + l_C \frac{\sigma_C^2}{n_C} = l_A \frac{p_A q_A}{n_A} + l_B \frac{p_B q_B}{n_B} + l_C \frac{p_C q_C}{n_C} = K.$$

for some prefixed constants l_A, l_B and l_C, where $n_A + n_B + n_C = n$, and Ψ_k, $k = A, B, C$, is a function of p_A, p_B and p_C such that Ψ_A is decreasing in p_A (for fixed p_B and p_C), and Ψ_A is positive. Similar interpretations hold for Ψ_B and Ψ_C.

Suppose $n_B/n_A = R_B$ and $n_C/n_A = R_C$, and hence

$$\pi_A = \frac{n_A}{n} = \frac{1}{1 + R_B + R_C}, \quad \pi_B = \frac{n_B}{n} = \frac{R_B}{1 + R_B + R_C}, \quad \pi_C = \frac{n_C}{n} = \frac{R_C}{1 + R_B + R_C}.$$

Clearly problem (11) reduces to minimizing (with respect to R_B and R_C)

$$\frac{n}{1 + R_B + R_C} (\Psi_A + R_B \Psi_B + R_C \Psi_C)$$

subject to

$$\frac{1 + R_B + R_C}{n} \left(l_A \sigma_A^2 + l_B \frac{\sigma_B^2}{R_B} + l_C \frac{\sigma_C^2}{R_C} \right) = K.$$

The solution for R_B and R_C can be obtained by differentiating

$$(\Psi_A + R_B \Psi_B + R_C \Psi_C) \left(l_A \sigma_A^2 + l_B \frac{\sigma_B^2}{R_B} + l_C \frac{\sigma_C^2}{R_C} \right),$$

keeping the constraint in mind, which gives

$$R_B = \frac{\sqrt{\Psi_A + R_C \Psi_C} \sqrt{l_B p_B q_B}}{\sqrt{\Psi_B} \sqrt{l_A p_A q_A + \frac{l_C p_C q_C}{R_C}}} = F_1(R_C),$$

$$R_C = \frac{\sqrt{\Psi_A + R_B \Psi_B} \sqrt{l_C p_C q_C}}{\sqrt{\Psi_C} \sqrt{l_A p_A q_A + \frac{l_B p_B q_B}{R_B}}} = F_2(R_B). \tag{12}$$

Note that, when $l_A = l_B = l_C$, the solution of (12) is simply

$$R_B = \sqrt{\frac{p_B q_B}{\Psi_B}} \Big/ \sqrt{\frac{p_A q_A}{\Psi_A}}, \quad R_C = \sqrt{\frac{p_C q_C}{\Psi_C}} \Big/ \sqrt{\frac{p_A q_A}{\Psi_A}},$$

which results in

$$\pi_j = \frac{\sqrt{\frac{p_A q_A}{\Psi_A}}}{\sqrt{\frac{p_A q_A}{\Psi_A}} + \sqrt{\frac{p_B q_B}{\Psi_B}} + \sqrt{\frac{p_C q_C}{\Psi_C}}},$$

for $j = A, B, C$. For $\Psi_j = p_j q_j^3$, we realise the allocation (10), and for $\Psi_j = q_j$ we obtain the allocation (9). Thus, for $l_A = l_B = l_C$, the optimal allocation can be directly extended from the corresponding two-treatment optimal allocation. But, the situation will be different when the l_j's are not the same.

3 Implementation and simulation

The implimentation of this optimal rule for unequal l_j's is as follows.

- The first m patients are treated with equal probability $1/3$ to each treatment. After m responses are available, we have sufficient data to obtain estimates of p_A, p_B and p_C.

- For the allocation of the $(i+1)$st patient, $i \geq m$, we calculate $\widehat{p}_{Ai}$, $\widehat{p}_{Bi}$ and $\widehat{p}_{Ci}$, the estimated proportions of successes to the corresponding treatments up to the first i patients. We treat these as true values at this stage, plug them into (12), and solve for R_B and R_C iteratively.

- To solve for $(R_{B,i+1}, R_{C,i+1})$, the (R_B, R_C)-values for the $(i+1)$st patient, one can take any reasonable value of R_B and R_C (say $R_B^{(0)}$ and $R_C^{(0)}$) as the starting values for the iteration for the $(i+1)$st patient. A reasonable choice may be $R_B^{(0)} = R_{B,i}$ and $R_C^{(0)} = R_{C,i}$, the R_B and R_C values for the ith patient. Let the values of R_B and R_C, after convergence, be $R_{B,i+1}$ and $R_{C,i+1}$. Then, we allocate the $(i+1)$st patient to the three treatments with probabilities $1/(1 + R_{B,i+1} + R_{C,i+1})$, $R_{B,i+1}/(1 + R_{B,i+1} + R_{C,i+1})$ and $R_{C,i+1}/(1 + R_{B,i+1} + R_{C,i+1})$, respectively.

Tables 1-2 give the π_j's for different p_j's (which might be estimates at some stage). Table 1 considers equal l_j-values, where the results of Table 2 are obtained assuming unequal l_js. We consider four designs for comparison; namely: (i) the RPW rule for three treatments, (ii) the Rosenberger et al (2001) optimal allocation for three treatments, (iii) our optimal design with $\Psi_j = p_j q_j^3$, and (iv) our optimal design with $\Psi_j = q_j$. It is observed that the limiting allocations of (i) and (iii) are the same for equal l_js, and those for (ii) and (iv) are also the same for equal l_js. But, when the l_js are different, the limiting allocation of the rules (iii) and (iv) change quite a bit, whereas those of (i) and (ii) do not change.

Keeping (7) and (8) in mind, the possible choices of Ψ_k can be

$$\Psi_k = p_k q_k^3, \; q_k,$$

for $k = A, B, C$. Other suitable choices of Ψ_k will provide other 'optimal' allocations. The convergence of the simultaneous equations (12) can be guaranteed as follows.

Table 1. Limiting allocation proportions for $k = 3$, $l_A = l_B = l_C$, for different values of (p_A, p_B, p_C). Design I: RPW rule for three treatments, Design II: Rosenberger et al.-type design for three treatments, Design III: optimal design with $\Psi_j = p_j q_j^3$, Design IV: optimal design with $\Psi_j = q_j$.

(p_A, p_B, p_C)	(π_A, π_B, π_C)	
	Design I≡Design III	Design II≡Design IV
(.8,.8,.8)	(.333,.333,.333)	(.333,.333,.333)
(.8,.8,.6)	(.400,.400,.200)	(.349,.349,.302)
(.8,.8,.4)	(.429,.429,.142)	(.369,.269,.262)
(.8,.8,.2)	(.444,.444,.111)	(.400,.400,.200)
(.8,.6,.6)	(.500,.250,.250)	(.366,.317,.317)
(.8,.6,.4)	(.545,.273,.182)	(.389,.386,.275)
(.8,.6,.2)	(.571,.286,.143)	(.423,.366,.211)
(.8,.4,.4)	(.600,.200,.200)	(.414,.293,.293)
(.8,.4,.2)	(.632,.210,.158)	(.453,.320,.227)
(.8,.2,.2)	(.666,.167,.167)	(.500,.250,.250)
(.6,.6,.6)	(.333,.333,.333)	(.333,.333,.333)
(.6,.6,.4)	(.375,.375,.250)	(.355,.355,.290)
(.6,.6,.2)	(.400,.400,.200)	(.388,.388,.224)
(.6,.4,.4)	(.428,.286,.286)	(.380,.310,.310)
(.6,.4,.2)	(.462,.308,.230)	(.418,.341,.241)
(.6,.2,.2)	(.500,.250,.250)	(.464,.268,.268)
(.4,.4,.4)	(.333,.333,.333)	(.333,.333,.333)
(.4,.4,.2)	(.364,.364,.272)	(.369,.369,.262)
(.4,.2,.2)	(.400,.300,.300)	(.414,.293,.293)
(.2,.2,.2)	(.400,.300,.300)	(.414,.293,.293)

Note that, here

$$\frac{\partial F_1}{\partial R_B} = 0, \ \frac{\partial F_2}{\partial R_C} = 0,$$

and

$$\frac{\partial F_1}{\partial R_C} = \frac{1}{2} \frac{\sqrt{l_B p_B q_B}\sqrt{\Psi_A + \Psi_C R_C}}{\sqrt{\Psi_B}\sqrt{l_A p_A q_A + \frac{l_C p_C q_C}{R_C}}} \left[\frac{\Psi_C}{\Psi_A + \Psi_C R_C} + \frac{l_C p_C q_C}{R_C^2 \left(l_A p_A q_A + \frac{l_C p_C q_C}{R_C}\right)} \right]$$

$$= \frac{1}{2} F_1(R_C) \times \left[\frac{\Psi_C}{\Psi_A + \Psi_C R_C} + \frac{l_C p_C q_C}{R_C^2 \left(l_A p_A q_A + \frac{l_C p_C q_C}{R_C}\right)} \right] = O(R_C^{-1/2}),$$

$$\frac{\partial F_2}{\partial R_B} = \frac{1}{2} \frac{\sqrt{l_C p_C q_C}\sqrt{\Psi_A + \Psi_B R_B}}{\sqrt{\Psi_C}\sqrt{l_A p_A q_A + \frac{l_B p_B q_B}{R_B}}} \left[\frac{\Psi_B}{\Psi_A + \Psi_B R_B} + \frac{l_B p_B q_B}{R_B^2 \left(l_A p_A q_A + \frac{l_B p_B q_B}{R_B}\right)} \right]$$

$$= \frac{1}{2} F_2(R_B) \times \left[\frac{\Psi_B}{\Psi_A + \Psi_B R_B} + \frac{l_B p_B q_B}{R_B^2 \left(l_A p_A q_A + \frac{l_B p_B q_B}{R_B}\right)} \right] = O(R_B^{-1/2}).$$

Table 2. Limiting allocation proportions for $k = 3$, $l_A = 1$, $l_B = 0.5$, $l_C = 0.25$, for different values of (p_A, p_B, p_C). Design I: RPW rule for three treatments (same as Table 1), Design II: Rosenberger et al.-type design for three treatments (same as Table 1), Design III: optimal design with $\Psi_j = p_j q_j^3$, Design IV: optimal design with $\Psi_j = q_j$.

(p_A, p_B, p_C)	(π_A, π_B, π_C)	
	Design III	Design IV
(.8,.8,.8)	(.453,.320,.227)	(.453,.320,.227)
(.8,.8,.6)	(.511,.361,.128)	(.467,.330,.203)
(.8,.8,.4)	(.534,.377,.089)	(.485,.343,.172)
(.8,.8,.2)	(.546,.386,.068)	(.511,.361,.128)
(.8,.6,.6)	(.624,.220,.156)	(.489,.299,.212)
(.8,.6,.4)	(.658,.233,.109)	(.509,.311,.180)
(.8,.6,.2)	(.676,.239,.085)	(.537,.329,.134)
(.8,.4,.4)	(.713,.168,.119)	(.540,.270,.190)
(.8,.4,.2)	(.735,.173,.092)	(.571,.286,.143)
(.8,.2,.2)	(.768,.136,.096)	(.624,.220,.156)
(.6,.6,.6)	(.453,.320,.227)	(.453,.320,.227)
(.6,.6,.4)	(.490,.347,.163)	(.473,.334,.193)
(.6,.6,.2)	(.511,.361,.128)	(.501,.354,.145)
(.6,.4,.4)	(.554,.261,.185)	(.504,.291,.205)
(.6,.4,.2)	(.581,.274,.145)	(.536,.309,.155)
(.6,.2,.2)	(.624,.220,.156)	(.589,.241,.170)
(.4,.4,.4)	(.453,.320,.227)	(.453,.320,.227)
(.4,.4,.2)	(.480,.340,.180)	(.485,.343,.172)
(.4,.2,.2)	(.525,.278,.197)	(.540,.270,.190)
(.2,.2,.2)	(.453,.320,.227)	(.453,.320,.227)

One can choose the initial value $(R_B^{(0)}, R_C^{(0)})$ large enough such that

$$\left| \left(\frac{\partial F_1}{\partial R_C} \right)_{R_B^{(0)}, R_C^{(0)}} \right| < 1, \quad \left| \left(\frac{\partial F_2}{\partial R_B} \right)_{R_B^{(0)}, R_C^{(0)}} \right| < 1,$$

and consequently the convergence of the simultaneous iteration procedure is guaranteed (see Scarborough (1966) , Ch. XII, p. 301).

4 Conclusions

The present paper describes optimal response-adaptive designs for three-treatments. A detailed research on such designs for general k (≥ 2) treatments with one constraint and with the issue of power, expected failure, etc., and illustration with some real data is under investigation. We overlook these issues in this short article for the sake of brevity.

We have the following generalizations in mind. First, one may think of finding optimal designs with more than one constraint. Then, optimal designs in the presence of covariates is of interest. We hope to pursue some of these issues in future communications.

Acknowledgement. The authors wish to thank two anonymous referees for their careful reading and helpful suggestions. S. Mandal is supported by a research grant from the Natural Sciences and Engineering Research Council of Canada.

References

Bartlett R, Roloff D, Cornell R, Andrews A, Dillon P, Zwischenberger J (1985) Extracorporeal circulation in neonatal respiratory failure: a prospective randomized trial. Pediatrics 76:479–487

Biswas A, Dewanji A (2004) Inference for a rpw-type clinical trial with repeated monitoring for the treatment of rheumatoid arthritis. Biometrical Journal 46:769–779

Durham S, Flournoy N, Li W (1998) A sequential design for maximizing the probability of a favorable response. Canadian Journal of Statistics 26:479–495

Hayre L (1979) Two-population sequential tests with three hypotheses. Biometrika 66:465–474

Ivanova A (2003) A play-the-winner type urn model with reduced variability. Metrika 58:1–13

Ivanova A, Rosenberger W, Durham S, Flournoy N (2000) A birth and death urn for randomized clinical trials: asymptotic methods. Sankhya B 62:104–118

Jennison C, Turnbull B (2000) Group Sequential Methods with Applications to Clinical Trials. Chapman and Hall/CRC, Boca Raton, Florida

Rosenberger W, Lachin J (2002) Randomization in Clinical Trials: Theory and Practice. Wiley, New York

Rosenberger W, Stallard N, Ivanova A, Harper C, Ricks M (2001) Optimal adaptive designs for binary response trials. Biometrics 57:173–177

Rout C, Rocke D, Levin J, Gouw's E, Reddy D (1993) A reevaluation of the role of crystalloid preload in the prevention of hypotension associated with spinal anesthesia for elective cesarean section. Anesthesiology 79:262–269

Scarborough J (1966) Numerical Mathematical Analysis, Sixth edition. Oxford and IBH Publishing Co. Pvt. Ltd., New Delhi

Tamura R, Faries D, Andersen J, Heiligenstein J (1994) A case study of an adaptive clinical trial in the treatment of out-patients with depressive disorder. Journal of the American Statistical Association 89:768–776

Wei L (1979) The generalized polya's urn design for sequential medical trials. Annals of Statistics 7:291–296

Wei L, Durham S (1978) The randomized play-the-winner rule in medical trials. Journal of the American Statistical Association 73:840–843

Zelen M (1969) Play the winner rule and the controlled clinical trial. Journal of the American Statistical Association 64:131–146

One-Half Fractions of a 2^3 Experiment for the Logistic Model

Roberto Dorta-Guerra[1], Enrique González-Dávila[1], and Josep Ginebra[2]

[1] Departamento de Estadística, Investigación Operativa y Computación, Universidad de La Laguna, c/Astrofísico Francisco Sánchez, 38271 La Laguna, Tenerife, Spain. `rodorta@ull.es`, `egonzale@ull.es`

[2] Departament d'Estadística i Investigació Operativa, Universitat Politècnica de Catalunya, Avgda. Diagonal 647, 08028 Barcelona, Spain. `josep.ginebra@upc.edu`

Summary. D-optimal experiments for binary response data have been extensively studied in recent years. On the other hand two-level fractional factorials are often used as screening designs at the preliminary stage of an investigation when the outcome is continuous. We explore the performance of the one-half two-level experiments for a logistic model with three factors, and show that the conventional wisdom about this kind of experiment does not apply when the response is binomial.

Key words: binary data, local D-optimality, two-level designs, one-half two-level designs

1 Introduction

When the response of interest can be conveniently modeled through first order normal linear regression models, two-level factorial experiments are either optimal or close to optimal among all experiments with the same sample size, for a broad class of experimental regions and for most sensible optimal design criteria, including the determinant of the information matrix. (e.g. see Pukelsheim (1973)). Furthermore, when 'fractioning' two-level factorial experiments, "spatially" balanced allocations treating all factors symmetrically by assigning the same number of experimental combinations to each of their levels and the same number of runs on each experimental combination, leading to orthogonal design matrices, are always at least as good as allocations leading to non-orthogonal design matrices with the same number of runs. Also, it is well known that under normal response models, two complementary fractions of a full two-level factorial are statistically equivalent. None of this holds for binary response models.

The design of experiments literature most often uses optimal design criteria based on the Fisher information matrix, $I(\beta)$. In particular, the D-optimal criteria of maximizing the determinant of $I(\beta)$ is the most widely studied one. Under binary response models, the Fisher information matrix depends on the parameters of the model, which much complicates a lot the construction of an optimal design. To get round this difficulty the local optimality approach assumes that the value of the parameters are known. The D-optimal designs obtained in this way are called local D-optimal designs.

Many authors have considered the construction of local D-optimal designs for logistic models, which are the usual models for binary response data. For example, Abdelbasit and Plackett (1983) discuss the construction of local D-optimal designs for a binary response with one explanatory variable, Sitter and Torsney (1995) deal with the case of two design variables, and Ford et al (1992) reduce the optimal design of experiments for generalised linear models, including binary response models, to a canonical form. The results in these, and in many other papers indicate that local D-optimal designs are not two-level experiments.

In this paper we study the performance of four point designs that fall within the class of two-level experiments with three factors, all under the logistic model for binary responses. In particular, we search for the experiments centered at a given point and maximizing the determinant of $I(\beta)$, assuming no restriction on the experimental region. We provide a set of easy-to-use tables that determine the best fractional experiment for each case, and we show how the standard 2^{3-1} fractions used for normal response models are not the best option for logistic models.

2 Logistic models

Often, quality improvement experiments on manufacturing processes asses quality through pass or fail inspection of the articles produced. For example, Bisgaard and Fuller (1995) consider the case of a grinding process, where one is interested in the effect of blade size, centering, leveling and speed on the presence or absence of undesirable marks on steel samples.

In binary response experiments with three design variables or factors, n_i articles are tested at levels $x_i = (x_{1i}, x_{2i}, x_{3i})$ of the design variables for $i = 1, \ldots, q$, (or n_i subjects are administered dose levels $x_i = (x_{1i}, x_{2i}, x_{3i})$, in dose response experiments), and the outcome is binary. Usually the total number of subjects, $n = \sum_{i=1}^{q} n_i$, is specified from the start and one assumes that the number of successes on the n_i subjects under x_i, y_i, are conditionally independent binomial random variables, $y_i | x_i, \beta \sim \text{Binomial}(n_i, p(x_i; \beta))$.

Under the main effects logistic regression model for binary responses, one assumes that

$$p(x_i; \beta) = \frac{1}{1 + exp(\beta_0 + \beta_1 x_{1i} + \beta_2 x_{2i} + \beta_3 x_{3i})} = F(z_i)$$

where $z_i = \beta_0 + \beta_1 x_{1i} + \beta_2 x_{2i} + \beta_3 x_{3i}$.

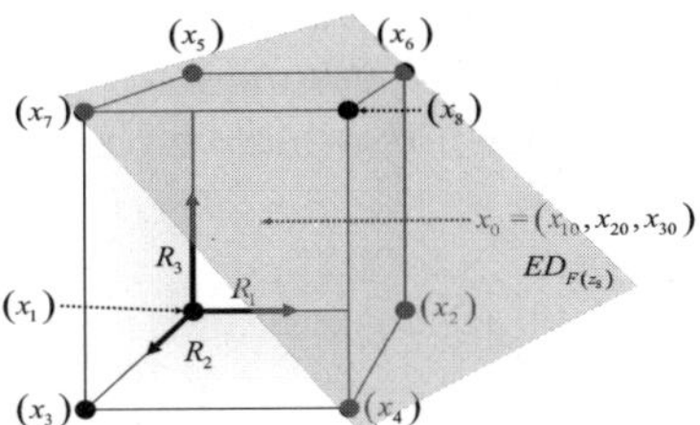

Fig. 1. Two-level three-factor experiment. The plane denoted by $ED_{F(z_8)}$ is the effective doses which contains point x_8. All the exprimental points are supported on planes parallel to $ED_{F(z_8)}$.

The contour levels of the surface, $p(x; \beta)$, are planes, and therefore the set of design points $x_i = (x_{1i}, x_{2i}, x_{3i})$ such that $p(x_i; \beta) = F(z)$ for a given z, which define the equal dose set $ED_{F(z)}$, are also planes. The Fisher information matrix for the experiment supported on $x_1, \ldots, x_q$ in R^3, is

$$I(\beta) = n \sum_{i}^{q} \lambda_i h(z_i)(1, x_{1i}, x_{2i}, x_{3i})^{'}(1, x_{1i}, x_{2i}, x_{3i}) \qquad (1)$$

where $\lambda_i = n_i/n$ are the weights of the support points and where $h(z_i) = F'(z_i)^2/(F(z_i)(1 - F(z_i))) = e^{z_i}/(1 + e^{z_i})^2$ (see Sitter and Torsney (1995)).

3 Local D-optimal one-half fraction of 2^3 experiments

In this paper, we restrict attention to one-half fractions of 2^3 experiments, which are experiments supported on 4 of the 2^3 vertices of a design as illustrated in Fig. 1. That is, we consider experiments supported on points x_i of the form $x_i = (x_{10} + a_{1i}R_1, x_{20} + a_{2i}R_2, x_{30} + a_{3i}R_3)$, where the a_{ji}'s are either -1 or $+1$, R_1, R_2, R_3 are the half ranges, $x_0 = (x_{10}, x_{20}, x_{30})$ is the center point of the 2^3 experiment which is on the $ED_{F(z_0)}$ plane with $z_0 = \beta_0 + \beta_1 x_{10} + \beta_2 x_{20} + \beta_3 x_{30}$ and $z_i = z_0 + a_{1i}\beta_1 R_1 + a_{2i}\beta_2 R_2 + a_{3i}\beta_3 R_3$.

Let A be the 8×4 standardized design matrix, whose i-th row is the vector $(1|a_i) = (1, a_{1i}, a_{2i}, a_{3i})$, that is

$$A = \begin{pmatrix} 1 & -1 & -1 & -1 \\ 1 & +1 & -1 & -1 \\ 1 & -1 & +1 & -1 \\ 1 & +1 & +1 & -1 \\ 1 & -1 & -1 & +1 \\ 1 & +1 & -1 & +1 \\ 1 & -1 & +1 & +1 \\ 1 & +1 & +1 & +1 \end{pmatrix} \begin{matrix} (1) \\ (2) \\ (3) \\ (4) \\ (5) \\ (6) \\ (7) \\ (8) \end{matrix}$$

and let A_{i_1,i_2,i_3,i_4} be the 4×4 submatrix of A formed by its rows i_1, i_2, i_3, i_4.

Dorta-Guerra et al (2005) evaluated the determinant of $I(\beta)$ under a generalized linear model for a two-level experiment. From this result it is easy to see that the determinant of $I(\beta)$ under the main effects logistic model described above is given by the following expression:

$$(\beta_1\beta_2\beta_3)^2 det(I(\beta))/n^4 = t_1^2 t_2^2 t_3^2 \sum_{\{i_1,i_2,i_3,i_4\}} c_{i_1,i_2,i_3,i_4} \lambda_{i_1} \lambda_{i_2} \lambda_{i_3} \lambda_{i_4} h(z_{i_1})h(z_{i_2})h(z_{i_3})h(z_{i_4}) \tag{2}$$

where $t_i = \beta_i R_i$, $z_i = z_0 + a_{1i}t_1 + a_{2i}t_2 + a_{3i}t_3$ and the summation is over all 70 possible combinations of (i_1,i_2,i_3,i_4) obtained as subsets of $\{1,2,3,4,5,6,7,8\}$ of size 4, and where $c_{1,4,6,7} = c_{2,3,5,8} = 4^4$, $c_{1,2,3,4} = c_{1,2,5,6} = c_{1,2,7,8} = c_{1,3,5,7} = c_{1,3,6,8} = c_{1,4,5,8} = c_{2,3,6,7} = c_{2,4,5,7} = c_{2,4,6,8} = c_{3,4,5,6} = c_{3,4,7,8} = c_{5,6,7,8} = 0$, and $c_{i_1,i_2,i_3,i_4} = 4^3$ for all other 56 terms. Finally $c_{i_1,i_2,i_3,i_4} = (det(A_{i_1,i_2,i_3,i_4}))^2$.

Observe that (2) is a function of z_0, (t_1,t_2,t_3) and $(\lambda_1,\ldots,\lambda_8)$. Notice that all two-level experiments centered on a point on the $ED_{F(z_0)}$ with the same t_i's and λ_j's have the same value of $(\beta_1\beta_2\beta_3)^2 det(I(\beta))/n^4$. Once we have the optimal t_i's we can also determine the optimal z_i's and automatically the $p(x_i; \beta)$'s through the $F(z_i)$'s. In the optimization process for fixed $F(z_0)$ we compute the optimal t_i's and λ_j's which maximize (2) using the Quasi-Newton method. It is worth noting that optimal values for t_i do not depend on the parameters of the model and therefore, neither do the $p(x_i; \beta)$'s depend on such parameters. Moreover, to calculate the experimental points, it is necessary to get an initial estimation of the parameters to obtain the half ranges $R_i = t_i/\beta_i$.

Our goal is to characterize the seventy possible one-half fractions of a 2^3 experiment. In fact, it turns that one-half fraction is the local D-optimal experiment among the class of all two-level experiments, including full two-level factorial experiments (see Dorta-Guerra et al (2005)). Taking into account that the model has four parameters and that this experiment has four points, the optimal λ_j's are equal to 1/4. In the next section we show that we can construct five classes of experiments taking into account the determinant of their information matrix. For example, when $\beta_1 > 0$, $\beta_2 > 0$ and $\beta_3 > 0$, the local D-optimal one-half fraction when it is centered above the 50 percentile is the one supported on the points x_2, x_3, x_4 and x_5. This fraction is denoted by (2345), and its corresponding standardized sub-matrix and z_i's are

$$A_{2,3,4,5} = \begin{pmatrix} 1 & +1 & -1 & -1 \\ 1 & -1 & +1 & -1 \\ 1 & +1 & +1 & -1 \\ 1 & -1 & -1 & +1 \end{pmatrix} \begin{aligned} &\Rightarrow z_2 = z_0 + \beta_1 R_1 - \beta_2 R_2 - \beta_3 R_3 \\ &\Rightarrow z_3 = z_0 - \beta_1 R_1 + \beta_2 R_2 - \beta_3 R_3 \\ &\Rightarrow z_4 = z_0 + \beta_1 R_1 + \beta_2 R_2 - \beta_3 R_3 \\ &\Rightarrow z_5 = z_0 - \beta_1 R_1 - \beta_2 R_2 + \beta_3 R_3 \end{aligned}$$

On the other hand, if we change the sign of the first parameter, that is, $\beta_1 < 0$, $\beta_2 > 0$ and $\beta_3 > 0$, the local D-optimal one-half fraction is the one supported at x_1, x_3, x_4 and x_6, which is labeled as (1346) and has as a standardized sub-matrix and z_i's

As can be seen, the z values are equal and hence the determinant for these two experiments is the same. It is easy to verify that $A_{1,3,4,6}$ can be obtained

$$A_{1,3,4,6} = \begin{pmatrix} 1 & -1 & -1 & -1 \\ 1 & -1 & +1 & -1 \\ 1 & +1 & +1 & -1 \\ 1 & +1 & -1 & +1 \end{pmatrix} \begin{matrix} \Rightarrow z_1 = z_0 - (-|\beta_1|)R_1 - \beta_2 R_2 - \beta_3 R_3 \\ \Rightarrow z_3 = z_0 - (-|\beta_1|)R_1 + \beta_2 R_2 - \beta_3 R_3 \\ \Rightarrow z_4 = z_0 + (-|\beta_1|)R_1 + \beta_2 R_2 - \beta_3 R_3 \\ \Rightarrow z_6 = z_0 + (-|\beta_1|)R_1 - \beta_2 R_2 + \beta_3 R_3 \end{matrix}$$

by multiplying the second column of $A_{2,3,4,5}$ by (-1), after reordering the rows. If a change of sign takes place for β_2 instead of β_1, we would have to multiply the third column of $A_{2,3,4,5}$ by (-1) instead of the second column, and so on. This implies that one can always reduce the complexity of the problem, by considering without loss of generality that the parameters of the model are all positive to start with. It should be clear from the above argument that the fraction (2345), depicted in Fig. 2 with label $\beta_1 > 0$, and the fraction (1346), depicted in Fig. 2 with label $\beta_1 < 0$, locate their four optimal support points on the same ED_F planes, and both lead to the same determinant of $I(\beta)$.

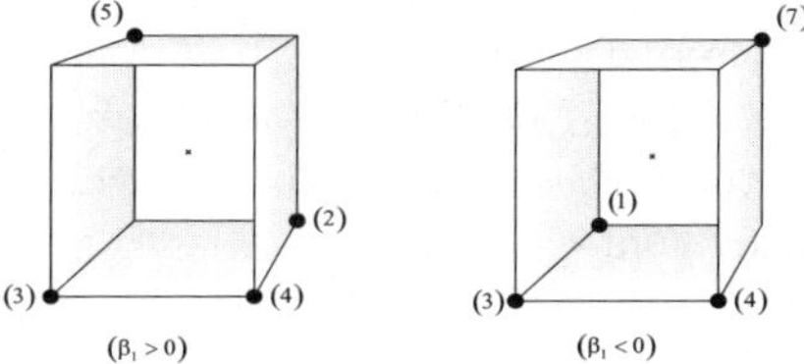

Fig. 2. Local D-optimal one half-fraction for the main effects three factor logistic model when $\beta_2 > 0$, $\beta_3 > 0$ and where the sign of β_1 is either positive or negative.

In addition, if columns 2 and 4 of the matrix $A_{2,3,4,5}$ are switched we obtain the $A_{2,3,5,7}$ matrix (sorting the indices in ascending order). Note that the optimal z's are equal for both experiments, so we have two experiments. So the same determinant, that is, the experiments are equivalent. Consequently, when β_1, β_2 and β_3 are positive we can use either experiment (2345) or (2357). Therefore, for given parameters, we can ensure that reordering the last three columns of a standardized sub-matrix we obtain all the equivalent experiments.

4 Classes of one-half fractions of 2^3 experiments

The seventy possible one-half fractions of a 2^3 experiment can be divided into five different classes according to their geometric configuration. In this section, we compute the local D-optimal experiment in each one of these five classes of one-half fractions. We also present tables which allow one to obtain the D-optimal configurations under all possible combinations of signs for β_1, β_2 and β_3, depending on whether the ED_F of the center point of the fraction is larger or smaller than 50%. The determinant of the D-optimal experiment in each of the five classes of one-half fractions is also given as a function of the $ED_{F(z_0)}$, where appropriate.

4.1 Class (a)

This class corresponds to the standard 2^{3-1} resolution III fraction of a full factorial for linear normal response models, which is the D-optimal fraction for normal models, and has the only geometric configuration which leads to an orthogonal standardized design matrix; its designs have two support points on each of the faces of the cube. This class is formed only by the fractions (1467) and (2358), and either one or the other of these two fractions will be local D-optimal in this class, depending on the signs of β_1, β_2 and β_3, and depending on whether $ED_{F(z_0)}$ is larger or smaller than 50%, in the way described in Table 1. For example, if we are interested in an experiment centered above 50% and all the parameters are positive the local D-optimal one-half experiment is formed by points (2358). In contrast, if the experiment is centered under 50% we should use the (1467) design. Notice that, contrary to the case of normal models, under the logistic model complementary fractions like (1467) and (2358) are not equivalent.

$ED_{F_{(z_0)}}$	$\beta_3 > 0$				$\beta_3 < 0$			
	$\beta_1 > 0$ $\beta_2 > 0$	$\beta_1 < 0$ $\beta_2 > 0$	$\beta_1 > 0$ $\beta_2 < 0$	$\beta_1 < 0$ $\beta_2 < 0$	$\beta_1 > 0$ $\beta_2 > 0$	$\beta_1 < 0$ $\beta_2 > 0$	$\beta_1 > 0$ $\beta_2 < 0$	$\beta_1 < 0$ $\beta_2 < 0$
$> 50\%$	(2358)	(1467)	(1467)	(2358)	(1467)	(2358)	(2358)	(1467)
$< 50\%$	(1467)	(2358)	(2358)	(1467)	(2358)	(1467)	(1467)	(2358)

	$ED_{F(z_0)}$	.10	.20	.30	.40	.50	.60	.70	.80	.90	.95
(1467)	$10^6 n^{-4}(\beta_1\beta_2\beta_3)^2 Det(I(\beta))$	271	533	614	565	447	307	175	73	15	3
(2358)	$10^6 n^{-4}(\beta_1\beta_2\beta_3)^2 Det(I(\beta))$	15	73	175	307	447	565	614	533	271	96

Table 1. Local D-optimal designs among all the fractions included in class (a).

4.2 Class (b)

This class includes all one-half fractions in which three of the four support points are on one of the faces of the cube, and the other one is on the opposite face of the cube, at the vertex connected to the vertex on the opposite face which does not have any support point. This class always includes the local D-optimal experiment among all one-half fractions. Therefore, in particular, it is obvious that the largest determinant attained through any fraction of in class (a) (which is the default one-half fraction under normal models) will always be smaller than the determinant of the best configuration included in class (b). In fact, we find the largest determinant in class (a) to be 78% smaller than in class (b).

$ED_{F(z_0)}$	$\beta_3 > 0$				$\beta_3 < 0$			
	$\beta_1 > 0$ $\beta_2 > 0$	$\beta_1 < 0$ $\beta_2 > 0$	$\beta_1 > 0$ $\beta_2 < 0$	$\beta_1 < 0$ $\beta_2 < 0$	$\beta_1 > 0$ $\beta_2 > 0$	$\beta_1 < 0$ $\beta_2 > 0$	$\beta_1 > 0$ $\beta_2 < 0$	$\beta_1 < 0$ $\beta_2 < 0$
$> 50\%$	(2345) (2357) (2356)	(1346) (1468) (1456)	(1247) (1457) (1478)	(1238) (2368) (2378)	(1678) (1367) (1267)	(2578) (2458) (1258)	(3568) (1358) (3458)	(4567) (2467) (3467)
$< 50\%$	(4567) (2467) (3467)	(3568) (1358) (3458)	(2578) (2458) (1258)	(1678) (1367) (1267)	(1238) (2368) (2378)	(1247) (1457) (1478)	(1346) (1468) (1456)	(2345) (2357) (2356)

	$ED_{F(z_0)}$	.05	.10	.20	.30	.40	.50	.60	.70	.80	.90
(2345)	$10^6 n^{-4}(\beta_1\beta_2\beta_3)^2 Det(I(\beta))$	1	12	123	489	1329	2909	5511	9345	14268	18495
(4567)	$10^6 n^{-4}(\beta_1\beta_2\beta_3)^2 Det(I(\beta))$	17332	18495	14268	9345	5511	2909	1329	489	123	12

Table 2. Local D-optimal designs among all the fractions included in class (b).

4.3 Class (c)

This class includes all the experiments with three support points on one face of the cube, with the fourth support point on the opposite face, at the vertex opposite to the vertex which corresponds to the support point 'between' the first three. The local D-optimal configurations among all those in this class can be constructed from the information in Table 3 for any combination of signs for β_i and position of the center point x_0 of the experiment.

$ED_{F(z_0)}$	$\beta_3 > 0$				$\beta_3 < 0$			
	$\beta_1 > 0$ $\beta_2 > 0$	$\beta_1 < 0$ $\beta_2 > 0$	$\beta_1 > 0$ $\beta_2 < 0$	$\beta_1 < 0$ $\beta_2 < 0$	$\beta_1 > 0$ $\beta_2 > 0$	$\beta_1 < 0$ $\beta_2 > 0$	$\beta_1 > 0$ $\beta_2 < 0$	$\beta_1 < 0$ $\beta_2 < 0$
$> 50\%$	(1235)	(1246)	(1347)	(2348)	(1567)	(2568)	(3578)	(4678)
$< 50\%$	(4678)	(3578)	(2568)	(1567)	(2348)	(1347)	(1246)	(1235)

	$ED_{F(z_0)}$	.05	.10	.20	.30	.40	.50	.60	.70	.80	.90
(1235)	$10^6 n^{-4}(\beta_1\beta_2\beta_3)^2 Det(I(\beta))$	0	0	1	9	35	112	310	804	2047	5329
(4678)	$10^6 n^{-4}(\beta_1\beta_2\beta_3)^2 Det(I(\beta))$	8496	5329	2047	804	310	112	35	9	1	0

Table 3. Local D-optimal designs among all the fractions included in class (c).

4.4 Class (d)

This class includes all the experiments with three support points on one of the faces of the cube, with the fourth support point located on the opposite face of the cube, opposite either of the first three design points which is not 'between' the other two. Contrary to the previous three classes of one-half fractions, in this class the determinant of the corresponding local D-optimal configuration does not depend on whether $ED_{F(z_0)}$ is larger or smaller than 50%.

4.5 Class (e)

This class is formed by those designs with four support points all falling on the same plane, and thus by the one-half fractions which do not allow estimation of

$\beta_3 > 0$				$\beta_3 < 0$			
$\beta_1 > 0$ $\beta_2 > 0$	$\beta_1 < 0$ $\beta_2 > 0$	$\beta_1 > 0$ $\beta_2 < 0$	$\beta_1 < 0$ $\beta_2 < 0$	$\beta_1 > 0$ $\beta_2 > 0$	$\beta_1 < 0$ $\beta_2 > 0$	$\beta_1 > 0$ $\beta_2 < 0$	$\beta_1 < 0$ $\beta_2 < 0$
(2346)	(1345)	(1248)	(1237)	(2678)	(1578)	(4568)	(3567)
(2347)	(1348)	(1245)	(1236)	(3678)	(4578)	(1568)	(2567)
(3567)	(4568)	(1578)	(2678)	(1237)	(1248)	(1345)	(2346)
(2456)	(1356)	(2478)	(1378)	(1268)	(1257)	(3468)	(3457)
(2567)	(1568)	(4578)	(3678)	(1236)	(1245)	(1348)	(2347)
(3457)	(3468)	(1257)	(1268)	(1378)	(2478)	(1356)	(2456)

	$ED_{F(z_0)}$	.05	.10	.20	.30	.40	.50	.60	.70	.80	.90
(2346)	$10^6 n^{-4}(\beta_1\beta_2\beta_3)^2 Det(I(\beta))$	196	541	1375	2163	2713	2909	2713	2163	1375	541

Table 4. Local D-optimal designs among all the fractions included in class (d).

one of the three parameters β_i. These fractions lead to a singular information matrix with a zero determinant. There are two sets of experiments with these characteristics; those with four points in the same face of the 2^3 experiment $\{(1234), (1357), (1256), (2468), (3478), (5678)\}$, and those with points on the diagonal planes of the 2^3 experiment $\{(1278), (1368), (1458), (2457), (2367), (3456)\}$.

5 Conclusions

In this paper we study the amount of information, measured through the determinant of $I(\beta)$, in one-half fractions of 2^3 experiments for the logistic model. The default 2^{3-1} fractions for normal response data are not the best ones for binary response data. We also catalogue all the possible two-level four point design configurations, and find the best experiments in each one of the 5 possible classes.

Acknowledgement. This work was partially supported by the Spanish MEC Proyecto MTM2006-09920.

References

Abdelbasit K, Plackett R (1983) Experimental design for binary data. J of the American Statistical Association 78:90–98

Bisgaard S, Fuller H (1995) Sample size estimates for 2^{k-p} designs with binary responses. J of Quality Technology 27:334–354

Dorta-Guerra R, González-Dávila E, Ginebra J (2005) Two-level experiments for binary response data. Universidad Politécnica de Cataluña, Barcelona

Ford I, Torsney B, Wu C (1992) The use of canonical form in the construction of locally optimal designs for non-linear problems. J of the Royal Statistical Society 54:569–583

Pukelsheim F (1973) Optimal Design of Experiments. Wiley, New York

Sitter R, Torsney B (1995) Optimal designs for binary response experiments with two design variables. Stat Sinica 5:405–419

Bayes Estimators of Covariance Parameters and the Influence of Designs

Younis Fathy[1] and Christine Müller[2]

[1] Fakultät V, Institut für Mathematik, Carl von Ossietzky University Oldenburg, Postfach 2503, 26111 Oldenburg, Germany.
`younis.fathy@mail.uni-oldenburg.de`
[2] Department of Mathematics and Computer Science, University of Kassel, 34109 Kassel, Germany. `cmueller@mathematik.uni-kassel.de`

Summary. It is assumed that the covariance matrix of N observations has the form $C_\theta = \sum_{r=1}^{R} \theta_r \, U_r$ where $U_1, \ldots, U_R$ are known covariance matrices and $\theta_1, \ldots, \theta_R$ are unknown parameters. Estimators for $\sum_{r=1}^{R} \theta_r \, b_r$ with known $b_1, \ldots, b_R$ are characterized which minimize the Bayes risk within all invariant quadratic unbiased estimators. In this characterization, the matrix A, which determines the quadratic form of the estimator, is given by a linear equation system which is not of full rank. It is shown that some solutions of the equation system prove to be asymmetric matrices A. Therefore, sufficient conditions are presented which ensures symmetry of the matrix A. Given this result, the influence of designs on the Bayes risk is studied.

Key words: Bayes invariant quadratic unbiased estimator, quadratic form, time dependence, spatial covariance, one and two dimensional designs

1 Introduction

We assume that an observation $z(x) \in \Re$ at the experimental condition $x \in \Re^q$ is given by

$$z(x) = f(x)^\top \beta + e(x),$$

where $f : \Re^k \longrightarrow \Re^p$ is a known regression function, $\beta \in \Re^p$ an unknown parameter vector and $e(x)$ is measurement error with expectation equal to 0. Several observations $z_1(x_1), \ldots, z_N(x_N)$ at $x_1, \ldots, x_N$ are given by the vector $Z = (z_1(x_1), \ldots, z_N(x_N))^\top$ and the design matrix $F = (f(x_1), \ldots, f(x_N))^\top$ so that we have the linear model $Z = F\beta + E$.

In many linear models it is assumed that the covariance matrix of the error vector $E = (e(x_1), \ldots, e(x_N))^\top$ is the identity matrix times an unknown variance parameter. This means in particular that normally distributed observations $z_1(x_1), \ldots, z_N(x_N)$ are stochastically independent. But if the experimental conditions are time points or points in space it is more likely that points which are closer together provide similar observations than points which are further apart. Then the errors $e(x_1), \ldots, e(x_N)$ and the observations $z_1(x_1), \ldots, z_N(x_N)$ are no longer uncorrelated. In general it is possible for E to have an arbitrary covariance matrix.

Since a general covariance matrix will have more parameters than available observations, some assumptions are needed for the covariance structure of E. Here we will consider the case that the covariance matrix is given by a linear combination of known covariance matrices $U_1, \ldots, U_R$ so that

$$C_\theta = \sum_{r=1}^{R} \theta_r \, U_r,$$

where only $\theta = (\theta_1, \ldots, \theta_R)^\top$ is unknown. Hence we consider a mixed linear model with variance components $\theta_1, \ldots, \theta_R$. In many cases one matrix U_r will be the identity matrix and another matrix will have components $U_r(x_i, x_j)$ which are a decreasing function of the distance between x_i and x_j. Then the identity matrix represents the effect which is known as the nugget effect in spatial statistics (see e.g. Cressie (1993)).

There are several possibilities for estimating the variance components $\theta_1, \ldots, \theta_R$ (see e.g. Rao and Kleffe (1988) or Koch (1999)). If a linear combination $\alpha = \sum_{r=1}^{R} b_r \theta_r$ with given $b = (b_1, \ldots, b_R)^\top$ is to be estimated, then this can be done by an estimator given by the quadratic from $\hat{\alpha}_0(Z) = Z^\top Q Z$. Such estimators should be invariant with respect to linear transformations of the linear model, i.e. they should satisfy $\hat{\alpha}_0(Z + F\beta) = \hat{\alpha}_0(Z)$ for any $\beta \in \Re^p$. This invariance property is in particular satisfied if the estimator has the form $\hat{\alpha}_0(Z) = Z^\top M A M Z$ where $M = I - F(F^\top F)^- F^\top$ is the projection matrix. Then one can also work with observation vectors $Y = M Z$ with expectation 0 and covariance matrix

$$\tilde{C}_\theta = M C_\theta M^\top = \sum_{r=1}^{R} \theta_r \, M U_r \, M^\top = \sum_{r=1}^{R} \theta_r \, V_r.$$

The aim is then to determine the matrix A in the quadratic form $\hat{\alpha}(Y) = Y^\top A Y$. Assuming an a-priori distribution for the unknown parameters $\theta_1, \ldots, \theta_R$, in Section 2 we present a characterisation of the matrix A of the estimator which minimizes the Bayes risk within all invariant quadratic unbiased estimators. It turns out that the matrix A is given by an equation $G O = W$ where the vector O contains the matrix A in vectorized form and the matrix G and the vector W depend on $V_1, \ldots, V_R$ and b, respectively. Thereby the form of A depends very much on which g-inverse of G is used to solve the equation. It is shown that sometimes a solution is obtained such that the matrix A is not symmetric. It is desirable that A be symmetric since quadratic forms based on a symmetric matrix have good inference properties, their distributions usually being of the χ^2 family. One possible solution is to symmetrise

A by replacing it by $(A + A^\top)/2$. Another possibility is to find symmetric solutions A satisfying the equation $GO = W$. Section 3 provides a condition which ensures this. Having a method by which to calculate a reasonable estimate $\hat{\alpha}(Y) = Y^\top A Y$, we study in Section 4 the influence of the design on the Bayes risk of this estimator by examples.

2 Bayes invariant quadratic unbiased estimators (BAIQUE)

Unbiased quadratic estimators $\hat{\alpha}(Y) = Y^\top A Y$ for $\alpha = \sum_{r=1}^{R} b_r \theta_r = b^\top \theta$ are characterized by Rao (1973) as follows.

Lemma 1. *The quadratic estimator $Y^\top A Y$ is an unbiased estimator for the linear function $\alpha = b^\top \theta$ if and only if*

$$tr AV_r = b_r \quad for \quad r = 1, \ldots, R. \tag{1}$$

Let p_Θ be a prior distribution for the vector of parameters $(\theta_1, \theta_2, \ldots, \theta_R)$ which has second order moments of the form

$$E(\Theta_i \Theta_j) = \int \theta_i \theta_j \, p_\Theta(\theta) d\theta = C_{ij} \geq 0; \quad i, j = 1, \ldots, R.$$

We assume here that the matrix of these second order moments is positive semidefinite so that we can write

$$\mathcal{C} = (C_{ij})_{i,j=1,\ldots,R} = S\, S^\top = (\sum_{r=1}^{R} s_{ir} s_{jr})_{i,j=1,\ldots,R}. \tag{2}$$

The Bayes risk of the estimator $\hat{\alpha}(Y) = Y^\top A Y$ is given by

$$r(\hat{\alpha}) = E(E_\Theta(\hat{\alpha}(Y) - \alpha)^2) = \int E_\theta(\hat{\alpha}(Y) - \alpha)^2 p_\Theta(\theta) d\theta.$$

Definition 1. *(Gnot and Kleffe (1983)) A quadratic form $\hat{\alpha}(Y) = Y^\top A Y$ is called a Bayes invariant quadratic unbiased estimator (BAIQUE) if it minimizes $E(E_\Theta(\hat{\alpha}(Y) - \alpha)^2)$ subject to all invariant unbiased quadratic estimators.*

Theorem 1. *(Compare Fathy and Qassim (2002))*
The quadratic form $Y^\top A Y$ is BAIQUE if and only if the matrix A satisfies

$$\begin{pmatrix} vecV_1\ vecV_2\ \ldots\ vecV_R & \vdots & T \\ \ldots\ \ \ldots\ \ \ldots\ \ \ldots & \vdots & \ldots \\ 0 \quad 0 \ \ldots \ 0 & \vdots & (vecV_1)^\top \\ 0 \quad 0 \ \ldots \ 0 & \vdots & (vecV_2)^\top \\ \vdots\ \ \vdots\ \ \vdots\ \ \vdots & \vdots & \vdots \\ 0 \quad 0 \ \ldots \ 0 & \vdots & (vecV_R)^\top \end{pmatrix} \begin{pmatrix} \lambda_1 \\ \lambda_2 \\ \vdots \\ \lambda_R \\ \ldots \\ vecA \end{pmatrix} = \begin{pmatrix} 0 \\ \vdots \\ 0 \\ \ldots \\ b_1 \\ \vdots \\ b_R \end{pmatrix} \tag{3}$$

where $\lambda_1, \ldots, \lambda_R$ are Lagrange multipliers corresponding to the constraint (1), vec stands for the vec operator, $T = \sum_{r=1}^{R} T_r \otimes T_r$, $T_r = \sum_{i=1}^{R} s_{ir} V_i$ with s_{ir} given by (2).

Proof. The Bayes risk for unbiased $\hat{\alpha}$ satisfies

$$r(\hat{\alpha}) = E(Var_\Theta(Y^\top AY)) = E(2tr A\tilde{C}_\Theta A\tilde{C}_\Theta) = 2(\sum_{i}^{R}\sum_{j}^{R} E(\Theta_i\Theta_j)tr AV_i AV_j)$$

$$= 2(\sum_{i}^{R}\sum_{j}^{R} C_{ij} tr AV_i AV_j) = 2\sum_{i=1}^{R}\sum_{j=1}^{R}\sum_{r=1}^{R} s_{ir}s_{jr} tr AV_i AV_j = 2\sum_{r=1}^{R} tr AT_r AT_r.$$

Since

$$\frac{\partial \sum_{r=1}^{R} tr AT_r AT_r}{\partial A} = 2\sum_{r=1}^{R} T_r AT_r, \qquad \frac{\partial(tr AV_r - b_r)}{\partial A} = V_r,$$

there exists Lagrange multipliers $\lambda_1, \ldots, \lambda_R$ such that

$$(\sum_{r=1}^{R} T_r \otimes T_r)vec A + \sum_{r=1}^{R} \lambda_r vec V_r = 0.$$

This implies the assertion together with the constraint (1).

Example (Asymmetric A)
Let consider the one dimensional design of observations at $1, 3, 5$ with the regression function $f(x) = 1$. Assume that the covariance model is

$$Var(Z) = \theta_1 \exp(-D) + \theta_2 I_{3\times 3}$$

where the matrix $D = (h_{ij})_{i,j=1,2,3}$ represents the matrix of Euclidean distances and $exp(-D)$ stands for the matrix with components $exp(-h_{ij})$. So we find

$$D = (h_{ij})_{i,j=1,2,3} = \begin{pmatrix} 0 & 2 & 4 \\ 2 & 0 & 2 \\ 4 & 2 & 0 \end{pmatrix}.$$

Suppose a uniform prior information on the parameters is given by

$$p_1(\theta_1) = 0.25; \ \ 1 \le \theta_1 \le 5, \qquad p_2(\theta_2) = 0.33; \ \ 0 \le \theta_2 \le 3,$$

and Θ_1 and Θ_2 are independent. Then equation (3) can be written as $GO = W$ with $G \in \Re^{11\times 11}$, $O \in \Re^{11}$, $W \in \Re^{11}$. A solution O, if it exists, is given by $O = G^- W$ where G^- stands for a generalized inverse of G. For the case $b = (1,1)^\top$, one obtains from O the Lagrange multipliers $\lambda_1 = -15.3284$, $\lambda_2 = 12.8534$ and the asymmetric matrix A given by

$$A = \begin{pmatrix} 0.6217 & 0.2478 & -0.8696 \\ 0.1217 & 0.1304 & -0.2522 \\ -0.7435 & 0 & 0 \end{pmatrix}.$$

3 Sufficient conditions for a symmetric matrix A

In this section we assume $R = 2$, $U_1 = J$ and $U_2 = I$, where J is an arbitrary covariance matrix and $I = I_{N \times N}$ denotes the identity matrix. Then we denote by $U = MJM$ and $V = MIM = M$ the given matrices in the linear representation of the covariance matrix of Y. Let be $u = vecU$ and $v = vecV$ and T be defined as in Theorem 1. Moreover, we assume that the prior satisfies $E(\Theta\Theta^\top) = I_{2\times 2}$, so that $T = U \bigotimes U + V \bigotimes V$. Then equation (3) can be written as $GO = W$ where

$$
G = \begin{pmatrix} u & v & \vdots & T \\ \cdots\cdots\cdots\cdots\cdots \\ 0 & 0 & \vdots & u^\top \\ & & \vdots & \\ 0 & 0 & \vdots & v^\top \end{pmatrix}, \quad W = \begin{pmatrix} 0 \\ \vdots \\ 0 \\ \cdots \\ b_1 \\ b_2 \end{pmatrix}, \quad O = \begin{pmatrix} \lambda_1 \\ \lambda_2 \\ \cdots \\ vecA \end{pmatrix}.
$$

Theorem 2. *Let $O = (G^\top G)^- G^\top W$ satisfy $GO = W$, where $(G^\top G)^-$ denotes the Moore-Penrose inverse of $G^\top G$, and let be $M = KK^\top$ with $K^\top K = I_{l\times l}$, where l is the rank of M. If the maximum eigenvalue of $K^\top J K$ is less than 1 and vecA is given by O, then A is a symmetric matrix.*

For proof of Theorem 2, we need the following lemmas.

Lemma 2. *Let $M = KK^\top \in R^{p\times p}$ be a symmetric idempotent matrix of rank r, where $K^\top K = I \in R^{r\times r}$ is the identity matrix, and let $B \in R^{p\times p}$ be a nonsingular symmetric matrix. Then the Moore-Penrose inverse of $(MBM)^2$ is given by $((MBM)^2)^- = K(K^\top BK)^{-2}K^\top = K[(K^\top BK)^{-1}]^2 K^\top$.*

Lemma 3. *See Kincaid and Cheney (1991), p.172-173.*
If A is an $n \times n$ matrix such that the maximum eigenvalue of $A^\top A$ is less than 1, then $I - A$ is invertible, and

$$
(I - A)^{-1} = \sum_{k=0}^{\infty} A^k.
$$

Lemma 4. *See Mirsky (1990), p.337-338. Let $\phi_p, \psi_p, \chi_p : \Re^{p\times p} \to \Re^{p\times p}$ be defined by*

$$
\phi_p(z) = \sum_{m=0}^{\infty} a_m z^m, \quad \psi_p(z) = \sum_{m=0}^{\infty} b_m z^m, \quad \chi_p(z) = \sum_{m=0}^{\infty} c_m z^m,
$$

with $c_m = a_0 b_m + a_1 b_{m-1} + \ldots + a_m b_0$ for $m = 0, 1, 2, \ldots$. Assume that ϕ_1, ψ_1, χ_1 are convergent for $z \in \Re$ with $|z| < \rho$ and suppose that $\phi_1(z)\psi_1(z) = \chi_1(z)$ for $|z| < \rho$. If all characteristic roots of $A \in \Re^{p\times p}$ are less than ρ, then

$$\phi_p(A)\psi_p(A) = \chi_p(A).$$

Proof of Theorem 2. At first note that

$$O = (G^\top G)^- G^\top W = \begin{pmatrix} -(D^- BC^-)L \\ (C^- + C^- B^\top D^- BC^-)L \end{pmatrix},$$

where

$$D = A - BC^- B^\top, \quad A = \begin{pmatrix} u^\top u & u^\top v \\ v^\top u & v^\top v \end{pmatrix}, \quad B_{2\times p^2} = \begin{pmatrix} u^\top T \\ v^\top T \end{pmatrix},$$

$C = T^2 + uu^\top + vv^\top$ and $L = b_1 u + b_2 v$. Only $(C^- + C^- B^\top D^- BC^-)L$ is important for determination of A. An extension of the Sherman-Morrison-Woodbury lemma (see e.g. Henderson and Searle (1981)) implies

$$C^- = \frac{cH^- - H^- uu^\top H^-}{c} - \frac{(cH^- - H^- uu^\top H^-)vv^\top (cH^- - H^- uu^\top H^-)}{c^2 d},$$

where $H = T^2$, $c = 1 + u^\top H^- u$ and $d = 1 + v^\top (H + uu^\top)^- v$. Using this, we obtain

$$C^- L = H^- (\eta_1 u + \eta_2 v),$$

and thus after some calculations

$$vecA = (C^- + C^- B^\top D^- BC^-)L = H^- (\zeta_1 u + \zeta_2 v + \xi_1 Tu + \xi_2 Tv)$$

where η_1, η_2, ζ_1, ζ_2, ξ_1, ξ_2, $\in \Re$. To determine the Moore-Penrose inverse of $H = T^2$, we use the fact that the maximum eigenvalue of $K^\top JK \otimes K^\top JK$ is less than 1 if the maximum eigenvalue of $K^\top JK$ is less than 1. Hence, since $T = M \otimes M(I \otimes I + J \otimes J)M \otimes M$, Lemma 2 and Lemma 3 provide

$$(T^2)^- = K \otimes K \left(\left(I \otimes I + K^\top JK \otimes K^\top JK \right)^{-1} \right)^2 K^\top \otimes K^\top$$

$$= K \otimes K \left(\sum_{k=0}^{\infty} (-1)^k (K^\top JK)^k \otimes (K^\top JK)^k \right)$$

$$\left(\sum_{l=0}^{\infty} (-1)^l (K^\top JK)^l \otimes (K^\top JK)^l \right) K^\top \otimes K^\top.$$

Since $\rho = 1$ satisfies the condition of Lemma 4 for $\phi_1(z) = \psi_1(z) = \sum_{k=0}^{\infty} (-1)^k z^k$ and $\chi_1(z) = \sum_{k=0}^{\infty} (-1)^k (k+1) z^k$, the order of the summation can be exchanged so that we obtain

$$H^- = (T^2)^- = \sum_{k=0}^{\infty} \sum_{l=0}^{\infty} (-1)^{k+l} (K(K^\top JK)^{k+l} K^\top \otimes K(K^\top JK)^{k+l} K^\top).$$

The assertion follows now from the fact that for any symmetric matrices Q and P, we have $Q \otimes Q \, vecP = vec(QPQ)$.

4 The influence of the design on the Bayes risk

Let $b = (1, 1)$ and let the covariance matrix J be given by $J = \exp(-D)$ where D is the matrix of distances between the design points, i.e.

$$D = (\|x_i - x_j\|)_{i,j=1,\ldots,N}.$$

For all considered designs the maximum eigenvalue of $K^\top J K$ is less than 1, so that the matrix A, in the BAIQUE $\hat{a}$, is symmetric if A is determined via $O = (G^\top G)^- G^\top W$. The Bayes risk $r(\hat{a}) = 2\,tr(A\,M\,J\,M\,A\,M\,J\,M) + 2\,tr(A\,M\,A\,M)$ of the BAIQUE is given for the following one and two dimensional simple designs. More examples can be found in Fathy (2006).

4.1 One dimensional designs

Table 1 provides the Bayes risk for different four point designs on $[0, 1]$ for the stationary model with $f(x) = 1$ and for the linear regression model with $f(x) = (1, x)^\top$. It can be seen that the influence of the design on the Bayes

Table 1. Bayes risks for one dimensional designs

Design points	Bayes risk for $f(x) = 1$	Bayes risk for $f(x) = (1, x)^\top$
0, 0.3, 0.7, 1	2.3145	386.93
0.1, 0.25, 0.85, 1	0.8856	455.77
0.1, 0.4, 0.7, 1	3.9366	213.51
0.1, 0.49, 0.61, 1	4.6978	29.48
0.4, 0.5, 0.6, 1	2.9779	279.05
0.7, 0.8, 0.9, 1	34.0702	219.41

risk depends very much on the model: the best design for the stationary model is the worst for the model with a linear trend and, if the last design is excluded from consideration, vice versa.

4.2 Two dimensional designs

Table 2 provides the Bayes risk for different four point designs on $[0, 15] \times [0, 10]$ for the model with linear trend so that $f((x(1), x(2)) = (1, x(1), x(2))^\top$. Thereby different forms of designs are compared which satisfy $x_1(1) + x_1(2) + x_2(1) + x_2(2) + x_3(1) + x_3(2) + x_4(1) + x_4(2) = 40$ for the design points $x_i = (x_i(1), x_i(2))$. The distance between design points is measured by L_1, L_2 and L_∞ norms. It turns out that the different distance measures provide similar Bayes risks and that the triangular design always has the smallest risk followed by the trapezoid design. This may be caused by the fact that in both designs two design points are rather close to each other, which is, as in the one dimensional case, a favorite property for estimating $\theta_1 + \theta_2$.

Table 2. Bayes risks for two dimensional designs

Design form	Design points	L_1	L_2	L_∞
square	$(1,1), (1,9), (9,1), (9,9)$	3.2325	3.8362	3.3503
rectangular	$(1,1), (1,6), (12,1), (12,6)$	3.3801	3.3805	3.3812
triangular	$(1,6), (6,1), (6,6), (8,6)$	0.5364	0.5341	0.5352
kite	$(2,2.5), (6,1), (6,10), (10,2.5)$	3.4884	2.6369	2.3821
trapezoid	$(1,5), (3,3), (9,3), (11,5)$	2.7287	1.4195	0.5505

References

Cressie N (1993) Statistics for Spatial Data. Wiley, New York

Fathy Y (2006) Bayes quadratic unbiased estimator of spatial covariance parameters. PhD thesis, University of Kassel, Germany

Fathy Y, Qassim M (2002) Bayesian estimation of spatial covariance function with two and three parameters. Al-Rafedain Journal of Science 13:43–50

Gnot S, Kleffe J (1983) Quadratic estimation in mixed linear models with two variance components. Journal of Statistical Planning and Inference 8:267–279

Henderson H, Searle S (1981) On deriving the inverse of a sum of matrices. Siam Review 23:53–60

Kincaid D, Cheney W (1991) Numerical Analysis. Wadsworth Inc., California

Koch K (1999) Parameter Estimation and Hypothesis Testing in Linear Models. Springer, Berlin

Mirsky L (1990) An Introduction to Linear Algebra. Clarendon Press, Oxford

Rao C (1972) Estimation of variance and covariance components in linear models. Journal of the American Statistical Association 67:112–115

Rao C, Kleffe J (1988) Estimation of Variance Components and Application. Elsevier, Amsterdam

Optimum Design for Correlated Fields via Covariance Kernel Expansions

Valerii V. Fedorov[1] and Werner G. Müller[2]

[1] GlaxoSmithKline, 1250 So Collegeville Rd, PO Box 5089, Collegeville PA
19426-0989, U.S.A.
`valeri.v.fedorov@gsk.com`
[2] Department of Applied Statistics (IFAS), Johannes-Kepler-University,
Freistädter Straße 315, 4040 Linz, Austria
`werner.mueller@jku.at`

Summary. In this paper we consider optimal design of experiments for correlated
observations. We approximate the error component of the process by an eigenvector
expansion of the corresponding covariance function. Furthermore we study the limiting behavior of an additional white noise as a regularization tool. The approach is
illustrated by some typical examples.

Key words: correlated errors, random field, regression experiment

1 Introduction

Let $\mathfrak{X}$ denote the design space, corresponding to a finite set of potential trials.
We can observe a random field

$$y_j(x_i) = \eta(x_i, \beta) + \varepsilon_j(x_i), \tag{1}$$

where $\eta(x, \beta)$ is the response function at $x \in \mathfrak{X}$ containing q unknown parameters $\beta = (\beta_1, \ldots, \beta_q)^{\mathrm{T}} \in I\!\!R^q$. Let us further assume that the random noise
$\varepsilon(x)$ consists of two independent components

$$\varepsilon(x) = u(x) + e(x),$$

such that for both $E[u] = E[e] = 0$ and thus $E[\varepsilon] = 0$, and that

$$\mathrm{Cov}[e(x), e(x')] = \sigma^2(x)\delta_{x,x'},$$

and

$$\mathrm{Cov}[u(x), u(x')] = k(x, x'),$$

the latter - the so-called covariance kernel - being a known function; $\delta_{x,x'}$ denotes the Kronecker-symbol. Component $e_j(x)$ can be viewed as an observational error, which can be reduced by placing, for example, r_i geiger counters at location x_i and $e'_j(x_i) = r_i^{-1} \sum_{l=1}^{r_i} e_{jl}(x_i)$ and $\mathrm{Var}[e'_j(x_i)] = \sigma^2(x_i)/r_i$. Component $u_j(x)$ is a random process which describes the deviation on, say a particular day j from the local average $\eta(x, \beta)$ and can thus not be replicated.

Hence $\mathrm{Cov}[\varepsilon(x), \varepsilon(x')] = \sigma^2(x)\delta_{x,x'} + k(x, x')$, where $x, x' \in \mathfrak{X}$. This setup has many potential applications, but it is extremely relevant in environmental studies, which exhibit, for instance spatial or temporal data or both (cf. Müller (forthcoming 2007)). While $e(x)$ represents a so called nugget effect, $u(x)$ describes local (temperature, humidity, etc.) fluctuation, which are usually correlated at relatively short distances.

We will in the following approximate the correlated component $u(x)$ by Mercer's eigenfunction expansion (cf. Mercer (1909)) of the respective kernel $k(x, x')$, thereby allowing embedding our problem into standard convex design theory. This idea has already been suggested in Fedorov (1996) and Fedorov and Flanagan (1997), but the approach has never been fully developed, implemented and tested. Furthermore we will relate our results to alternative methods for the limiting cases $\sigma^2 \to \infty$ (independence) and $\sigma^2 \to 0$ (random fields).

2 Expansion of the covariance kernel

The approach is based on the fact that the error component $u(x)$ in the random process can be represented by the infinite expansion

$$u(x) = \sum_{l=1}^{\infty} \gamma_l \varphi_l(x),$$

and correspondingly $k(x, x') = \sum_{l=1}^{\infty} \lambda_l \varphi_l(x) \varphi_l(x')$ where the γ_l are specific independent random values with $E[\gamma_l] = 0$ and $\mathrm{Cov}[\gamma_l, \gamma_{l'}] = \lambda_l \delta_{l,l'} = \Lambda_{l,l'}$, and the $\varphi_l(x)$ and λ_l are the eigenfunctions and eigenvalues, respectively, of the covariance kernel $k(x, x')$, given through $\lambda_l \varphi_l(x) = \int_{\mathfrak{X}} k(x, x') \varphi_l(x') dx'$. Usually this fact is referred to as Mercer's theorem (cf. Mercer (1909)).

If $\{\lambda_l\}$ diminishes rapidly then the random process (1) can be approximated:

$$y_j(x) = \eta(x, \beta) + \sum_{l=1}^{p} \gamma_{lj} \varphi_l(x) + e(x), \tag{2}$$

where y_j can be seen as an observation on day j and γ_{lj} is sampled from $N(0, \lambda_l)$.

On an intuitive level the choice of p should depend upon the locality of $\eta(x, \beta)$, $x \in \mathfrak{X}$ and the variance $\sigma^2(x)$ of the measurement error. The presentation can now be regarded as a mixed effects model (a special form of a random coefficient regression model, cf. eg. Pinheiro and Bates (2000)) as long as it admits replications, where $\theta = \{\beta_1, \ldots, \beta_q, \gamma_1, \ldots, \gamma_p\}$ with

$$E[\theta] = \bar{\theta} = \{\beta, 0_p\},$$

and

$$\mathrm{Cov}[\theta, \theta'] = \begin{pmatrix} 0_{q \times q} & 0_{q \times p} \\ 0_{p \times q} & \Lambda_{p \times p} \end{pmatrix}.$$

The covariance of the total error $\varepsilon(x)$ can thus be written as $\mathrm{Cov}[\varepsilon(x), \varepsilon(x')] = \sigma^2(x)\delta_{x,x'} + \varphi^T(x)\Lambda\varphi(x')$, where $\varphi^T(x) = \{\varphi_1(x), \ldots \varphi_p(x)\}$. Often replications can be understood as observations of the corresponding number of closely allocated sensors, meters, etc..

Random coefficient models can be embedded into standard convex design theory (cf.eg. Gladitz and Pilz (1982)), thus allowing the use of powerful design tools such as equivalence theorems and first order gradient algorithms. There remains the issue, however, of which design criterion is relevant for a specific situation and to resolve this there seem to be at least three plausible options:

a) The main interest of the researcher is in the "trend" parameters β and the component $u(x)$ is regarded solely as a nuisance. This problem appears, for instance, if one is interested in weather patterns in the region $\mathcal{X}$.

b) The emphasis is on the prediction of individual instances of the process, thus the information on γ_l must also enter a design criterion, a problem which occurs in weather prediction.

c) One desires prediction of the average process, i.e. $\eta(x, \beta)$ at a given set of $x \in \mathcal{Z} \neq \mathcal{X}$.

In this paper we will, for clarity of exposition, consider in detail only case a), but treatment of the other cases is similar. Furthermore, to eventually avoid the arbitrary choice of the order of approximation p, we will want to reformulate results in terms of the original covariance kernel $k(x, x')$ rather than its approximation.

3 Design for the estimation of trend

For simplification we now and in the following consider a linearisation $f(x) = \partial \eta(x, \beta)/\partial \beta|_{\beta = \beta_0}$ of the response around a prior guess of the parameter β_0, eventually leading to so-called locally optimum designs.

Let us further assume that we observe the random field $y(x)$ at n distinct points $x_1, \ldots, x_n$, that is observations are generated according to

$$y_{ij} = \beta^T f(x_i) + \sum_{l=1}^{p} \gamma_{lj}\varphi_l(x_i) + e_{ij}, \quad i = 1, \ldots, n, \quad j = 1, \ldots, m,$$

with $r_1, \ldots, r_n$ repeated measurements respectively. We assume that for all j the collection $\{x_i, \frac{r_i}{R}\}_1^n = \xi_n$, with $R = \sum_{i=1}^n r_i$, (the design) is the same. Actually, we assume that $m = 1$; the reader can add a multiplier m^{-1} where needed. Also we then assume for the sake of simplicity $\sigma^2(x) = \sigma^2$. Thus from now on we admit the possibility of repeated observations, which will reflect microscale variations. For instance, in the case of designing a spatial network, these replications can stem from observations from very closely neighboring measurement sites.

In this setup the best linear unbiased estimator of the trend parameter β is

$$\hat{\beta} = \{FV^{-1}F^T\}^{-1}FV^{-1}\bar{y},$$

with

$$F = \{f(x_1), \ldots, f(x_n)\},$$
$$\Phi = \{\varphi(x_1), \ldots, \varphi(x_n)\},$$
$$V = \sigma^2\Omega + \Phi^T\Lambda\Phi, \text{ and}$$
$$\bar{y}^T = (\bar{y}_1, \ldots, \bar{y}_n), \ \bar{y}_i = \frac{1}{r_i}\sum_{l=1}^{r_i} y_{i1}, \ \Omega_{ii'} = \frac{\delta_{ii'}}{r_i}.$$

The asymptotic covariance matrix of the estimator $\hat{\beta}$ is $M^{-1}(\xi_n) = m^{-1}\{FV^{-1}F^T\}^{-1}$ and thus an optimum design ξ_n^* must seek to satisfy

$$\xi_n^* = \arg\min_{\xi_n}\Psi\{M(\xi_n)\}, \tag{3}$$

for a reasonably chosen design criterion, say $\Psi\{M\} = -\log\det M$.

Let us now consider the best linear unbiased estimator $\hat{\theta}^T = (\hat{\beta}^T, \hat{\gamma}^T)$ in the full (random coefficient) model. Its covariance matrix is

$$D(\xi_n) = \begin{pmatrix} D_{ff} & D_{f\varphi} \\ D_{f\varphi}^T & D_{\varphi\varphi} \end{pmatrix} = \begin{pmatrix} FWF^T & FW\Phi^T \\ \Phi WF^T & \Phi W\Phi^T + \Lambda^{-1} \end{pmatrix}^{-1},$$

where $W_{ii'} = \frac{R}{\sigma^2}\delta_{ii'}\xi(x_i)$, $\xi(x_i)$ being a design measure at x_i.

It is easy to show (Frobenius formula) that

$$D_{ff}(\xi) \propto M^{-1}(\xi_n)$$

and thus the criterion (3) is equivalent to subset D-optimality (D_S-optimality) in a random coefficient regression model. It now follows directly from standard design theory (cf. Theorem 2.7.1 in Fedorov (1972)), that a necessary and sufficient condition for an approximate design ξ^* to optimize (3) is

$$\phi(x, \xi^*) = \phi_\theta(x, \xi^*) - \varphi^T(x)\left[\Lambda^{-1} + \Phi W\Phi^T\right]^{-1}\varphi(x)$$
$$\leq \operatorname{tr} D(\xi^*)M(\xi^*) - \operatorname{tr}\left[\Lambda^{-1} + \Phi W\Phi^T\right]^{-1}\Phi W\Phi^T, \tag{4}$$

with $\phi_\theta(x, \xi^*) = (f(x), \varphi(x))^T D(\xi^*)(f(x), \varphi(x))$. There exists an optimal design with no more than $q(2p + q + 1)/2 + 1$ design points.

Note that this equivalence condition effectively reflects the design problem discussed in section 3 of Wynn (2004), which embeds our problem further into the maximum entropy framework. It is essential for the development of numerical algorithms and analysis of their properties. We have used the first order exchange algorithm, which at its s iteration adds some mass to point

$$x_s^+ = \arg\max_{x \in \mathcal{X}} \phi(x, \xi_s)$$

and substracts some at point

$$x_s^- = \arg\min_{x \in \text{supp}\,\xi_s} \phi(x, \xi_s).$$

Details of the algorithm and its properties (convergence) can be found in Fedorov (1972).

The so-called sensitivity function $\phi(x, \xi)$ can thus be routinely employed in such a standard design optimization algorithm. However, since the original problem is formulated in terms of the covariance kernel a preferred formulation is

$$\phi(x, \xi) = \mathsf{f}^T(x, \xi) D_{\mathsf{ff}}(\xi) \mathsf{f}(x, \xi) \tag{5}$$

with

$$\mathsf{f}^T(x, \xi) = f^T(x) - k^T(x, \xi) K^{-1}(\xi) \left(W + K^{-1}(\xi)\right)^{-1} W F^T,$$
$$D_{\mathsf{ff}}(\xi) = \left\{ F \left[W - W \left(W + K^{-1}(\xi)\right)^{-1} W \right] F^T \right\}^{-1},$$
$$k^T(x, \xi) = \{k(x, x_1), \ldots, k(x, x_n)\}, \text{ and}$$
$$[K(\xi)]_{ii'} = k(x_i, x_i') \text{ for all } x_i, x_i' \in \text{supp}\,\xi.$$

Here we use the fact that for sufficiently large p and after neglecting remainder terms one can (cf. Fedorov and Flanagan (1997)) adopt the approximation $K \simeq \Phi^T \Lambda \Phi$. The advantage of this formulation is, that a practitioner solely has to specify the response function and the covariance kernel and needs not to bother with the implicit eigenvalue expansion. A derivation of this presentation can be found in the Appendix.

4 Examples

To illustrate the technique presented above we have performed a standard one-point correction design algorithm based on the sensitivity function (4) on two typical examples. Calculations were performed on a 101 design point grid with a uniform measure as the initial design and stopped after 10000 iterations. Asymmetries are due to numerical inaccuracies and the finite number of iterations and possible slight shifts are due to the finiteness of the grid. As a criterion we used D-optimality, where $\Psi\{M\} = -\log\det M$.

4.1 Näther's Case

Firstly consider a simple linear regression model, i.e. $f^T(x_i) = (1, x_i)$, on $\mathfrak{X} = [-1, 1]$ with an error covariance kernel given by

$$k(x, x') = \begin{cases} 1 - |x - x'| & \text{for } |x - x'| < 1 \\ 0 & \text{for } |x - x'| \geq 1. \end{cases}$$

This example gained some prominence in the design literature for correlated errors, since it linearly relates the response function to the covariance kernel, which allows for a direct proof of uniform optimality of a three point design concentrated on $\{-1, 0, 1\}$, see Näther (1985). Moreover note, that it served as a motivating case for the considerations in Müller and Pàzman (2003), where also regulatory noise was employed. In contrast, in the present paper, this has been made dependent upon the design measure itself.

It is thus easy to guess what a reasonable design algorithm should yield for the extreme settings of large σ (independence; $\sigma > 10^2$ was used for the computations) and small σ (dependence due to k; here $\sigma = 10^{-2}$). In the former case, as expected, the algorithm yields the design measure equally distributed between the extremal points -1 and 1, whereas in the latter case it yields the measure displayed in Figure 1, which adds some (unfortunately barely visible, but from the sensitivity function easily deducible) very small measure at the center, which highly corresponds to the computational results of Müller and Pàzman (2003) displayed in their Fig.1. In all our figures the dashed line represents a rescaled sensitivity function, the solid line the design measure. Note that for intermediate choices of σ a respective proportion of the central weight is distributed to the extremal points.

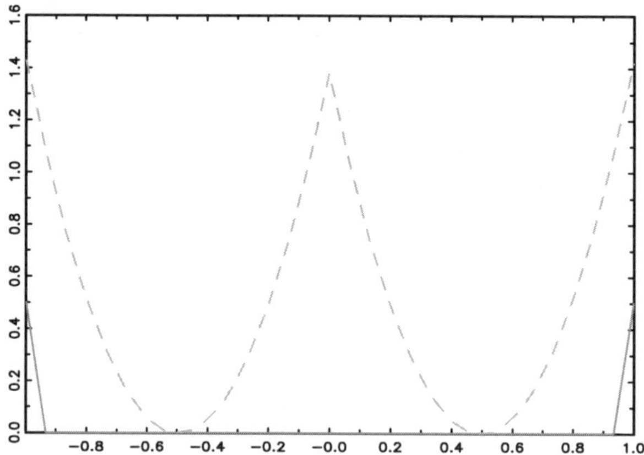

Fig. 1. Design measure (solid line) and rescaled sensitivity function (dashed line) for Näther's case on a 101-point grid; horizontal x, vertical $\xi(x)$.

4.2 Third order polynomial and poisson kernel

Next, assume the response function to be a third order polynomial, i.e. $f^T(x_i) = (1, x_i, x_i^2, x_i^3)$ and that the error covariance structure is described by the Poisson kernel

$$k(x, x') = \frac{1 - \zeta^2}{1 - 2\zeta \cos \pi(x - x') + \zeta^2},$$

where $0 \leq \zeta \leq 1$ is a shape parameter and the design region is $\mathfrak{X} = [0, 1]$. We have chosen $\zeta = 0.2$ for all the numerical examples in this subsection.

The optimum design in the uncorrelated case follows from Theorem 2.3.3. in Fedorov (1972) as the roots of the polynomial $[1 - (2x - 1)^2][3(2x - 1)^2 - 1]$, which yields the points 0,0.211,0.789,1. This is also the result that our numerical algorithm yields (approximately) for any large σ (here even $\sigma > 10^1$), see Figure 2.

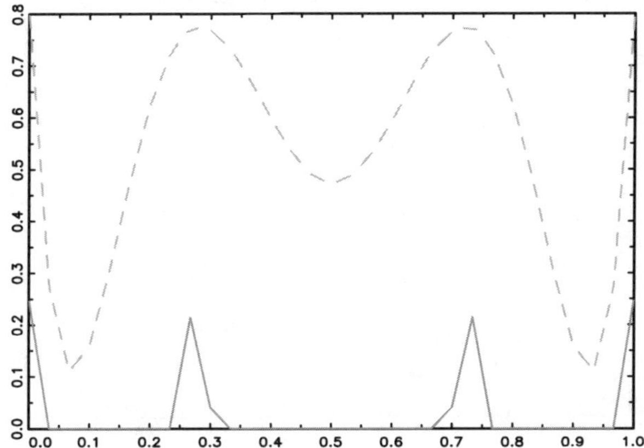

Fig. 2. Design measure (solid line) and rescaled sensitivity function (dashed line) for third order polynomial (under independence) on a 101-point grid; hor. x, ver. ξ.

The situation is very different for the case of letting σ decrease, i.e. approaching the 'purely' correlated case. In the beginning the two inner points move outwards to merge with the extremal points. The measure around these two points is now much more spread out (see Figure 3 for $\sigma = 10^{-2}$). Further decreasing σ now shifts the collapsed points toward the center (see Figure 4 for $\sigma = 10^{-30}$) to end up with a design with an extremly flat sensitivity function over a wide range of the region.

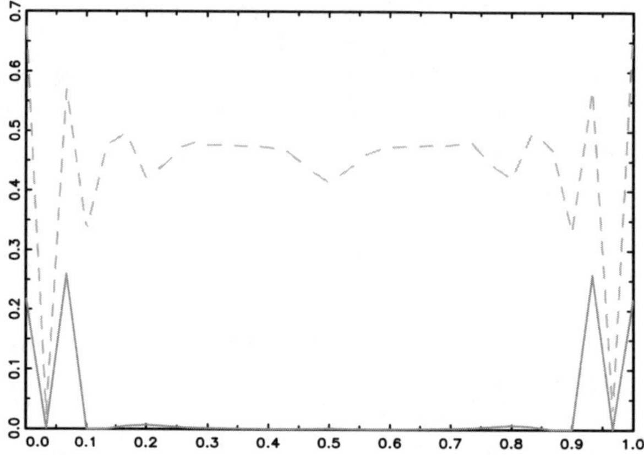

Fig. 3. Design measure (solid line) and rescaled sensitivity function (dashed line) for third order polynomial ($\sigma = 10^{-}2$) on a 101-point grid; hor. x, ver. ξ.

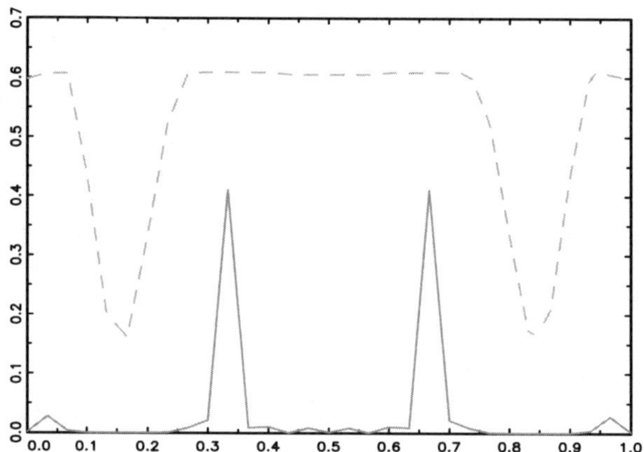

Fig. 4. Design measure (solid line) and rescaled sensitivity function (dashed line) for third order polynomial ($\sigma = 10^{-30}$) on a 101-point grid; hor. x, ver. ξ.

Appendix

For the derivation of ϕ_θ we require the entries D_{ff}, $D_{f\varphi}$ and $D_{\varphi\varphi}$. For brevity we omit arguments. We have

$$D_{ff} = \left\{ FWF^T - FW\Phi^T \left(\Phi W \Phi^T + \Lambda^{-1} \right)^{-1} \Phi W F^T \right\}^{-1}$$

and hence

$$D_{ff}^{-1} = FWF^T - FW\Phi^T \left(\Phi W\Phi^T + \Lambda^{-1}\right)^{-1} \Phi WF^T$$

$$\simeq FWF^T - FW\Phi^T \left(\Lambda - \Lambda\Phi(W^{-1} + K)^{-1}\Phi^T\Lambda\right)\Phi WF^T$$

$$\simeq F\left[W - W(W + K^{-1})^{-1}W\right]F^T.$$

Note that $K \simeq \Phi^T\Lambda\Phi$. Also we have
$$D_{f\varphi} = -D_{ff}FW\Phi\left(\Phi W\Phi^T + \Lambda^{-1}\right)^{-1} \text{ and}$$
$$D_{\varphi\varphi} = \left(\Phi W\Phi^T + \Lambda^{-1}\right)^{-1} + \left(\Phi W\Phi^T + \Lambda^{-1}\right)^{-1}\Phi WFD_{ff}FW\Phi\left(\Phi W\Phi^T + \Lambda^{-1}\right)^{-1}$$

Now we can write

$$\phi_\theta(x,\xi) = (f(x),\varphi(x))^T \begin{pmatrix} D_{ff} & D_{f\varphi} \\ D_{f\varphi}^T & D_{\varphi\varphi} \end{pmatrix} (f(x),\varphi(x))$$

$$= f^T(x)D_{ff}f(x) + 2f^T(x)D_{f\varphi}\varphi(x) + \varphi^T(x)D_{\varphi\varphi}\varphi(x)$$

$$= f^T(x)D_{ff}f(x) - 2f^T(x)D_{ff}FW\Phi\left(\Phi W\Phi^T + \Lambda^{-1}\right)^{-1}\varphi(x)$$

$$+\varphi^T(x)\left(\Phi W\Phi^T + \Lambda^{-1}\right)^{-1}\varphi(x)$$

$$+\varphi^T(x)\left(\Phi W\Phi^T + \Lambda^{-1}\right)^{-1}\Phi WFD_{ff}FW\Phi\left(\Phi W\Phi^T + \Lambda^{-1}\right)^{-1}\varphi(x).$$

Note that the one but last summand will be subtracted in the definition of the sensitivity function so that we can subsume

$$\phi(x,\xi) = \left(f^T(x) - \varphi^T(x)\left(\Phi W\Phi^T + \Lambda^{-1}\right)^{-1}\Phi WF\right)D_{\mathsf{ff}} \times$$

$$\times \left(f^T(x) - \varphi^T(x)\left(\Phi W\Phi^T + \Lambda^{-1}\right)^{-1}\Phi WF\right)^T$$

$$= \mathsf{f}^T(x)D_{\mathsf{ff}}\mathsf{f}(x).$$

So, it remains to revert $\mathsf{f}(x)$ to a presentation in terms of the original covariance function. For that purpose we require the presentation

$$\left(\Phi W\Phi^T + \Lambda^{-1}\right)^{-1} \simeq \Lambda - \Lambda\Phi\left(K + W^{-1}\right)^{-1}\Phi^T\Lambda$$

Then

$$\mathsf{f}^T(x) = f^T(x) - \varphi^T(x)\left(\Phi W\Phi^T + \Lambda^{-1}\right)^{-1}\Phi WF^T$$

$$\simeq f^T(x) - \varphi^T(x)\left(\Lambda - \Lambda\Phi\left(K + W^{-1}\right)^{-1}\Phi^T\Lambda\right)\Phi WF^T$$

$$\simeq f^T(x) - \left(k^T(x) - k^T(x)\left(K + W^{-1}\right)^{-1}K\right)WF^T,$$

using the convenient notation $k^T(x) \simeq \varphi^T(x)\Lambda\Phi$.
And from

$$k^T(x) - k^T(x)\left(K + W^{-1}\right)^{-1}K = k^T(x)K^{-1}\left(W + K^{-1}\right)^{-1}$$

the sensitivity function given in (4) easily follows.

Acknowledgement. We are most grateful to the helpful remarks of the referees, which made us aware of how to improve the presentation.

References

Fedorov V (1972) Theory of Optimal Experiments. Academic Press, New York

Fedorov V (1996) Handbook of Statistics, Volume 13, Elsevier, Amsterdam, chap Design of spatial experiments: model fitting and prediction

Fedorov V, Flanagan D (1997) Optimal monitoring network design based on mercer's expansion of covariance kernel. Journal of Combinatorics, Information & System Sciences 23:237–250

Gladitz J, Pilz J (1982) Construction of optimal designs in random coefficient models. Mathematische Operationsforschung und Statistik, Series Statistics 13:371–385

Mercer J (1909) Functions of positive and negative type and their connection with the theory of integral equations. Philosophical Transactions Royal Society London A 209:415–446

Müller WG (forthcoming 2007) Collecting Spatial Data, 3rd edition. Springer Verlag, Heidelberg

Müller WG, Pàzman A (2003) Measures for designs in experiments with correlated errors. Biometrika 90:423–434

Näther W (1985) Exact designs for regression models with correlated errors. Statistics 16:479–484

Pinheiro JC, Bates D (2000) Mixed-Effects Models in S and S-PLUS. Springer Verlag, New York

Wynn H (2004) mODa7 - Advances in Model-Oriented Design and Analysis, Physica, Heidelberg, chap Maximum entropy sampling and general equivalence theory

Generalized Probit Model in Design of Dose Finding Experiments

Valerii V. Fedorov and Yuehui Wu

Research Statistical Unit, SQS, R&D, GlaxoSmithKline, Collegeville, P.O.Box
5089, Collegeville, PA, 19426 U.S.A.
`Valeri.V.Fedorov@gsk.com`   `Yuehui.2.Wu@gsk.com`

Summary. In clinical studies, continuous endpoints are very commonly seen. However, either for ease of interpretation or to simplify the reporting process, some continuous endpoints are often reported and (unfortunately) analyzed as binary or ordinal responses. We emphasize the usefulness of differentiation between response and utility functions and develop tools to build locally optimal designs for corresponding models. It is also shown that dichotomization of responses may lead to significant loss in statistical precision. We consider an example with two responses and one utility function. The generalization to a larger number of responses and utility functions is straightforward.

Key words: dichotomized and continuous responses, multivariate probit model, optimal design, utility function

1 Generalized probit model

1.1 Background model and notations

In clinical trials, there are often multiple endpoints to a treatment. Exclusively for the sake of simplicity, we assume there are two endpoints and let Z denote these two continuous responses. We further assume that

$$Z \sim \mathcal{N}(\eta, \Sigma), \tag{1}$$

where η is a vector of means and Σ is the variance-covariance matrix. We assume that the first component corresponds to toxicity and the second to efficacy. Generalization to higher dimension looks straightforward but leads to unavoidably complicated notation and more intensive computing. Potentially both η and Σ may depend on some covariates like doses of various compounds(drugs), age, sex, etc. In this paper the only covariate is drug dose, $x \in \mathcal{X}$, and the matrix Σ is constant within $\mathcal{X}$.

Responses $Z^T = (Z_1, Z_2)$ can be observed either directly or only some functions of them can be observed (and partial loss of information is to be expected). A popular choice is a dichotomization of Z (compare with Ashford and Sowden (1970)):

$$Y_k = \begin{cases} 1, & \text{if } Z_k \le c_k \\ 0, & \text{otherwise} \end{cases}, \qquad k = 1, 2, \tag{2}$$

i.e. Y_k is binomially distributed with the parameter

$$P(Y_k = 1) = \Phi_{.k}(v_k^*; R) = \int_{-\infty}^{\infty} dv_{k'} \int_{-\infty}^{v_k^*} dv_k \phi(v_1, v_2, R), \tag{3}$$

where $\phi(v_1, v_2, R)$ is the density function of the standard bivariate normal distribution, $v = (\text{diag}\Sigma)^{-1/2}(z - \eta)$, $v_k^* = (c_k - \eta_k)/\sigma_k$, $\sigma_k^2 = \Sigma_{kk}$, and R is the correlation matrix corresponding to Σ. Similarly,

$$P(Y_1 = 1, Y_2 = 1) = \Phi_{12}(v_1^*, v_2^*; R) = \int_{-\infty}^{v_1^*} dv_1 \int_{-\infty}^{v_2^*} dv_2 \phi(v_1, v_2, R). \tag{4}$$

Responses Y_k defined by (2) can be replaced by multilevel ordinal, or by a mixture of continuous, binary, ordinal responses, etc. For instance, one can define $Y_k = L_k^T Z$, where L_k is a given vector or

$$Y_k = \begin{cases} 2, & \text{if } Z_k \le d_{k2} \\ 1, & \text{if } d_{k2} < Z_k \ge d_{k1} \\ 0, & \text{otherwise.} \end{cases}$$

In general Components of a response vector are dependent if Σ is not diagonal. This feature is useful in modelling stochastic dependence of various responses to treatment; see Dragalin et al (2005), and Dragalin et al (2006).

Often, while responses Z can be observed, a regulatory agency is interested only in the function

$$p(x) = P(Y_1 = 1, Y_2 = 0|x) = P(Y_1 = 1|x) - P(Y_1 = 1, Y_2 = 1|x), \tag{5}$$

where Y_1 and Y_2 are toxicity and efficacy values respectively.

Unfortunately there are examples when practitioners use dichotomized versions of the observed data for analysis (not for reporting). We show that this may lead to a substantial loss of information. Thus it is worth distinguishing between response functions and utility functions, understanding that sometimes they may coincide; distinguishing as in Dragalin et al (2006).

For illustration purposes, in Figure 1 we show three cases: two continuous responses; both responses dichotomized; efficacy continuous, toxicity dichotomized. For the first two cases, the utility function is based on dichotomized responses and in the last case utility is the conditional ($Y_1 = 1$, i.e. no toxicity) mean of efficacy. Note that

$$E(Y_2|Y_1 = 1) = \frac{\theta_2^T f_2(x)\Phi_{.1}((c_1 - \theta_1^T f_1(x))/\sigma_1) - \sigma_2\rho\phi((c_1 - \theta_1^T f_1(x))/\sigma_1)}{\Phi_{.1}(v_1^*)}.$$

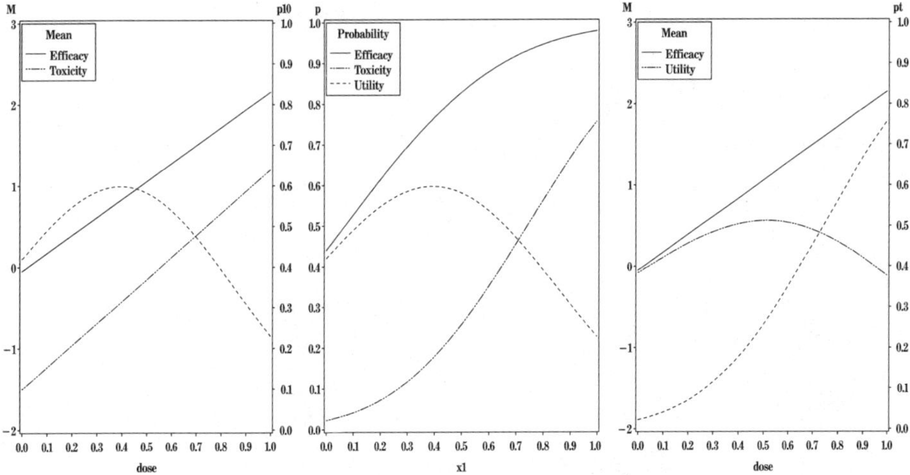

Fig. 1. Examples of various combinations of continuous/dichotomized responses and utility function.

2 Optimal design

2.1 Notation

Let $\xi = \{x_i, \lambda_i\}_1^N, x_i \in \mathcal{X}, \sum_{i=1}^{N} \lambda_i = 1$ and

$$M(\xi, \boldsymbol{\Theta}) = \sum_{i=1}^{N} \lambda_i \mu(x_i, \Theta), \tag{6}$$

where $\mu(x_i, \Theta)$ is the information matrix of a single observation at dose x (Dragalin et al (2006)), and Θ is a collection of unknown parameters. Our main goal is to build and to compare locally optimal continuous designs

$$\xi^*(\Theta) = \arg\min_{\xi} \Psi[M(\xi, \Theta)], \tag{7}$$

for various scenarios of dichotomization and different optimality criteria.

2.2 Information matrices for a single observation

Calculation of the information matrix for a single observation is a crucial step in optimal design construction. For the sake of simplicity we assume that Σ is known, i.e. does not contain unknown parameters. Otherwise a technique similar to Fedorov et al (2001) should be applied.

When both Y_1 and Y_2 are continuous, and $\eta_k(x, \Theta) = \theta_k f(x), k = 1, 2$ and $\Theta^T = (\theta_1^T, \theta_2^T)$ then

$$\mu(x,\Theta) = F(x)\Sigma^{-1}F^T(x) = \Sigma^{-1}\bigotimes f(x)f^T(x),$$

where $F(x) = \begin{pmatrix} f(x) & 0 \\ 0 & f(x) \end{pmatrix}$.

When both responses are dichotomized, the information matrix is:

$$\mu(x,\Theta) = \mathbf{C}_1\mathbf{C}_2(\mathbf{P} - \mathbf{p}\mathbf{p}^T)^{-1}\mathbf{C}_2^T\mathbf{C}_1^T,$$

with

$$\mathbf{C}_1 = \begin{pmatrix} \phi(\frac{c_1 - \theta_1^T f(x)}{\sigma_1}) & \mathbf{0} \\ \mathbf{0} & \phi(\frac{c_2 - \theta_2^T f(x)}{\sigma_2}) \end{pmatrix} \bigotimes f(x),$$

$$\mathbf{C}_2 = \begin{pmatrix} \Phi(u_1) & 1 - \Phi(u_1) & -\Phi(u_1) \\ \Phi(u_2) & -\Phi(u_2) & 1 - \Phi(u_2) \end{pmatrix},$$

$$u_k = \frac{(c_{k'} - \theta_{k'}^T f(x))/\sigma_{k'} - \rho(c_k - \theta_k^T f(x))/\sigma_k}{\sqrt{1 - \rho^2}},$$

$$\mathbf{P} = \begin{pmatrix} P(Y_1 = 1, Y_2 = 1) & 0 & 0 \\ 0 & P(Y_1 = 1, Y_2 = 0) & 0 \\ 0 & 0 & P(Y_1 = 0, Y_2 = 1) \end{pmatrix}, \quad \mathbf{p} = \text{diag } \mathbf{P}.$$

Note that we set $1 \leq \text{rank}\{\mu(x,\Theta)\} \leq 2$ and therefore the optimal designs may have less support points than $\dim(\theta_1) + \dim(\theta_2)$ as one might expect. We do not discuss the mixed case (only one response is dichotomized) because we failed to derive a closed form expression for $\mu(x,\Theta)$.

2.3 Locally optimal designs

All computation were done under the assumption that the parameters of the "normal" background model are: $\theta = (-1.5, 2.7, -0.05, 2.2)$.

If one is interested only in $x^*(\Theta) = \arg\min_x p\,(x;\Theta)$ then $\Psi[M(\xi,\Theta)] = L^T(\Theta)M^{-1}(\xi,\Theta)L(\Theta)$, where $L^T(\Theta) = \partial x^*/\partial\Theta$.

Since $x^*(\Theta)$ depends on all unknown parameters, we may conclude that it is expedient to build $\xi_D^* = \arg\max_\xi |M(\xi,\Theta)|$.

Locally D-optimal designs for estimating Θ using continuous responses consist of two boundary points with equal weights and stays the same for all parameter values (since the model is linear with respect to unknown parameters), while the design region is $0 \leq x \leq 1$ and there is no dependence on the unknown parameters (compare with Dragalin et al (2006)).

The L-optimal design for continuous responses has the same support points 0 and 1, but with weights 0.6 and 0.4 respectively. Deviation of weights from 0.5 leads to very minor changes in both D and L criteria.

For dichotomized responses and the same utility function, the D-optimal design is $\xi_D^* = \{0, 0.4; 0.35, 0.5; 0.95, 0.1\}$ and the L-optimal design is $\xi_L^* = \{0, 0.5; 0.85, 0.5\}$. For practical reason, the weights in the designs have been rounded but this does not greatly affect the properties of the design.

Table 1. Determinant of variance-covariance matrix of $\hat{\Theta}$ for sample size 200

Model	Continuous case, design		Dichotomized case, design	
	L-optimal	D-optimal	L-optimal	D-optimal
Continuous	0.049	0.045	0.086	0.068
Binary	15.19	14.00	13.68	13.04

Table 2. Standard error of $\hat{x}^*$ for sample size 200

Model	Continuous case, design		Dichotomized case, design	
	L-optimal	D-optimal	L-optimal	D-optimal
Continuous	0.028	0.029	0.029	0.030
Binary	0.039	0.039	0.038	0.039

Table 1 lists the determinant of the variance-covariance matrix of $\hat{\Theta}$ for different models and designs. The D-optimal design always provides the smallest value for the corresponding model and it is obvious that the continuous response model provides more accurate parameter estimates, i.e. a smaller value of the determinant of variance-covariance matrix. If one uses dichotomized responses, 270% more subjects are needed to achieve the same precision in parameter estimation as using continuous responses. Table 2 lists the theoretical standard deviation, for 200 subjects, of the estimated target dose under different designs and two models using continuous or binary outcomes. As expected, under the same model, the corresponding L-optimal design provides the most accurate estimates of target dose x^* in terms of minimizing the standard deviation of $\hat{x}^*$. However, under the bivariate regression model, all of the standard errors of the estimated target dose are around 0.03 which is smaller than the standard errors obtained under the bivariate probit model (around 0.04), showing that using dichotomized responses leads to loss of accuracy in the target dose estimator. In other words, to attain the same accuracy with dichotomized responses as compared to continuous responses requires 78% more subjects.

3 Simulation

To compare the performance for locating the maximum utility between using continuous and dichotomized responses, we generate 200 continuous observations according to four designs: L and D optimal designs under the bivariate probit model and bivariate regression model introduced in section 2, respectively. We then fit a bivariate regression model to obtain parameter estimates, calculate the corresponding $p(Y_1 \leq c_1, Y_2 \geq c_2)$ and locate the target doses. We then fit a bivariate probit model using the same 200 observations with cutoff values $c_1 = 0.5$ and $c_2 = 0.1$ and estimate the target dose which maximizes p_{10}. We repeat the procedure 1000 times to compare the distributions of the estimated target doses under the two models. Note that at dose zero the toxicity rate is close to 0 and at the highest dose one, the efficacy rate is close to 1. In the simulated data, it is likely that there is no non-efficacious response at dose 1 or no toxicity response at dose 0. This consequently leads to singularities in the parameter estimation procedure. To fix this problem, we introduce regularization of the likelihood function.

The simulation results support the theoretical results in Table 1. Due to space limitations, we only plot the results of the L-optimal design under the bivariate regression model and the bivariate probit model in Figure 2. The graph on the left illustrates the results when both responses are binary and the graph on the right shows the results when both responses are continuous. The spread of the histogram under the bivariate regression model is substantially smaller than under the bivariate probit model. All the results suggest that using continuous responses provides more accurate estimates of the target dose. We suggest use of the L-optimal design. However, if this is difficult to obtain, the D-optimal design should be a good alternative.

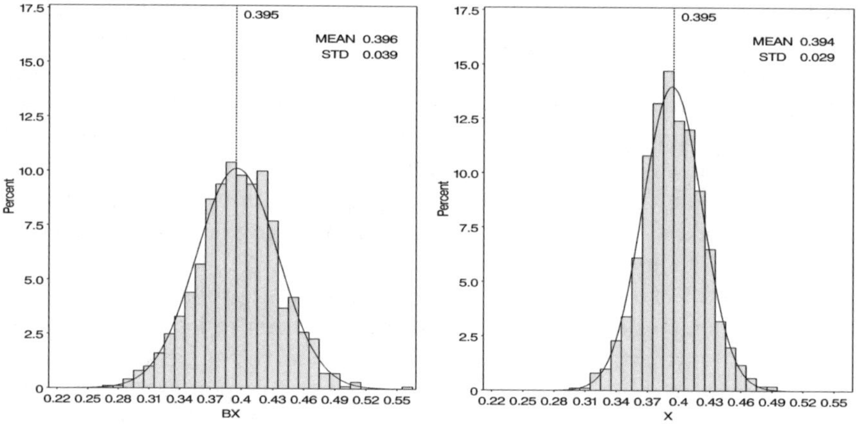

Fig. 2. Histogram for $\hat{x}^*$ from simulations. Left panel: Histogram of $\hat{x}^*$ according to L-optimal design under the bivariate probit model. Right panel: Histogram of $\hat{x}^*$ according to L-optimal design under the bivariate regression model.

4 Conclusion

The models discussed, which are based on a multivariate normal distribution, are particular cases of what is called a generalized probit model (probably we abuse established terminology but failed to find something better) and they are parsimonious and flexible. They allow a reasonable description of cases which can be met in dose finding practice. We discussed only locally optimal designs with the clear understanding that further research on adaptive and Bayesian designs similar to Dragalin and Fedorov (2006) and Dragalin et al (2006) should follow. The result on wastage of information, if one resorts to dichotomization, can be viewed as a spin off to the above.

References

Ashford JR, Sowden RR (1970) Multi-variate probit analysis. Biometrics 26:535–546

Dragalin V, Fedorov V (2006) Adaptive designs for dose-finding based on efficacy-toxicity response. Journal of Statistical Planning and Inference 136:1800–1823

Dragalin V, Fedorov V, Wu Y (2005) Optimal designs for bivariate probit model. GSK Technical Report 2005-07

Dragalin V, Fedorov V, Wu Y (2006) Adaptive designs for selecting drug combinations based on efficacy-toxicity response. Journal of Statistical Planning and Inference

Fedorov V, R G, Leonov S (2001) Optimal design for multiple responses with variance depending on unknown parameters. GSK Technical Report 2001-03

Optimal Design of Bell Experiments

Richard D. Gill[1] and Philipp Pluch[2]

[1] Mathematical Institute, Leiden University, 2300 RA Leiden, Netherlands
   gill@math.leidenuniv.nl
[2] Department of Statistics, Klagenfurt University, University Street 65-67, 9020
   Klagenfurt, Austria philipp.pluch@uni-klu.ac.at

Summary. In this paper, we discuss quantum nonlocality experiments and show
how to optimize them in respect of their experimental setup. The usage of statistical
tools from missing data and maximum likelihood are crucial. An aim of this paper
is to bring this kind of theory to the statistical community.

Key words: Bell inequalities, Aspect experiment, counterfactuals, hidden variables, missing data

1 Introduction

During the last century quantum theory developed to a successful theory
which describes the physical reality of the mesoscopic and microscopic world.
Quantum mechanics was developed with the aim of describing atoms and
explaining detected spectral lines in a measurement apparatus. It took more
than eighty years from its discovery before it was possible to experimentally
determine and visualize the most fundamental object in quantum mechanics,
the wave function. At this point the statistical community also began to take
part in developments.

There are two papers from the last century, of interest from an interpretational
point of view: Einstein et al (1935) and Bell (1964). In the first, the authors
argued that the theory of quantum mechanics is not a complete theory of
nature. So if we are able to formulate a complete theory (including hidden
variables), we are able to make precise predictions. In their opinion quantum
mechanics is incomplete if it is true. Their main assumption is nowadays
described as *local realism*. Bell showed that the most general local hidden
variable theory could not reproduce some correlations that arise in quantum
mechanics. So local realism is false if quantum mechanics is true. Clauser et al
(1969), Greenberger et al (1989) and Hardy (1993) proposed experiments to
try to establish one of the above statements. In the literature there also exists
a large number of proofs of Bell's theorem stating that quantum mechanics is
incompatible with local realistic theories of Nature.

2 Bell-type experiments

In the following we study the sets of all possible joint probability distributions of the outcomes of a *Bell-type experiment*, under two sets of assumptions, corresponding respectively to local realism and to quantum physics. Bell's theorem can also be interpreted as stating that the set of LR (local realism) probability laws is strictly contained in the QM (quantum mechanics) set.

We now want to give a description of a $p \times q \times r$ Bell experiment, where $p \geq 2$, $q \geq 2$ and $r \geq 2$ are fixed integers. The experiment involves a source and a number p of *parties*, usually called *Alice*, *Bob*, and so on in the physical literature. The source sends entangled[3] photons to Alice and each other party. Before the photons arrive at the parties, each of the parties commits him or herself to use of a particular measurement device out of some set of possible measurement setups. Let us suppose that each party can choose one of q measurements.

When the photons arrive, each of the parties measure with their chosen setting. We suppose that the possible outcomes for each of the parties can be classified into one of r different outcome categories. Given that Alice chose setting a, Bob b, and so on, there is some joint probability $p(x, y, \ldots | a, b, \ldots)$ that Alice will then observe outcome x, Bob y, $\ldots$. We suppose that the parties chose their settings a, b, $\ldots$, at random from some joint distribution with probabilities $\sigma(a, b, \ldots)$; $a, b, \cdots = 1, \ldots, q$. Altogether, one run of the whole experiment has outcome $(a, b, \ldots; x, y, \ldots)$ with probability $p(a, b, \ldots; x, y, \ldots) = \sigma(a, b, \ldots) p(x, y, \ldots | a, b, \ldots)$.

One can consider "unbalanced" experiments with possibly different numbers of measurements per party, different numbers of outcomes per party's measurement. More complicated multi-stage measurement strategies are also of interest. We stick here to the basic "balanced" designs, for ease of exposition.

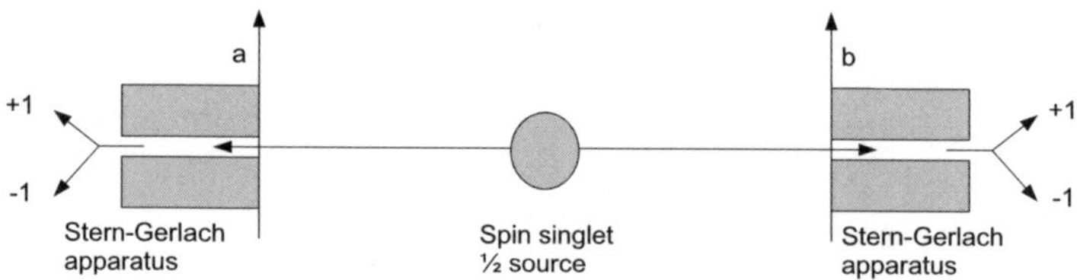

Fig. 1. Schematic Bell experiment

In fig.(1) we give an example of a $2 \times 2 \times 2$ Bell experiment. The outcomes X,

---

[3] Quantum entanglement is a fundamental concept in quantum physics in which the quantum states of two or more objects have to be described with reference to each other, even if the individual objects are spatially separated. This leads to correlations between observable physical properties of the system.

Y are conventionally coded ± 1. Alice and Bob both choose between settings numbered or labelled $1, 2$ for the setting of the measurement A, B; they can switch between $a = 1$ or 2 and $b = 1$ or 2 on Alice's and Bob's apparatus. The random variable A denotes the measurement setting of Alice and the random variable B denotes the measurement setting of Bob, both taking values in $\{1, 2\}$. The experimenter decides on the distribution σ of (A, B),

$$\sigma(a, b) = \mathbf{P}\{A = a, B = b\}. \tag{1}$$

The random variable X denotes the measurement outcome of Alice and Y the outcome of Bob's measurement. The joint distribution of (X, Y) depends on the chosen setting pair $(a, b) \in \{1, 2\}^2$. The state of the entangled qubit[4] Ψ together with the measurement settings determines four conditional distributions of (X, Y) given (A, B),

$$Q(x, y | a, b; \Psi) = \mathbf{P}^{\Psi}_{a,b}\{X = x, Y = y\} \tag{2}$$

The joint distribution of (A, B, X, Y) for a single trial of the experiment reads as

$$Q(a, b, x, y; \Psi, \sigma) = \sigma(a, b) Q^{\Psi}_{ab}(X = x, Y = y). \tag{3}$$

We explain how the quantum probabilities are computed in a moment.
Local realistic theories are characterized by the possibility of jointly modelling the outcomes given all possible settings. In this case we have four binary random variables (X_1, X_2, Y_1, Y_2). For given $a \in \{1, 2\}$, $X_a \in \{-1, 1\}$ denotes the outcome which Alice would have observed if her measurement setting was a. Take π as the probability distribution for (X_1, X_2, Y_1, Y_2) which can be thought of as an arbitrary 16-dimensional probability vector. Then $\pi \in \Pi$ determines four distributions of (X, Y) given (A, B) given by

$$\mathbf{P}^{\pi}_{a,b}(X = x, Y = y) = P(x, y | a, b; \pi) = \sum_{\substack{x_1, x_2, y_1, y_2 \in \{\pm 1\} \\ x_a = x, y_b = y}} \pi_{x_1, x_2, y_1, y_2} \tag{4}$$

A nonlocality proof is to find a state Ψ and measurements that violate local realism, i.e. there exists no π such that $P(., . | a, b; \pi) = Q(., . | a, b; \Psi) \forall (a, b) \in \{1, 2\}^2$.

The classical polytope

Local realism means the following:

> "Measurements which were not done also have outcomes; and both actual and potential measurement outcomes are independent of the measurement settings actually used by all the parties".

---

[4] is a unit of quantum information similar to a bit in classical information theory. It is described by a state vector in a two-level quantum mechanical system.

For ease of notation let us consider a two party experiment. As before, X_1, $\ldots$, X_q and Y_1, $\ldots$, Y_q denote the counterfactual outcomes of each of Alice's and Bob's possible q measurements (taking values in $\{1, \ldots, r\}$. We may think of these in statistical terms as missing data; in physical terms they are called hidden variables. Denote by A and B Alice's and Bob's random settings, each taking values in $\{1, \ldots, q\}$. The actual outcomes obse rved by Alice and Bob are therefore $X = X_A$ and $Y = Y_B$. The data arising from one run of the experiment, A, B, X, Y, has joint probability distribution function

$$p(a, b; x, y) = \sigma(a, b)\pi(X_a = x, Y_b = y).$$

Now the joint probability distribution of X_a and Y_b can be arbitrary, but in any case it is a mixture of all possible degenerate distributions of these variables. Consequently, for a fixed setting distribution σ, the joint distribution of A, B, X, Y is also a mixture of the possible distributions corresponding to degenerate (deterministic) hidden variables. Since there are only finitely many degenerate distributions when p, q and r are fixed, we see that

> under local realism and freedom, the joint probability laws of the observable data lie in a convex polytope, whose vertices correspond to degenerate hidden variables.

We call this polytope the classical polytope.

The quantum body

The basic rule for computation of probabilities in quantum physics is called Born's law: take the squared lengths of the projections of the state vector into a collection of orthogonal subspaces corresponding to the different possible outcomes. Let $\mathcal{H}$ and $\mathcal{K}$ denote two complex Hilbert spaces. We take a unit vector $|\Psi\rangle$ in $\mathcal{H} \otimes \mathcal{K}$. For each a, let L_x^a, $x = 0, \ldots, r - 1$, denote orthogonal closed subspaces of $\mathcal{H}$, together spanning all of $\mathcal{H}$. Similarly, let M_y^b denote the elements of q collections of decompositions of $\mathcal{K}$ into orthogonal subspaces. Finally, define $Q^\Psi(x, y|a, b) = \|\Pi_{L_x^a} \otimes \Pi_{M_y^b}|\Psi\rangle\|^2$, where Π denotes orthogonal projection into a closed subspace. This defines a collection of joint probability distributions of X and Y, indexed by (a, b). Note that the quantum probabilities depend not just on Ψ but also on the collections of subspaces indicated above; when we write Ψ as a parameter of these probability distributions, we are really thinking of the whole collection of state and subspaces.
We note following facts:

> The collection of all possible quantum probability laws of A, B, X, Y (for fixed setting distribution σ) forms a closed convex body containing the local polytope.

The no-signalling polytope:

The two convex bodies so far defined are forced to live in a lower dimensional affine subspace, by the basic normalization properties of probability distributions: $\sum_{x,y} P(a,b|x,y) = \sigma(a,b)$ for all a, b. Moreover, probabilities are necessarily non negative, so this restricts us further to some convex polytope. However, physics (locality) implies another collection of equality constraints, putting us into a still smaller affine subspace. These constraints are called the no-signalling constraints: $\sum_y p(a,b;x,y)$ should be independent of b for each a and x, and vice versa. It is easy to check that both the local realist probability laws, and the quantum probability laws, satisfy no-signalling. Quantum mechanics is certainly a local theory as far as manifest (as opposed to hidden) variables are concerned.

The set of probability laws satisfying no-signalling is therefore another convex polytope in a low dimensional affine subspace; it contains the quantum body, which in turn contains the classical polytope.

3 GHZ paradox

van Dam et al (2005) proved Bell's theorem in the $3 \times 2 \times 2$ case. Under local realism we can introduce hidden variables X_1, X_2, Y_1, Y_2, Z_1, Z_2, standing for the counterfactual outcomes of a three party experiment. The experimenters' settings are (assigned by) setting 1 or 2. These variables are binary, their possible outcomes are ± 1. Now note that

$$(X_1 Y_2 Z_2).(X_2 Y_1 Z_2).(X_2 Y_2 Z_1) = (X_1 Y_1 Z_1).$$

Thus, if the setting patterns $(1,2,2)$, $(2,1,2)$ and $(2,2,1)$ *always* result in X, Y and Z with $XYZ = +1$, it will also be the case that the setting patte rn $(1,1,1)$ *always* results in X, Y and Z with $XYZ = +1$.
Next define the 2×2 matrices

$$\sigma_1 = \begin{pmatrix} 0 & 1 \\ 1 & 0 \end{pmatrix}, \quad \sigma_2 = \begin{pmatrix} 1 & 0 \\ 0 & -1 \end{pmatrix}.$$

One can easily check that $\sigma_1 \sigma_2 = -\sigma_2 \sigma_1$, (anticommutation), $\sigma_1^2 = \sigma_2^2 = \mathbf{1}$, the 2×2 identity matrix, since σ_1 and σ_2 are both Hermitean with eigenvalues ± 1.
Now define matrices $X_1 = \sigma_1 \otimes \mathbf{1} \otimes \mathbf{1}$, $X_2 = \sigma_2 \otimes \mathbf{1} \otimes \mathbf{1}$, $Y_1 = \mathbf{1} \otimes \sigma_1 \otimes \mathbf{1}$, $Y_2 = \mathbf{1} \otimes \sigma_2 \otimes \mathbf{1}$, $Z_1 = \mathbf{1} \otimes \mathbf{1} \otimes \sigma_1$, $Z_2 = \mathbf{1} \otimes \mathbf{1} \otimes \sigma_2$. It is now easy to check that

$$(X_1 Y_2 Z_2).(X_2 Y_1 Z_2).(X_2 Y_2 Z_1) = -(X_1 Y_1 Z_1),$$

and that $(X_1 Y_2 Z_2)$, $(X_2 Y_1 Z_2)$, $(X_2 Y_2 Z_1)$ and $(X_1 Y_1 Z_1)$ commute with one another.

Since these four 8×8 Hermitean matrices commute they can be simultaneously diagonalized. Some further elementary considerations lead one to conclude the existence of a simultaneous eigenvector $|\psi\rangle$ common to all four, with eigenvalues $+1, +1, +1, -1$ respectively. We take this to be the state $|\psi\rangle$, with the three Hilbert spaces all equal to $\mathbb{C}^2$. We take the two orthogonal subspaces for the 1 and 2 measurements of the parties all to be the two eigenspaces of σ_1 and σ_2 respectively. This generates quantum probabilties such that the setting patterns $(1,2,2)$, $(2,1,2)$ and $(2,2,1)$ *always* result in X, Y and Z with $XYZ = +1$, while the setting pattern $(1,1,1)$ *always* results in X, Y and Z with $XYZ = -1$.

Thus we have shown that a vector of quantum probabilities exists, which cannot possibly occur under local realism. Since the classical polytope is closed, the corresponding quantum law must be strictly outside the classical polytope. It therefore violates a generalized Bell inequality corresponding to some face of the classical polytope, outside of which it must therefore lie.

4 GHZ experiment

How should one design good Bell experiments, and what is the connection of all this physics with mathematical statistics? Indeed there are many connections — as already alluded to; the hidden variables of a local realist theory are simply the missing data of a nonparametric missing data problem.

In the laboratory one creates the state Ψ by a source of entangled photons, and the measurement devices of Alice and Bob by assemblages of polarization filters, beam splitters and photodetectors implementing thereby the measurements corresponding to the subspaces L_a^x, etc. One also settles on a joint setting probability π. One repeats the experiment many times, hoping to indeed observe a quantum probability law lying outside the classical polytope, i.e., violating a Bell inequality. The famous Aspect et al (1982) experiment implemented this program in the $2 \times 2 \times 2$ case, violating the so-called CHSH inequality (which we will describe later) by a large number of standard deviations. What is being done here is statistical hypothesis testing, where the null hypotheses is local realism, the alternative is quantum mechanics; the alternative being true by design of the experimenter and validity of quantum mechanics.

5 How to compare different experiments

In the laboratory one will prefer an experiment under which the distance from the quantum physical reality is far from the nearest local realistic or classical description. Physicists have invested a lot of research into Euclidean distance for this purpose; but it is not clear what this distance means operationally, and whether it is comparable over experiments of different types. Moreover the Euclidean distance is altered by taking different setting distributions π. It shows that Euclidean distance is closely related to *noise resistance*, a kind of robustness to experimental imperfection. As one mixes the quantum probability distribution more and more with completely random, uniform outcomes, corresponding to pure noise in the photodetectors, the quantum probability distribution shrinks towards the center of the classical polytope, at some point passing through one of its faces. The amount of noise which can be allowed while still admitting violation of local realism is directly related to Euclidean distance.

6 Kullback-Leibler divergence and statistical strength

van Dam et al (2005) proposed use of relative entropy,

$$D(q:p) = \sum_{abxy} q(abxy) \log_2(q(abxy)/p(abxy)),$$

where q now stands for the "true" probability distribution under some quantum description of reality, and p stands for a local realist probability distribution. They evaluate $\sup_q \inf_p D(q:p)$ where the supremum is taken over parameters at the disposal of the experimenter (the quantum state $|\psi\rangle$, the measurement projectors, the set ting distribution π); while the infimum is taken over probability distributions of outcomes given settings allowed by local realism. (Thus the q and p in the above supremum and infimum roles stand for something different from the probability laws q and p lying in the quantum body and classical polytope respectively; hopefully this abuse of notation may be excused.)

They argue that this relative entropy gives direct information about the number of trials of the experiment required to give a desired level of confidence in the conclusion of the experiment. Two experiements which differ by a factor 2 are such that the one with the smaller divergence needs to be repeated twice as often as the other in order to give an equally convincing rejection of local realism.

Moreover, optimizing over different sets of quantum parameters leads to various measures of "strength of non-locality". For instance, one can ask what is the best experiment based on a given entangled state $|\psi\rangle$? Exper iments of different format can be compared with one another, possibly discounting the relative entropies according to the numbers of quantum systems involved in the different experiments in an obvious way. They proved that the interior infimum is basically the computation of a nonparametric maximum likelihood estimator in a missing data problem; so statistical methods can be used.

Let Z be an arbitrary finite set. For a distribution Q over Z, $Q(z)$ denotes the probability of event $\{z\}$. For two (arbitrary) distributions Q and P defined over Z, the Kullback-Leibler (KL) divergence from Q to P is defined as

$$D(Q\|P) = \sum_{z \in Z} Q(z) \log \frac{Q(z)}{P(z)} \tag{5}$$

where the logarithm is to base 2. We use the conventions that, for $y > 0$, $y \log 0 := \infty$, and $0 \log 0 := \lim_{y \to 0} y \log y = 0$. As above we have fixed our experiment and σ; from a Bayesian data analyst's point of view we will use the KL distance as

$$\frac{-\log(P(\mathcal{P}^\sigma | \text{data}))}{N} \approx \inf_\pi D(Q^{\sigma\Psi} \| P^{\sigma\Psi}). \tag{6}$$

In the frequentist setup we consider the p-value of the best test statistics of $H_0 : \mathcal{P}^\sigma$ versus $H_1 : \mathcal{Q}^\sigma$ and derive

$$\frac{-\log(p(\text{data}))}{N} \approx \inf_\pi D(Q^{\sigma\Psi} \| P^{\sigma\Psi}). \tag{7}$$

This measure of statistical strength can be performed for different experimental setups in a quantum experiment and is used for evaluating the performance of a quantum experiment. For details on this we refer to van Dam et al (2005).

7 Conclusions

Bell experiments offer a rich field involving many statistical ideas, beautiful mathematics, and offering some exciting challenges. Moreover it is a hot topic in quantum information and quantum optics. Much remains to be done.

References

Aspect A, Dalibard J, G R (1982) Experimental test of bell's inequalities using time-varying analyzers. Phys Rev Lett 49:1804–1807

Bell J (1964) On the einstein-podolsky-rosen paradox. Physics 1:195–200

Clauser J, Horne M, Shimony A, Holt R (1969) Proposed experiment to test local hidden-variable theories. Phys Rev Lett 23:880–884

van Dam W, Gill RD, Grünwald PD (2005) The statistical strength of nonlocality proofs. IEEE – Trans Inf Theory 51:2812–2835

Einstein A, Podolsky B, Rosen N (1935) Can quantum-mechanical description of physical reality be considered complete? Phys Rev 47:777–780

Greenberger D, Horne M, Zeilinger A (1989) Going beyond bell's theorem. In: Kafatos M (ed) Bell's Theorem, Quantum Theory and Conceptions of the Universe, Kluwer, Dordrecht

Hardy L (1993) Nonlocality for two particles without inequalities for almost all entangled states. Phys Rev Lett 71:1665–1668

A Comparison of Efficient Designs for Choices Between Two Options

Heiko Großmann[1], Heinz Holling[2], Ulrike Graßhoff[3], and Rainer Schwabe[3]

[1] School of Mathematical Sciences, Queen Mary, University of London, Mile End Road, London E1 4NS, United Kingdom `h.grossmann@qmul.ac.uk`
[2] Psychologisches Institut IV, Westfälische Wilhelms-Universität Münster, Fliednerstr. 21, 48149 Münster, Germany `holling@psy.uni-muenster.de`
[3] Institut für Mathematische Stochastik, Otto-von-Guericke-Universität Magdeburg, PF 4120, 39016 Magdeburg, Germany `ulrike.grasshoff@mathematik.uni-magdeburg.de`, `rainer.schwabe@mathematik.uni-magdeburg.de`

Summary. Optimal designs for choice experiments with choice sets of size two are frequently derived under the assumption that all model parameters in a multinomial logit model are equal to zero. In this case, optimal designs for linear paired comparisons are also optimal for the choice model. It is shown that the methods for constructing linear paired comparison designs often require a considerably smaller number of choice sets when the parameters of primary interest are main effects.

Key words: choice experiments, optimal design, paired comparisons

1 Introduction

The question why people prefer some products or services over others is easy to ask but usually not easy to answer. Generally, a number of factors or attributes enter into the evaluation of the available options and an advantage of, say, a mobile phone in terms of display size may well be offset by its tiny keypad. Choice experiments are by now widely used to learn how the various attributes influence decisions, and applications in marketing, health economics and other fields abound. Excellent descriptions of these experiments, their background, underlying models and analysis can be found in textbooks such as Louviere et al (2000) and Train (2003).

A typical choice experiment consists of a series of choice tasks. Each task offers a choice set of options and asks the respondent to select, for example, the most attractive alternative. The choice sets presented are generated according to an experimental design which specifies attribute levels for the options in each set. Originating with the work of Louviere and Woodworth (1983) the optimal and efficient design of choice experiments has attracted considerable attention. Most of these developments have been summarized by Großmann et al (2002) and Louviere et al (2004) and new results continue to appear (Sándor and Wedel (2005); Kessels et al (2006)).

Models for choice data are nonlinear in the unknown model parameters. For the common alphabetic optimality criteria it is therefore not possible to find designs that are optimal regardless of the parameter values. Yet, optimal designs that do not depend on the parameters can be derived when it is assumed that all model parameters are equal to zero or, equivalently, that within each choice set each option is chosen with the same probability. More precisely, the optimal design problem for the choice model is then equivalent to the corresponding problem for an approximating linear model (see e.g. Großmann et al, 2002). Under the above assumption optimal choice designs have been derived in a series of papers by researchers at the University of Technology, Sydney (Street et al (2001); Burgess and Street (2003); Street and Burgess (2004a); Burgess and Street (2005)). For the practically important case of choices between two options these designs can be contrasted with optimal designs for linear paired comparison models (Graßhoff et al (2003, 2004)) which are also optimal for choice experiments when the model parameters are assumed to be zero.

This note aims to provide guidance for the design of choice experiments with choice sets of size two by comparing the above design approaches. In particular, we will be consider which designs are available when the attributes in an experiment have varied numbers of levels and how many choice sets are required. After a discussion of choice and linear paired comparison models the main part of the paper is devoted to designs for main effects models. Subsequently, we very briefly comment on the few design approaches available to date for models which include interaction terms.

2 Models

Suppose there are K attributes that are assumed to drive the preferences for the options in a product category. Although some of these may be quantitative in nature, usually only a finite set $\mathcal{X}_k = \{1, \ldots, v_k\}$ of levels is investigated for each attribute k in a choice experiment. The options are then represented by combinations of attribute levels. In what follows, we restrict ourselves to choices between two options. The first option in each choice set is denoted by $\mathbf{s} = (s_1, \ldots, s_K)$ and the second one by $\mathbf{t} = (t_1, \ldots, t_K)$, which are both elements of $\mathcal{X}_1 \times \ldots \times \mathcal{X}_K$.

Many choice models can be derived from random utility theory (see e.g. Train, 2003). To this end, it is assumed that the utility $U(\mathbf{s})$ obtained from each option $\mathbf{s}$ can be additively decomposed into a deterministic part $V(\mathbf{s})$ and a random part ε. Furthermore, it is assumed that a decision maker always chooses the option with the highest utility. Thus if Y denotes the binary random variable that equals 1 when $\mathbf{s}$ is chosen and 0 otherwise, this amounts to $P(Y = 1) = P(U(\mathbf{s}) > U(\mathbf{t}))$ for any two options $\mathbf{s}$ and $\mathbf{t}$. The additional assumption that all ε variables are independent identically distributed according to the Gumbel distribution gives rise to the multinomial logit (MNL) model which is most commonly used in practice and for which the choice probabilities take the form $P(Y = 1) = e^{V(\mathbf{s})}/(e^{V(\mathbf{s})} + e^{V(\mathbf{t})})$.

The deterministic part of the utility function is related to the attributes by a minimal linear parametrization $V = \mathbf{f}'\boldsymbol{\beta}$, where $\mathbf{f} = (f_1, \ldots, f_p)'$ is a vector of p known regression functions and $\boldsymbol{\beta}$ contains the unknown model parameters. The choice probabilities in the MNL model can then be written as $P(Y = 1) = e^{(\mathbf{f}(\mathbf{s})-\mathbf{f}(\mathbf{t}))'\boldsymbol{\beta}}/(1 + e^{(\mathbf{f}(\mathbf{s})-\mathbf{f}(\mathbf{t}))'\boldsymbol{\beta}})$ from which it is easily seen that for choices between two options the MNL model is equivalent to logistic regression with predictors given by the components of $(\mathbf{f}(\mathbf{s}) - \mathbf{f}(\mathbf{t}))'$.

Linear paired comparison models were developed for situations in which again two options are compared at a time, but where the response variable is (at least approximately) continuous. Such variables occur, for example, when respondents are asked to state how much they prefer one option over the other. In this case, the evaluation of the pair $(\mathbf{s}, \mathbf{t})$ can be described by the model equation $\tilde{Y} = (\mathbf{f}(\mathbf{s}) - \mathbf{f}(\mathbf{t}))'\tilde{\boldsymbol{\beta}} + \tilde{\epsilon}$ where $\tilde{\boldsymbol{\beta}}$ denotes the unknown parameter vector and the random variable $\tilde{\epsilon}$ has mean zero. Observations corresponding to different pairs are assumed to be uncorrelated with constant variance σ^2. A positive value of $\tilde{Y}$ indicates the degree to which $\mathbf{s}$ is preferred over $\mathbf{t}$ and similarly a negative value of $\tilde{Y}$ corresponds to the opposite preference. As an aside, it is worth noting that the linear paired comparison model can be used to estimate parameter vectors $\tilde{\boldsymbol{\beta}}$ for individual respondents, whereas the MNL model usually requires a group-level analysis.

An exact design ξ_N for a choice experiment specifies the N choice sets that are presented for evaluation. When each choice set consists of two options, ξ_N can be represented as an N-tuple of pairs, that is $\xi_N = ((\mathbf{s}_1, \mathbf{t}_1), \ldots, (\mathbf{s}_N, \mathbf{t}_N))$. Notice that not all pairs specified by ξ_N need to be different. Moreover, note that here the choice sets are represented by ordered pairs so that two pairs $(\mathbf{s}_i, \mathbf{t}_i)$ and $(\mathbf{s}_j, \mathbf{t}_j)$ with $\mathbf{s}_j = \mathbf{t}_i$ and $\mathbf{t}_j = \mathbf{s}_i$ are regarded as different.

Exact designs for the linear paired comparison model are specified in exactly the same manner. The common optimality criteria for measuring the quality of a design $\xi_N = ((\mathbf{s}_1, \mathbf{t}_1), \ldots, (\mathbf{s}_N, \mathbf{t}_N))$ in this model are based on the normalized information matrix $\mathbf{M}(\xi_N) = \frac{1}{N}\mathbf{X}'\mathbf{X}$, where $\mathbf{X}$ is the $N \times p$ design matrix whose ith row is the vector $(\mathbf{f}(\mathbf{s}_i) - \mathbf{f}(\mathbf{t}_i))'$, $i = 1, \ldots, N$. Similarly, measures for the performance of ξ_N in the MNL model are usually functionals of the normalized information matrix $\mathbf{M}(\xi_N; \boldsymbol{\beta})$ in that model which depends on the unknown parameter vector $\boldsymbol{\beta}$. Yet if $\boldsymbol{\beta} = \mathbf{0}$ or if, equivalently, the choice probabilities are equal to $\frac{1}{2}$, it follows that $\mathbf{M}(\xi_N; \boldsymbol{\beta}) = \frac{1}{4}\mathbf{M}(\xi_N)$. The important consequence for the common optimality criteria such as the D-criterion is then that within the class $\Xi = \{\xi_N : N \in \mathbb{N}\}$ any optimal design for the linear paired comparison model is also optimal for the MNL model and vice versa, whenever $\boldsymbol{\beta} = \mathbf{0}$ is assumed. With regard to the definition of optimality in terms of the normalized information matrix we note that if $\xi^* \in \Xi$ is a design for which the criterion function based on the normalized information matrix attains its optimum, and if for a given number of choice sets N a design ξ_N with $\mathbf{M}(\xi_N) = \mathbf{M}(\xi^*)$ can be found, then this design is also optimal among all designs of size N when the criterion function is evaluated for the non-normalized information matrix $\mathbf{X}'\mathbf{X}$.

When optimal designs for the MNL model are derived under the above assumption it is worthwhile to note the following implications. First, the designs are generally not optimal, if $\boldsymbol{\beta} \neq \mathbf{0}$. Second, the MNL model is implicitly replaced by a simpler linear model and optimal designs for this linear model are also optimal for the MNL model. Finally, since optimal designs are not unique, optimal designs for the surrogate linear model can be more economical in that they require a smaller number of choice sets.

In what follows we present the first comparison of efficient designs for choices between two options that were derived under the assumption $\boldsymbol{\beta} = \mathbf{0}$ and optimal designs for the linear paired comparison model. In doing so we hope to provide orientation for practical applications.

3 Designs for estimating main effects

Many choice experiments focus on the mean effects of the various attribute levels on the responses assuming that attribute interactions are negligible. The deterministic part of the utility obtained from option $\mathbf{s}$ is then taken to be the sum $V(\mathbf{s}) = \sum_{k=1}^{K} \beta_{k,s_k}$ where β_{k,s_k} is interpreted as the utility of the level s_k of attribute k.

Such experiments can be described by a main effects model with a minimal parametrization derived from suitable identifiability conditions. Here we adopt the standard conditions $\sum_{i=1}^{v_k} \beta_{k,i} = 0$ for every $k = 1, \ldots, K$ which implies that the attribute levels in the vector $\mathbf{f}$ are effects–coded. More precisely, we have $\mathbf{f} = (\mathbf{f}_1', \ldots, \mathbf{f}_K')'$ with components $\mathbf{f}_k$, $k = 1, \ldots, K$, where $\mathbf{f}_k(i)$ is the ith unit vector of length $p_k = v_k - 1$ for $i = 1, \ldots, v_k - 1$ and $\mathbf{f}_k(i) = -\mathbf{1}_{p_k}$ for $i = v_k$. As usual, $\mathbf{1}_a$ denotes a column vector of length a with all entries equal to 1. The parameter vector in the MNL model can be partitioned accordingly as $\boldsymbol{\beta} = (\boldsymbol{\beta}_1', \ldots, \boldsymbol{\beta}_K')'$ with $\boldsymbol{\beta}_k = (\beta_{k,1}, \ldots, \beta_{k,p_k})'$ for $k = 1, \ldots, K$. In total, there are then $p = p_1 + \ldots + p_K$ model parameters.

Most of the optimal designs for choice and paired comparison experiments have been derived using the D-criterion which aims at maximizing the determinant of the normalized information matrix. When all attributes have the same number of levels $v_1 = \ldots = v_K = v$, optimal designs which require $\frac{1}{2}v^K(v-1)^K$ choice sets or pairs have been constructed by Graßhoff et al (2004). For the special case $v = 2$ the same designs were also derived by Street et al (2001) using a different approach. Generally, the number of responses required by these designs is much too large to be useful in practice. Constructions that yield smaller optimal designs were presented by Street and Burgess (2004a) for $v = 2$ and for general v by Graßhoff et al (2004) and Burgess and Street (2005).

The method described by Street and Burgess (2004a, p. 188) uses a regular fractional factorial of resolution *III* or higher for K two-level factors to generate the choice sets. Each row of the fractional factorial represents the first option in a pair. The corresponding second option is then obtained by taking the foldover of the first one. Using the coding 0 for the low and 1 for the high level of each factor, the implication is that for generating the second option every 0 in the first option is replaced by 1 and every 1 by 0. With this coding, the second options can equivalently be constructed by adding the vector $\mathbf{1}'_K$ to every row of the fractional factorial using modulo 2 arithmetic. If for any two options $\mathbf{s}$ and $\mathbf{t}$ both $(\mathbf{s}, \mathbf{t})$ and $(\mathbf{t}, \mathbf{s})$ are generated by this procedure, then only one of these pairs is used as a choice set in the final design. We note that in our notation the low level of each attribute would be represented by 1 and the high level by 2. The fractional factorial then defines the option $\mathbf{s}$ in every pair and the second option $\mathbf{t}$ is obtained by setting $t_k = 2$ when $s_k = 1$ and $t_k = 1$ when $s_k = 2$ for every k.

In general, the minimum number of choice sets required by this foldover construction is equal to 2^{K-m} where m is the largest number for which a regular two-level fractional factorial of resolution *III* or higher can be generated by means of m defining contrasts. If there exists a corresponding fractional factorial such that the m defining contrasts have even wordlength (see e.g. Wu and Hamada, 2000, p. 157), this number can be reduced further to 2^{K-m-1}.

A different technique, which is also applicable when the number of levels v is larger than two, uses Hadamard matrices to construct designs with a reduced number of pairs (Graßhoff et al, 2004, p. 366). As originally presented, this construction produces the design in the form of the design matrix. Alternatively, the method can be described as follows. For K factors with v levels each, consider the smallest Hadamard matrix of order $u \geq K$ and choose a $u \times K$ submatrix $\mathbf{H}$ of that matrix. Generate a column vector $\mathbf{a}$ whose components are the $v(v-1)/2$ ordered pairs corresponding to the subsets of size two of $\{1, \ldots, v\}$ listed in some arbitrary order, for example, $\mathbf{a} = ((1,2), (1,3), \ldots, (1,v), (2,3), \ldots, (v-1,v))'$. Denote by $\mathbf{a}^-$ the vector obtained by replacing each pair (i, j) in $\mathbf{a}$ with (j, i). Furthermore, replace every 1 in $\mathbf{H}$ with $\mathbf{a}$ and every -1 with $\mathbf{a}^-$ respectively. The resulting array has N rows and K columns where $N = \frac{1}{2}uv(v-1)$ and is denoted by $\mathbf{D}$. The element in row n and column k of $\mathbf{D}$ is an ordered pair $(x_{n,k}, y_{n,k})$. For $n = 1, \ldots, N$ set $\mathbf{s}_n = (x_{n,1}, \ldots, x_{n,K})$ and $\mathbf{t}_n = (y_{n,1}, \ldots, y_{n,K})$. The design ξ_N consisting of the pairs $(\mathbf{s}_n, \mathbf{t}_n)$ is then optimal.

For $v = 2$ the foldover construction and the Hadamard approach both require 4 choice sets when $K = 3$ or $K = 4$, and 8 sets when $5 \leq K \leq 8$. However, a comparison of the designs for arbitrary values of K is complicated. The reason is, that in order to determine the minimum number of choice sets for the foldover construction it is necessary to investigate whether there exists a regular fractional factorial of resolution *III* or higher generated by m (as defined above) defining contrasts of even wordlength. Generally, this investigation requires the use of some algorithm as, for example, the one presented by Laycock and Rowley (1995). In practice, implementations of such algorithms are rarely available, however, and fractional factorials are usually taken from textbooks or similar sources. Suppose then that for a given number of attributes K a regular fraction of resolution III or higher with the least possible number of runs has been identified in this way, which is subsequently used in the foldover construction. Suppose further that of the m corresponding defining contrasts at least one has odd wordlength. In this case, if there exists a regular fraction of resolution III or higher with the same number of runs that is generated by m defining contrasts which all have even wordlength, then the number of choice sets resulting from the foldover construction will generally be considerably larger than the number of sets required by the Hadamard approach.

For $v > 2$ the Hadamard approach can also be compared with a general construction for situations where the attributes can have different numbers of levels (Burgess and Street, 2005, p. 296). The optimal designs generated by this approach consist of at least $v_1 \times \ldots \times v_K$ choice sets. When $v_k = v$ for $k = 1, \ldots, K$, this lower bound is equal to v^K. By contrast, the Hadamard construction produces optimal designs with $\frac{1}{2}uv(v-1)$ sets, where $u \leq K+3$ for all practically relevant values of K. In fact, the inequality holds at least for every $K \leq 424$ since Hadamard matrices are known to exist for every multiple of 4 up to this value (e.g. Graßhoff et al, 2004, p. 366). Thus the number of choice sets grows only linearly in the number of factors, whereas this number increases exponentially for the construction proposed by Burgess and Street (2005).

For specific values of v some additional constructions are available which often produce even smaller optimal designs than the Hadamard approach. Corresponding results for $v = 3, 4, 5$ are presented by Graßhoff et al (2004). For example, these methods can be used to generate an optimal design for $K = 5$ and $v = 4$ with 24 choice sets, whereas 48 sets are required by the Hadamard approach as well as an optimal design considered by Street et al (2005, p. 467). Specific constructions sometimes even produce saturated designs; that is, the number of choice sets equals the number of model parameters.

Only a few results are available for the case of attributes with different numbers of levels $v_1, \ldots, v_K$. The method of Burgess and Street (2005, p. 296) is generally applicable, but the number of choice sets can quickly become very large. In fact, the number of sets required is generally a multiple of $v_1 \times \ldots \times v_K$. Another approach is similar in vein to the Hadamard construction and replaces the symbols in the columns of a mixed orthogonal array of strength two with pairs of attribute levels (Graßhoff et al, 2004, p. 368). This method is less widely applicable, but yields optimal designs with a comparatively small number of choice sets. Graßhoff et al (2004) illustrate the construction for six factors with three and one factor with four levels with a design consisting of 18 choice sets. By contrast, $2916 \times 3 = 8748$ sets are needed when the method of

Burgess and Street (2005) is used. Moreover, the method based on orthogonal arrays is particularly useful when some attributes in a choice experiment have two and all other attributes have three levels. For example, optimal designs with 36 choice sets are available for experiments with up to 11 two-level and up to 12 three-level attributes.

4 Interactions

Optimal designs for estimating main effects and two-factor interactions were derived by van Berkum (1987), Street et al (2001) and Graßhoff et al (2003). Generally, these designs are too large for practical applications. For the case of two-level attributes Street and Burgess (2004a) presented a generalization of the foldover construction for main effects models. The size of the corresponding designs is a multiple of the number of runs of a two-level regular fractional factorial of resolution V. When the attributes have different numbers of levels, results are only available for $K = 2, 3, 4$ attributes (Burgess and Street (2005)).

5 Concluding remarks

The rapidly growing number of applications of choice experiments in many fields has led to an increased interest in choice designs in recent years. One strand of research has focused on deriving optimal designs under the assumption that the parameters in the multinomial logit model are equal to zero. For choices between two options this assumption implies that optimal designs for linear paired comparison models are also optimal within the context of choice experiments. When the parameters of primary interest are main effects, these designs are particularly attractive since they require a comparatively small number of choice sets.

Further research is needed to derive optimal designs with practical numbers of choice sets for models with interaction terms. Moreover, there are other important problems, such as the construction of designs for partial profiles (Graßhoff et al (2003, 2004); Großmann et al (2006)) and for choice experiments containing a no-choice option (Street and Burgess (2004b)), which need further study.

Acknowledgement. This research was supported in part by the Deutsche Forschungsgemeinschaft (DFG) under grant HO 1286/2-3.

References

van Berkum E (1987) Optimal paired comparison designs for factorial and quadratic models. Journal of Statistical Planning and Inference 15:265–278

Burgess L, Street D (2003) Optimal designs for 2^k choice experiments. Communications in Statistics—Theory and Methods 32:2185–2206

Burgess L, Street D (2005) Optimal designs for choice experiments with asymmetric attributes. Journal of Statistical Planning and Inference 134:288–301

Graßhoff U, Großmann H, Holling H, Schwabe R (2003) Optimal paired comparison designs for first-order interactions. Statistics 37:373–386

Graßhoff U, Großmann H, Holling H, Schwabe R (2004) Optimal designs for main effects in linear paired comparison models. Journal of Statistical Planning and Inference 126:361–376

Großmann H, Holling H, Schwabe R (2002) Advances in optimum experimental design for conjoint analysis and discrete choice models. In: Franses P, Montgomery A (eds) Advances in Econometrics, vol 16, Econometric Models in Marketing, JAI Press, Amsterdam, pp 93–117

Großmann H, Holling H, Graßhoff U, Schwabe R (2006) Optimal designs for asymmetric linear paired comparisons with a profile strength constraint. Metrika 64:109–119

Kessels R, Goos P, Vandebroek M (2006) A comparison of criteria to design efficient choice experiments. Journal of Marketing Research 43:409–419

Laycock P, Rowley P (1995) A method for generating and labelling all regular fractions or blocks for q^{n-m}-designs. Journal of the Royal Statistical Society B 57:191–204

Louviere J, Woodworth G (1983) Design and analysis of simulated consumer choice or allocation experiments: An approach based on aggregate data. Journal of Marketing Research 20:350–367

Louviere J, Hensher D, Swait J (2000) Stated Choice Methods: Analysis and Application. Cambridge University Press, Cambridge

Louviere J, Street D, Burgess L (2004) A 20+ years' retrospective on choice experiments. In: Wind Y, Green P (eds) Market Research and Modeling: Progress and Prospects, Kluwer Academic Publishers, Boston, MA, pp 201–214

Sándor Z, Wedel M (2005) Heterogeneous conjoint choice designs. Journal of Marketing Research 42:210–218

Street D, Burgess L (2004a) Optimal and near-optimal pairs for the estimation of effects in 2-level choice experiments. Journal of Statistical Planning and Inference 118:185–199

Street D, Burgess L (2004b) Optimal stated preference choice experiments when all choice sets contain a specific option. Statistical Methodology 1:37–45

Street D, Bunch D, Moore B (2001) Optimal designs for 2^k paired comparison experiments. Communications in Statistics—Theory and Methods 30:2149–2171

Street D, Burgess L, Louviere J (2005) Quick and easy choice sets: Constructing optimal and nearly optimal stated choice experiments. International Journal of Research in Marketing 22:459–470

Train K (2003) Discrete Choice Methods with Simulation. Cambridge University Press, Cambridge

Wu C, Hamada M (2000) Experiments: Planning, Analysis, and Parameter Design Optimization. Wiley, New York

D-optimal Designs for Logistic Regression in Two Variables

Linda M. Haines[1], Gaëtan Kabera[2], Principal Ndlovu[2] and Timothy E. O'Brien[3]

[1] Department of Statistical Sciences, University of Cape Town, Rondebosch 7700, South Africa. lhaines@stats.uct.ac.za
[2] School of Statistics and Actuarial Science, University of KwaZulu-Natal, Pietermaritzburg 3200, South Africa.
201291190@ukzn.ac.za   ndlovup@ukzn.ac.za
[3] Department of Mathematics and Statistics, Loyola University Chicago, 6525 N. Sheridan Road, Chicago, Illinois 60626, U.S.A.   teobrien@gmail.com

Summary. In this paper locally D-optimal designs for the logistic regression model with two explanatory variables, both constrained to be greater than or equal to zero, and no interaction term are considered. The setting relates to dose-response experiments with doses, and not log doses, of two drugs. It is shown that there are two patterns of D-optimal design, one based on 3 and the other on 4 points of support, and that these depend on whether or not the intercept parameter β_0 is greater than or equal to a cut-off value of -1.5434. The global optimality of the designs over a range of β_0 values is demonstrated numerically and proved algebraically for the special case of the cut-off value of β_0.

Key words: D-optimality, logistic regression in two variables

1 Introduction

Logistic regression models with two or more explanatory variables are widely used in practice, as for example in dose-response experiments involving two or more drugs. There has however been only sporadic interest in optimal designs for such models, with the papers of Sitter and Torsney (1995), Atkinson and Haines (1996), Jia and Myers (2001), Torsney and Gunduz (2001) and Atkinson (2006) and the thesis of Kupchak (2000) providing valuable insights into the underlying problems. In the present study a simple setting, that of the logistic regression model in two explanatory variables with no interaction term, is considered. The variables are taken to be doses, and not log doses, of two drugs and are thus constrained to be greater than or equal to zero. The aim of the study is to construct locally D-optimal designs, and in so doing to identify patterns in the designs that may depend on the values of the parameters in the model, and in addition to demonstrate the global optimality of these designs both numerically and algebraically.

2 Preliminaries

Consider the logistic dose-response model defined by

$$\text{logit}(p) = \beta_0 + \beta_1 d_1 + \beta_2 d_2$$

where p is the probability of success, β_0, β_1 and β_2 are unknown parameters and d_1 and d_2 are doses, not log doses, of two drugs such that $d_1 \geq 0$ and $d_2 \geq 0$. Responses are assumed to increase with dose for both drugs and the parameters β_1 and β_2 are thus taken to be greater than 0. In addition, from a practical point of view, the response at the control $d_1 = d_2 = 0$ is assumed to be less than 50% and the intercept parameter β_0 is accordingly taken to be less than 0. Note that, without loss of generality, the model can be expressed in terms of the scaled doses $z_1 = \beta_1 d_1$ and $z_2 = \beta_2 d_2$ as

$$\text{logit}(p) = \beta_0 + z_1 + z_2 \text{ with } z_1 \geq 0 \text{ and } z_2 \geq 0. \tag{1}$$

Then the information matrix for the parameters $\beta = (\beta_0, \beta_1, \beta_2)$ at a single observation $z = (z_1, z_2)$ is given by

$$M(\beta; z) = g(z)g(z)^T = \frac{e^u}{(1 + e^u)^2} \begin{bmatrix} 1 & z_1 & z_2 \\ z_1 & z_1^2 & z_1 z_2 \\ z_2 & z_1 z_2 & z_2^2 \end{bmatrix}$$

where $g(z) = \dfrac{e^{\frac{u}{2}}}{(1 + e^u)}(1, z_1, z_2)$ and $u = \beta_0 + z_1 + z_2$.

Consider now an approximate design which puts weights w_i on the distinct points $z_i = (z_{1i}, z_{2i})$ for $i = 1, \ldots r$, expressed as

$$\xi = \left\{ \begin{array}{ccc} (z_{11}, z_{21}), & \ldots, & (z_{1r}, z_{2r}) \\ w_1, & \ldots, & w_r \end{array} \right\} \quad \text{where } 0 < w_i < 1 \text{ and } \sum_{i=1}^{r} w_i = 1.$$

Then the information matrix for the parameters β at the design ξ is given by

$$M(\beta; \xi) = \sum_{i=1}^{r} w_i g(z_i) g(z_i)^T.$$ In the present study locally D-optimal designs,

that is designs which maximize the determinant of the information matrix at best guesses of the unknown parameters β_0, β_1 and β_2, are sought (Chernoff (1953)).

3 *D*-optimal designs

3.1 Designs based on 4 points

Sitter and Torsney (1995) and Jia and Myers (2001) considered the two-variable dose-response model, $\mathrm{logit}(p) = \beta_0 + z_1 + z_2$, with a design space based on the entire (z_1, z_2)-plane but constrained to lie within two parallel lines chosen arbitrarily and so as not to coincide with lines of constant logit. These authors constructed D-optimal designs comprising 4 equally weighted points of support located at the intersection of the parallel boundaries of the design space and lines of constant logit defined by $\beta_0 + z_1 + z_2 = \pm u$ with $u = 1.223$. It would thus seem sensible to adopt the form of these designs in the present case with $z_1 \geq 0$ and $z_2 \geq 0$ and to consider a 4-point design denoted by $\xi_f^\star$ and given by

$$\xi_f^\star = \left\{ \begin{array}{cccc} (-u-\beta_0, 0) & (0, -u-\beta_0) & (u-\beta_0, 0) & (0, u-\beta_0) \\ w & w & \frac{1}{2} - w & \frac{1}{2} - w \end{array} \right\}$$

with $0 < u \leq -\beta_0$. The support points of this design lie on the boundary of the design space on lines of constant, complementary u-values and the allocation of the weights is based on symmetry arguments. Note that the constraint on u ensures that the doses are positive. The determinant of the associated information matrix is given by

$$|M(\beta; \xi_f^\star)| = \frac{2e^{3u}u^2 w(1 - 2w)\{(u - \beta_0)^2 + 8\beta_0 uw\}}{(1 + e^u)^6}$$

and is maximized by setting its derivatives with respect to w and u to zero and solving the resultant equations simultaneously. Specifically, the optimal weight satisfies the quadratic equation

$$48\beta_0 uw^2 + 4(u^2 - 6\beta_0 u + \beta_0^2)w - (u - \beta_0)^2 = 0$$

together with the feasibility constraint $0 < w < \frac{1}{2}$ and is given uniquely by

$$w^\star = \frac{-u^2 + 6u\beta_0 - \beta_0^2 + \sqrt{u^2 + 14\beta_0 u + \beta_0^2}}{24\beta_0 u}.$$

It then follows that the optimal u value, denoted by $u^\star$, satisfies the transcendental equation

$$u^2(3 + 3e^u + 2u - 2ue^u) + \beta_0^2(1 + e^u + 2u - 2ue^u) + a(1 + e^u + u - ue^u) = 0 \quad (2)$$

where $a = \sqrt{u^4 + 14\beta_0^2 u^2 + \beta_0^4}$, together with the constraint $0 < u \leq -\beta_0$. Equation (2) cannot be solved explicitly, only numerically, but it is nevertheless instructive to examine the dependence of the optimal values of u and w on β_0. Values for $u^\star$ and $w^\star$ for selected values of β_0 are presented in Table 1.

Table 1. Values of $u^\star$ and $w^\star$ for selected β_0 for 4-point designs

β_0	-5	-4.5	-4	-3.5	-3	-2.5	-2	-1.55
$u^\star$	1.292	1.306	1.323	1.346	1.376	1.418	1.474	1.542
$w^\star$	0.1975	0.1934	0.1888	0.1838	0.1785	0.1731	0.1686	0.1667

Note that $u^\star$ decreases monotonically with β_0, that for a value of $\beta_0 = -10$ the probability of a success at the control $d_1 = d_2 = 0$ is very small (of the order of 4.5×10^{-5}) and that there is a cut-off value of β_0, approximately equal to -1.5434, above which the optimal doses $-u^\star - \beta_0$ become negative. This latter result is discussed in more detail in Sect. 3.3.

The global optimality or otherwise of the proposed D-optimal designs can be confirmed by invoking the appropriate Equivalence Theorem (see Atkinson and Donev (1992)) and, specifically, by proving that the directional derivative of the log of the determinant $|M(\beta; \xi)|^{-1}$ at $\xi_f^\star$ in the direction of $z = (z_1, z_2)$, written $\phi(\xi_f^\star, z, \beta)$, is greater than or equal to 0 over the design space. In fact 4-point designs of the form $\xi_f^\star$ were shown to be globally D-optimal *numerically* for a wide range of β_0 values less than -1.5434. As an example, consider $\beta_0 = -4$. The proposed D-optimal design is given by

$$\xi_f^\star = \left\{ \begin{array}{cccc} (2.677, 0) & (0, 2.677) & (5.323, 0) & (0, 5.323) \\ 0.1888 & 0.1888 & 0.3112 & 0.3112 \end{array} \right\}$$

and the directional derivative by

$$\phi(\xi_f^\star, z, \beta) = 3 - \frac{3.955e^{-4+z_1+z_2}(18.817 - 8z_1 - 8z_2 + z_1^2 + z_2^2 + 1.701z_1z_2)}{(1 + e^{-4+z_1+z_2})^2}.$$

A careful search of the values of $\phi(\xi_f^\star, z, \beta)$ over a fine grid of points $z = (z_1, z_2)$ in the region $[0, 10] \times [0, 10]$ indicated that the design $\xi_f^\star$ is indeed globally D-optimal and the 3-dimensional plot of $\phi(\xi_f^\star, z, \beta)$ against $z_1 \geq 0$ and $z_2 \geq 0$ given in Figure 1(a) illustrates this finding. An algebraic proof of the global D-optimality or otherwise of the proposed 4-point designs was somewhat elusive, the main problems being that the weights assigned to the support points are not equal and that the optimal u value cannot be determined explicitly. A strategy for the required proof is indicated later in the paper.

3.2 Designs based on 3 points

For values of $\beta_0 \geq -1.5434$, the 4-point designs described in the previous section are no longer feasible and it is appealing to consider candidate D-optimal designs which put equal weights on the three support points $(0, 0)$, $(u - \beta_0, 0)$ and $(0, u - \beta_0)$ where $u > \beta_0$. The determinant of the standardized information matrix for the parameters β at such a 3-point design, denoted by $\xi_t^\star$, is given by

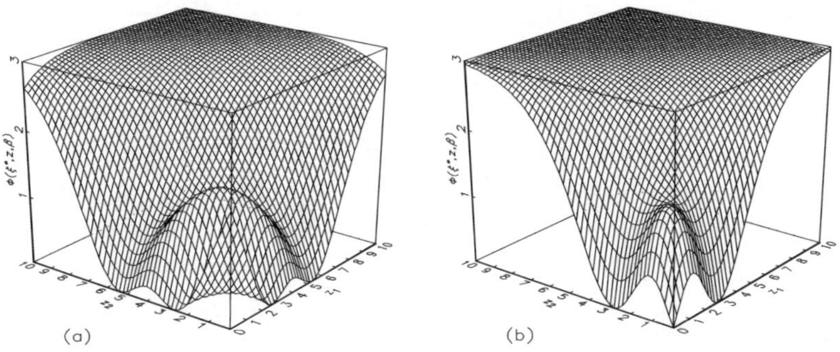

Fig. 1. Plots of the directional derivative $\phi(\xi, z, \beta)$ against z_1 and z_2 for model (1) with (a) $\beta_0 = -4$ and (b) $\beta_0 = -1$.

$$|M(\beta; \xi_t^\star)| = \frac{(u - \beta_0)^4 e^{\beta_0 + 2u}}{27(1 + e^{\beta_0})^2(1 + e^u)^4}$$

and the value of u maximizing this determinant satisfies

$$\frac{\partial |M(\beta; \xi_t^\star)|}{\partial u} = \frac{(u - \beta_0)^3(2 - \beta_0 + 2e^u + \beta_0 e^u + u - ue^u)}{27(1 + e^{\beta_0})^2(1 + e^u)^5} = 0.$$

The solution $u = \beta_0$ is not meaningful since the resultant design comprises the single point $(0, 0)$. Thus the value of u for which $|M(\beta; \xi_t^\star)|$ is a maximum satisfies the equation

$$2 - \beta_0 + 2e^u + \beta_0 e^u + u - ue^u = 0 \tag{3}$$

Numerical studies indicate that there is a unique solution to (3) for values of $u > \beta_0$, say $u^\star$, but this solution does not have an explicit form. Values of $u^\star$ for selected values of β_0 are presented in Table 2.

Table 2. Values of $u^\star$ for selected β_0 for 3-point designs

β_0	-1.5434	-1.5	-1.25	-1	-0.75	-0.5	-0.25	0
$u^\star$	1.5434	1.562	1.674	1.796	1.930	2.075	2.231	2.399

The global *D*-optimality or otherwise of the candidate designs can be confirmed by demonstrating that the directional derivative $\phi(\xi_t^\star, z, \beta)$ is greater than or equal to zero for all points z in the positive quadrant. This check was performed numerically for selected values of β_0 in the range -1.5434 to 0 using a fine grid of points in the region $[0, 10] \times [0, 10]$ as outlined for the 4-point designs of the previous section. For example, consider $\beta_0 = -1$. The proposed 3-point *D*-optimal design puts equal weights on the points $(0, 0), (0, 2.796)$ and $(2.796, 0)$. The directional derivative is given by

$$\phi(\xi_t^\star, z, \beta) = 3 - \frac{5.095e^{-1+z_1+z_2}\{2.995 - 2.142(z_1 + z_2) + 0.766z_1 z_2 + z_1^2 + z_2^2\}}{(1 + e^{-1+z_1+z_2})^2}$$

and the 3-dimensional plot of $\phi(\xi_t^\star, z, \beta)$ against $z_1 \geq 0$ and $z_2 \geq 0$ shown in Figure 1(b) indicates that the design is indeed globally optimal. For $\beta_0 \geq -1.5434$, confirming the global optimality or otherwise of 3-point designs of the form $\xi_t^\star$ algebraically is not straightforward however, since, in particular, the support points of the proposed designs are not associated with complementary u values.

3.3 A special case

The 4-point design introduced in Sect. 3.1 with optimal u value, $u^\star = -\beta_0$, reduces to the 3-point design which puts equal weights on the support points $(0,0), (-2\beta_0^\star, 0)$ and $(0, -2\beta_0^\star)$ where $\beta_0^\star$ satisfies the equation

$$1 + \beta_0 + e^{\beta_0} - \beta_0 e^{\beta_0} = 0 \tag{4}$$

for $\beta_0 < 0$. In other words $\beta_0^\star \approx -1.5434$ and the 3-point design of interest, denoted by $\xi_g^\star$, is given by $(0,0), (3.0868, 0)$ and $(0, 3.0868)$. Note that the support points are associated with the complementary u values, $\pm\beta_0^\star$. The design with $u^\star = -\beta_0 = -\beta_0^\star$ can be shown to be globally D-optimal as follows.

Theorem 1. *Consider the logistic regression model in two variables defined by (1) with $u^\star = \beta_0 = \beta_0^\star$. Then the 3-point design $\xi_g^\star$ which puts equal weights on the support points $(0,0), (-2\beta_0^\star, 0)$ and $(0, -2\beta_0^\star)$ is globally D-optimal.*

Proof. Assume that $\beta_0 = \beta_0^\star$. Then the directional derivative of $\ln |M(\beta; \xi)|$ at $\xi_g^\star$ in the direction of a single point $z = (z_1, z_2)$ is given by

$$\phi(\xi_g^\star, z, \beta) = 3 - 3\frac{e^{\beta_0+z1+z2}}{e^{\beta_0}} \frac{(1 + e^{\beta_0})^2}{(1 + e^{\beta_0+z1+z2})^2} \left\{ \frac{2\beta_0^2 + z_1^2 + z_2^2 + z_1 z_2 + 2\beta_0(z_1 + z_2)}{2\,\beta_0^2} \right\}.$$

Further, since $u_1 = \beta_0 + z_1 + z_2$ implies $z_2 = u_1 - \beta_0 - z_1$, the directional derivative can be reexpressed as

$$\phi(\xi_g^\star, z, \beta) = 3 - 3\frac{e^{u1}}{e^{\beta_0}} \frac{(1 + e^{\beta_0})^2}{(1 + e^{u_1})^2} \left\{ \frac{\beta_0^2 + u_1^2 + (\beta_0 - u_1)z_1 + z_1^2}{2\,\beta_0^2} \right\} \tag{5}$$

with $0 \leq z_1 \leq u_1 - \beta_0$. It now follows from the Equivalence Theorem for D-optimal designs that the design $\xi_g^\star$ is globally D-optimal provided the condition $\phi(\xi_g^\star, z, \beta) \geq 0$ holds. Consider u_1 fixed, i.e. consider points z on a line of constant logit. Then $\phi(\xi_g^\star, z, \beta)$ given by (5) is proportional to the quadratic function $f(z_1) = \beta_0^2 + u_1^2 + z_1(\beta_0 - u_1) + z_1^2$ which has a unique minimum at $z_1 = \frac{u_1 - \beta_0}{2}$. Therefore, the maxima of $f(z_1)$ within the design space are located at the boundary points $z_1 = 0$ and $z_1 = u_1 - \beta_0$. Thus the minima of the directional derivative $\phi(\xi_g^\star, z, \beta)$ for all points z in the positive quadrant occur on the boundaries $z_1 = 0$ and $z_2 = 0$. Now on setting $z_1 = 0$ or $z_1 = u_1 - \beta_0$ in (5), the inequality $\phi(\xi_g^\star, z, \beta) \geq 0$ reduces to

$$2 \frac{e^{\beta_0}}{e^{u_1}} \frac{(1 + e^{u_1})^2}{(1 + e^{\beta_0})^2} \geq \frac{(\beta_0^2 + u_1^2)}{\beta_0^2}. \tag{6}$$

This condition, together with the fact that β_0 satisfies equation (4) and thus $\beta_0 = \beta_0^\star$, is precisely the condition which emerges in invoking the appropriate directional derivative to prove the global optimality of the D-optimal design for a logistic regression model with one explanatory variable. Thus it follows immediately from that setting that condition (6) holds for all $u_1 \in \mathbb{R}$ and thus, in the present case, for all feasible $u_1 \geq \beta_0^\star$.

The framework of the above theorem can be used to devise a strategy for proving the global D-optimality of the candidate 3- and 4-point designs discussed in the earlier sections.

4 Conclusions

The main aim of the present study has been to construct locally D-optimal designs for the logistic regression model in two variables subject to the constraint that the values of the variables are greater than or equal to zero. In particular it is shown that the designs so constructed depend on the parameters β_1 and β_2 of model (1) only through the scaling of the two explanatory variables but that the basic pattern of the designs is determined by the intercept parameter β_0. Specifically, if $\beta_0 < \beta_0^\star$, where $\beta_0^\star$ satisfies equation (4) then the D-optimal design is based on 4, points of support located on complementary logit lines, whereas if $\beta_0^\star \leq \beta_0 \leq 0$ then the design comprises 3 points including a control. The global D-optimality of the designs for a wide range of β_0 values was demonstrated numerically but was only proved algebraically for the case with $\beta_0 = \beta_0^\star$. The broad strategy used in the proof for the latter setting, that is in Theorem 1, should be applicable to all D-optimal designs reported here. However the extension is not entirely straightforward and is currently being investigated.

There is much scope for further work. In particular there is a need to relate the D-optimal designs constructed here to the geometry of the design locus as elucidated in Sitter and Torsney (1995), to the "minimal" point designs developed in Torsney and Gunduz (2001) and to the results of Wang et al (2006) on Poisson regression. In fact the work reported here forms part of a larger study aimed at identifying patterns and taxonomies of designs for the logistic regression model in two variables both with and without an interaction term and with a range of constraints on the variables. Finally it would be interesting to extend the design construction to accommodate other criteria and, following Torsney and Gunduz (2001), to logistic regression models with more than two explanatory variables.

Acknowledgement. The authors would like to thank the University of Cape Town, the University of KwaZulu-Natal and the National Research Foundation, South Africa, for financial support.

References

Atkinson A (2006) Response Surface Methodology and Related Topics, World Scientific, New Jersey, chap 8: Generalized linear models and response transformation

Atkinson A, Donev A (1992) Optimum Experimental Designs. Clarendon Press, Oxford

Atkinson A, Haines L (1996) Handbook of Statistics, vol 13, Elsevier, Amsterdam, chap 11: Designs for nonlinear and generalized linear models

Chernoff H (1953) Locally optimal designs for estimating parameters. Ann Math Statist 24:586–602

Jia Y, Myers R (2001) Optimal experimental designs for two-variable logistic regression models. Preprint

Kupchak P (2000) Optimal designs for the detection of drug interaction. PhD thesis, University of Toronto, Toronto

Sitter R, Torsney B (1995) Optimal designs for binary response experiments with two design variables. Statistica Sinica 5:405–419

Torsney B, Gunduz N (2001) Optimum Design 2000, Kluwer, Dordrecht, chap 24: On optimal designs for high dimensional binary regression models

Wang Y, Myers R, Smith E, Ye K (2006) d-optimal designs for poisson regression models. J Statist Plann Inference 136:2831–2845

Design of Experiments for Extreme Value Distributions

Patrick J. Laycock[1] and Jesús López-Fidalgo[2]

[1] School of Mathematics, University of Manchester, Sackville St, Manchester M60
   1QD, UK, `pjlaycock@manchester.ac.uk`
[2] Department of Mathematics, University of Castilla la Mancha
   Avda. Camilo José Cela 3, 13071–Ciudad Real, Spain
   `jesus.lopezfidalgo@uclm.es`

Summary. In this paper experimental designs are considered for classic extreme value distribution models. A careful review of the literature provides some information matrices in order to study experimental designs. Regression models and their design implications are discussed for some situations involving extreme values. These include a constant variance and a constant coefficient of variation model plus an application in the context of strength of materials. Relative efficiencies calculated with respect to D–optimality are used to compare the designs given in this example.

Key words: generalised extreme value distribution, D–optimality, regression, Weibull distribution

1 Introduction

There are many situations where extreme values or extreme objectives might affect the design of experiments. In this paper we consider regression models where the dependent variable is an extreme value or has an implied extreme value distribution. Classic experimental designs, such as factorial designs, fractional and block designs, or response surface designs are typically constructed on the assumption of a linear regression model for the response variate with additive, finite variance, errors. More specifically, the usual model assumes that the data vector $\mathbf{y}$ is $N(\mathbf{X}\theta, \sigma^2 \mathbf{I})$ and we choose $\mathbf{X} = \{\mathbf{x}_1, \mathbf{x}_2, ..., \mathbf{x}_n\}^T$ so as to optimise the estimation of θ in some straightforward way. Such designs are typically optimised for conditions where ANOVA techniques are used, implying linearity, additivity and finite variance. Fortunately, many designs are known to be useful and relevant under wide variations of this model. See for example Silvey (1980), Atkinson and Donev (1992) or Fedorov and P. (1997). Our examples will concentrate on the strength or endurance of materials, where max-stable extreme value distributions form a natural alternative to the Normal family. These were introduced by Fisher and Tippett (1982) and some standard reference books are Leadbetter et al (1983) and the recent one of Coles (2001).

The layout of this paper is as follows. In Section 2 we study design implications for regression models where the dependent variable is a measured extreme. Constant variance models are examined in Section 2.1 then constant coefficient of variation models in Section 2.2 - illustrated by an extensive numerical study of a pitting corrosion data set. A model introduced by Leadbetter et al (1983) applicable to the strength of materials is studied in Section 3 and illustrated by an application to designs for examining the breaking strength of wide paper strips.

2 Designs for regression of extremes

It is a frequent occurrence in data collection for only the maxima (or equivalently, minima) of appropriately grouped sets of values to be observed, or recorded. It is then natural to assume, in the first instance and without specific alternatives, that a member of the generalized extreme value (GEV) family of distributions is applicable. See, for example Jenkinson (1955), Finley (1967) or Walden et al (1981). This family has a cumulative distribution function which can be expressed in the form (for maxima)

$$F(y) = \exp\{-[1 - \kappa(y - \delta)/\alpha]^{1/\kappa}\}, \ \alpha > 0, \ y \le \frac{\alpha}{\kappa} + \delta \tag{1}$$

with upper (else lower, depending on the sign of κ) bound $\xi = \delta + \alpha/\kappa$. We write GEV($\delta$, α, κ) for the distribution in (1). The parameters δ and α are location and scale parameters respectively whilst κ is an index, usually held constant. This third parameter effectively defines the particular member of the GEV family to be applied to the current data set. The distribution has mean function

$$\mu = \xi - \frac{\alpha}{\kappa}\Gamma(1 + \kappa), \ \text{provided } \kappa > -1 \tag{2}$$

and standard deviation

$$\sigma = \frac{\alpha}{\kappa}\left[\Gamma\left(1 + 2\kappa\right) - \Gamma\left(1 + \kappa\right)^2\right]^{\frac{1}{2}}, \ \text{provided } \kappa > -\frac{1}{2} \tag{3}$$

with appropriate limiting forms for the special case $\kappa = 0$, which is also known as the *Type I* extreme value distribution or *the* extreme value distribution or Gumbel's distribution; see Gumbel (1958). The GEV family has regular likelihood properties only in the range $(-\frac{1}{2} < \kappa < \frac{1}{2})$. Outside this range, the distribution becomes heavily skewed relative to an upper (or lower) boundary, which becomes the dominant fact characterizing the data.

2.1 Constant variance model

A natural regression model for this distribution is $y_i \sim GEV(\delta_i, \alpha, \kappa)$ with $\delta_i = \beta^{\mathbf{T}}\mathbf{x}_i$; see for example Reiss and Thomas (1997), page 102. With the parameters α and κ absorbed into the regression constant, this model produces linearity in the mean and constant variance for the regression of y on x, and hence satisfies the standard Gauss-Markov assumptions for ordinary least-squares. That is, we will have

$$E[y_i \mid \mathbf{x}_i] = \beta_0 + \beta^{\mathbf{T}}\mathbf{x}_i, \text{ with } Var[y_i \mid \mathbf{x}_i] = (\alpha/\kappa)^2 \left[\Gamma\left(1 + 2\kappa\right) - \Gamma\left(1 + \kappa\right)^2\right],$$

where $\beta_0 = (\alpha/\kappa)[1 - \Gamma(1 + \kappa)]$. Therefore all the usual experimental designs, as referred to above, can be justified for the estimation of β. Also $\widehat{\beta}_0$ and the residual variance can subsequently be used to provide moment estimators for α and κ. Full simultaneous maximum likelihood for β, α and κ produces a parameter-dependent information matrix, with the usual requirements for some form of prior information concerning the parameters. An example of such an analysis is offered in the next section. Alternative procedures might be based on the canonical forms described in Ford et al (1992).

2.2 Constant coefficient of variation model

It is commonplace in statistical practice that data is more likely to exhibit a constant coefficient of variation rather than a constant variance. For comments on this see Kendall and Stuart (1976), §37, or Aitchison and Brown (1957) and for the alternative generalized linear model approach, see McCullagh and Nelder (1989), pp22-23. An extreme value model exhibiting constant coefficient of variation is

$$y_i \sim GEV(\delta_i, \alpha_i, \kappa)$$

with link function η common to δ_i and α_i, so that

$$\delta_i = \delta_0\eta(\beta^{\mathbf{T}}\mathbf{x}_i), \ \alpha_i = \alpha_0\eta(\beta^{\mathbf{T}}\mathbf{x}_i), \text{ giving } \mu_i = \mu_0\eta(\beta^{\mathbf{T}}\mathbf{x}_i) \text{ and } \sigma_i = \sigma_0\eta(\beta^{\mathbf{T}}\mathbf{x}_i)$$

and hence a constant coefficient of variation, σ_0/μ_0, where μ_0, σ_0 are given by (2) and (3) respectively, using δ_0 and α_0. This model has been used and justified for pitting corrosion data by Laycock et al (1990) and extended for order statistics of pitting data by Scarf et al (1992). A convenient link here is

$$\eta(\beta^{\mathbf{T}}\mathbf{x}_i) = \exp(\beta^{\mathbf{T}}\mathbf{x}_i) = \prod_j z_{ji}^{\beta_j} \text{ with } z_{ji} = \exp(x_{ji}),$$

so that, setting $y_i^* = \log(y_i)$ we have

$$\mu_i^* = E[y_i^* \mid \mathbf{x}_i] \approx \beta_0^* + \beta^{\mathbf{T}}\mathbf{x}_i,$$

where $\beta_0^* = \log(\mu_0)$ and y_i^* has (approximately) constant variance, independent of β and $\mathbf{x}$. Therefore, with this link function, all the usual experimental designs for the linear regression model can be approximately justified for the estimation of β_0^* and β. Full maximum likelihood again produces a parameter-dependent information matrix, with the usual design requirements for some form of prior information concerning the parameters. We give an example below of a search for a locally optimum design.

3 Strength of materials

Leadbetter et al (1983), page 267, model the breaking strength of materials as a function of tested length, x, using extreme value distributions for minima. For the recommended *Type III* GEV or Weibull distribution, they set

$$\alpha_i = \alpha_0 x_i^{-\kappa} \qquad \text{and suggest } x_0 = 0$$

implying

$$\mu_i = (\alpha_0 x_i^{-\kappa}/\kappa)\Gamma(1+\kappa) \text{ and } \sigma_i = \frac{\alpha_0 x_i^{-\kappa}}{\kappa}\left[\Gamma(1+2\kappa) - \Gamma(1+\kappa)^2\right]^{\frac{1}{2}}$$

and hence constant coefficient of variation.

But taking logs here does not produce an (approximately) linear model, since the regression parameter is the highly non-linear shape parameter, κ. Their distribution function for a single observation, assuming $x_0 = 0$, is

$$F(y) = 1 - \exp\{-x[\kappa y/\alpha_0]^{1/\kappa}\}, \; x > 0, \; \alpha_0 > 0, \; y > 0,$$

where a convenient parameterization has been used here. If we fix the scale parameter at $\alpha_0 = 1$ and make use of the well known equality,

$$E\left[-\frac{\partial^2 \log f}{\partial \kappa^2}\right] = E\left[\left(\frac{\partial \log f}{\partial \kappa}\right)^2\right]$$

for evaluating Fisher's information function, I, valid in this situation with known bound, it can be shown that at $z_i = \exp(x_i)$

$$I_i = [(6+12(\gamma-1)\kappa+(6-12\gamma+6\gamma^2+\pi^2)\kappa^2+12\kappa(1+(\gamma-1)\kappa)z_i+6\kappa^2 z_i^2]/(6\kappa^4),$$

where γ is Euler's constant. This is a simple quadratic in z_i with positive coefficients; and since information is additive the total information on κ will be maximized by the 'extreme design' which places all the observations at the top end of any design interval $x \in [a, b]$. We were not able to derive a general expression for the information matrix in the case of joint estimation of α_0 and κ. So numerical integration was required in the following example.

Design	a	b	c	d	e	f	g	h
D-%efficiency	100	108	116	147	125	125	69	6
equivalent n	11	10.2	9.5	7.5	8.8	8.8	15.9	174

Table 1. Design efficiencies for strength of materials model

Leadbetter et al (1983), page 272, Example 14.2.1, give mean plots for the breaking strength, y, (in KNewtons per metre, of 5cm wide paper strips) for 'several' experiments in which the strips varied in length, x, from 8cm up to 10m. From their plots we inferred approximate values: $\alpha_0 = 2$ and $\kappa = 0.10$ and have then used these values in a numerical search for a locally 'best' design based on a maximum of 11 distinct observations. This is both a convenient choice and one which would typically be regarded as a large number of distinct levels for a designed experiment. A uniform design and seven alternatives were compared via the usual D-optimality. Their relative efficiencies in percentage terms - using the square-root of the determinant of the information matrix - compared to the uniform design, are given in Table 1. In the row labeled "equivalent n" the number of observations needed to get the same D–efficiency as design a) with 11 observations is computed; that is: the equivalent $n = 11 \times 100/D -$ %efficiency. The relative ordering of these designs proved to be almost exactly the same under A, E and c–optimality for the variance of $\widehat{\kappa}$ (not shown explicitly in the table).

The details for these designs are as follows (see Table 2 and Figure 1):

a) Uniform design over the design region with 11 levels and equal weights. The 'n' here corresponds to a design with one observation at each design point.
b) Uniform design on a log scale, also with 11 levels, giving an 8% increase in efficiency.
c) Bailey (1982) suggests $\{\sin(2\pi i/p), \ i = 1, 2..., p\}$, with appropriate scale and location, instead of the uniform design with p levels. This results in orthogonal polynomials becoming simple contrasts and with p prime ensures a canonical partitioning of the degrees of freedom for quantitative treatments. Here a design with p=11, a prime power, is used giving a further 8% increase in efficiency.
d) The classic D–optimal 2 point 'extreme design', giving a nominal 47% increase in efficiency over the 11 point uniform design, but with the usual weaknesses for model checking. This was also the best design for estimating κ.
e) A 3 point design, uniform on the log scale, offering a 25% improvement on the 1st design.
f) A 3 point design, with Bailey spacing on a log scale, also offering a 25% improvement.

Design, ξ	Support points and weights (below)									
a	8	107.2	206.4	305.6	404.8	504	603.2	702.4	801.6	900.8	1000
	$\frac{1}{11}$	$\frac{1}{11}$	$\frac{1}{11}$	$\frac{1}{11}$	$\frac{1}{11}$	$\frac{1}{11}$	$\frac{1}{11}$	$\frac{1}{11}$	$\frac{1}{11}$	$\frac{1}{11}$	$\frac{1}{11}$
b	8	13.0	21.0	34.1	55.2	89.4	145.0	234.9	380.7	617.0	1000
	$\frac{1}{11}$	$\frac{1}{11}$	$\frac{1}{11}$	$\frac{1}{11}$	$\frac{1}{11}$	$\frac{1}{11}$	$\frac{1}{11}$	$\frac{1}{11}$	$\frac{1}{11}$	$\frac{1}{11}$	$\frac{1}{11}$
c	8.2	10.0	14.4	24.2	45.3	89.4	176.6	329.9	554.5	804.0	975.7
	$\frac{1}{11}$	$\frac{1}{11}$	$\frac{1}{11}$	$\frac{1}{11}$	$\frac{1}{11}$	$\frac{1}{11}$	$\frac{1}{11}$	$\frac{1}{11}$	$\frac{1}{11}$	$\frac{1}{11}$	$\frac{1}{11}$
d	8	1000									
	$\frac{1}{2}$	$\frac{1}{2}$									
e	8	89.4	1000								
	$\frac{1}{3}$	$\frac{1}{3}$	$\frac{1}{3}$								
f	11.1	89.4	723.7								
	$\frac{1}{3}$	$\frac{1}{3}$	$\frac{1}{3}$								
g	8										
	1										
h	1000										
	1										

Table 2. Designs for constant coefficient of variation model

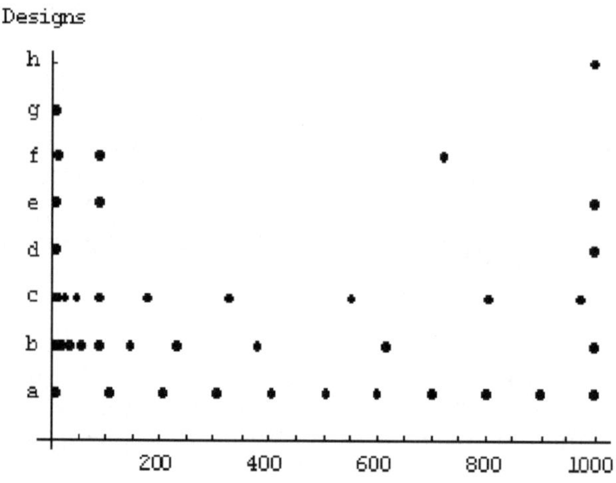

Fig. 1. Designs for Leadbetter's example (for each design all support points receive equal weights)

As it happens, a one point design is technically feasible for this nonlinear problem. This is because here both the mean and variance depend on the same two parameters. The situation can then be compared, for example, to estimating both the mean and (residual) variance for a normal straight line regression through the origin from multiple observations at a single point. Nevertheless, the one–point designs only give 69% efficiency when $x_\star = 8$ (design g), falling to 6% when $x_\star = 1000$ (design h).

It seems worthwhile to base designs on a log scale using the Bailey spacing with a reasonable number of design points, for the scope this gives in fully exploring this model in an efficient manner.

4 Conclusions

We have examined a variety of design situations where extreme value theory can be used to suggest good or optimal designs. In particular, some regression situations involving extremes can be shown to be suitable for the application of classic design theory, while others require search techniques adapted from other areas of nonlinear design theory. These searches have resulted in suggestions for a "best" design in the example. Situations involving extremes, but not examined in this paper, include designs suitable for the general class of 'heavy tailed' distributions, by which is meant distributions with tails heavier than the normal, distributions with infinite variance or the modern high throughput 'screening designs' using extreme value theory for Gaussian sequences.

Acknowledgement. Work partly supported by a Royal Society grant and the Spanish grants of Ministerio de Educación y Ciencia MTM2004-06641-C02-01 and Junta de Castilla y León SA125/04.

References

Aitchison J, Brown J (1957) The Lognormal Distribution. CUP, Cambridge

Atkinson A, Donev A (1992) Optimum Experimental Designs. Oxford Science Publications, Oxford

Bailey RA (1982) The decomposition of treatment degrees of freedom in quantitative factorial experiments. J Roy Stat Soc, B 44:63–70

Coles S (2001) An Introduction to Statistical Modeling of Extreme Values Series. Springer, New York

Fedorov V, P H (1997) Model-Oriented Design of Experiments. Springer, New York

Finley HF (1967) An extreme value statistical analysis of maximum pit depths and time to first perforation. Corrosion 23:83–87

Fisher RA, Tippett LHC (1982) Limiting forms of the frequency distributions of the largest or smallest member of a sample. Proc Camb Phil Soc 24:180–190

Ford I, Torsney B, Wu C (1992) The use of a canonical form in the construction of locally optimal designs for non-linear problems. J Roy Stat Soc, B 54:569–583

Gumbel E (1958) Statistics of Extremes. Columbia University Press, New York

Jenkinson AF (1955) The frequency distribution of the annual maximum (or minimum) values of meteorological elements. Quart J Roy Met Soc 81:158–165

Kendall M, Stuart A (1976) The Advanced Theory of Statistics, Vol 3, Design and Analysis and Time Series. Hafner, 3rd edn., New York

Laycock P, Cottis R, Scarf P (1990) Extrapolation of extreme pit depths in space and time. J Electrochem Soc 137:64–69

Leadbetter M, Lindgren G, Rootzen H (1983) Extremes and Related Properties of Random Sequences and Processes. Springer, New York

McCullagh P, Nelder J (1989) Generalized Linear Models. Chapman and Hall, 2nd edn., London

Reiss RD, Thomas M (1997) Statistical Analysis of Extreme Values. Birkhauser Verlag, Basel

Scarf P, Laycock P, Cottis R (1992) Extrapolation of extreme pit depths in space and time, using the r deepest depths. J Electrochem Soc 139:2621–2627

Silvey S (1980) Optimal design. Chapman and Hall, London

Walden A, Prescott P, Webber N (1981) Some important considerations in the analysis of annual maximum sea levels. Coastal Engineering 4:335–342

A Model Selection Algorithm for Mixture Experiments Including Process Variables

Hugo Maruri-Aguilar[1] and Eva Riccomagno[1,2]

[1] Department of Statistics, The University of Warwick, Gibbet Hill Road, Warwick CV4 7AL, U.K.   `H.Maruri-Aguilar@warwick.ac.uk`
[2] Department of Mathematics, Polytechnic of Turin, Corso Duca degli Abruzzi, 24, 10129 Torino, Italy   `Eva.Riccomagno@polito.it`

Summary. Experiments with mixture and process variables are often constructed as the cross product of a mixture and a factorial design. Often it is not possible to implement all the runs of the cross product design, or the cross product model is too large to be of practical interest.

We propose a methodology to select a model with a given number of terms and minimal condition number. The search methodology is based on weighted term orderings and can be extended to consider other statistical criteria.

Key words: mixture-amount designs, term orderings, condition number

1 Models for mixture experiments with process variables

When a mixture experiment cannot be performed under uniform conditions or when the responses depend on factors other than the mixture components, like the total mixture amount or some process variables, then the cross product of a standard mixture design and a full factorial design in the non mixture factors is often recommended (see e.g. (Cornell, 2002, Ch. 7) and Prescott (2004)). But product designs might have a larger number of runs than desired and only a subset of the design is implemented.

Let $x = (x_1, \ldots, x_q) \in \mathbb{R}^q$ be the mixture components and $z = (z_1, \ldots, z_k) \in \mathbb{R}^k$ the process variables. The x_i, $i = 1, \ldots, q$ are to be interpreted as proportions, possibly scaled, of a total amount and it might be one of the z_j, which is often denoted by m. We assume that a design $\mathcal{D}$ is a finite set of points in $\mathbb{R}^{q+k}$, usually the mixture components are listed first, and that there are no replicated runs. The projection of $\mathcal{D}$ over the x-space is $\mathcal{D}_x$ and $\mathcal{D}_z$ is the projection over the process variable space. Both $\mathcal{D}_x$ and $\mathcal{D}_z$ admit replicates. For small values of q and k the full product design, $\mathcal{D}_x \times \mathcal{D}_z$, is recommended in the literature, where $\mathcal{D}_x$ is a simplex lattice design and $\mathcal{D}_z$ is a full factorial design (Cornell, 2002, Ch. 7).

In the literature various models for studying the combined effect of the x and z factors are proposed. Often they have a fairly regular structure derived from a standard cubic or quadratic model for factorial designs (g below) and a Scheffé quadratic or cubic polynomial model, in a relevant parametrization, in the mixture components (f). In Draper and Pukelsheim (1998) for pure mixture designs Kronecker type models are recommended of degree two or three. Typically proposed models are additive regression type models like $y(x, z) = f(x) + g(z)$, or complete cross product models of the type $y(x, z) = f(x)g(z)$, or combinations of these such as $y(x, z) = f(x) + g(z) + \sum_{i=1}^{q} \sum_{j=1}^{k} f_{ij}(x_i, z_j)$ (see e.g.(Cornell, 2002, §7.10)), where the f_{ij} comprises products of terms in f and g. For a mixture amount experiment (Cornell, 2002, §7.5 and page 405) a mixture amount model of the form $f_0(x) + m f_1(x) + \ldots + m^p f_p(x)$ is suggested where

$$f_p(x) = \sum_i \gamma_i^{(p)} x_i + \sum_{i<j} \gamma_{ij}^{(p)} x_i x_j + \ldots + \sum_{i_1 < \ldots < i_l} \gamma_{1,\ldots,l}^{(p)} x_{i_1} \ldots x_{i_l},$$

$l \leq q$, p is a positive integer and the $\gamma^{(p)}$ are regression parameters. A component amount model has a smaller number of terms and takes the form $f(a_1, \ldots, a_q)$ for $a_i = x_i m$, $i = 1, \ldots, q$ and a suitable polynomial function f.

2 Homogeneous representation of a mixture experiment

In algebraic statistics an indeterminate x_i is identified with the i-th factor in the experiment and the design, $\mathcal{D}$, is described and defined by the set of polynomials in the x_i's vanishing at all the design points. This infinite set of polynomials is called the design ideal, Ideal($\mathcal{D}$). The ideal generated by the polynomials $f_1, \ldots, f_v$ is defined as $\langle f_1, \ldots, f_v \rangle = \{\sum_{i=1}^{v} s_i f_i : s_i$ being polynomials$\}$.

Example 1. A $\{q, m\}$ *simplex lattice design* (see Scheffé (1958)) is the intersection of the simplex in $\mathbb{R}^q$ and the full factorial design in q factors and with the uniformly spaced levels $\{0, 1/m, \ldots, 1\}$. Thus the ideal of the $\{q, m\}$ simplex lattice design is

$$\langle \prod_{j=0}^{m} (x_1 - j/m), \ldots, \prod_{j=0}^{m} (x_q - j/m), \sum_{i=1}^{q} x_i - 1 \rangle$$

That is, the first q polynomials give the full factorial design and the last one is the simplex condition which selects the points of the full factorial whose components sum to one.

In Maruri-Aguilar et al (2006) it is shown that for a pure mixture design, i.e., $\mathcal{D} = \mathcal{D}_x$, an alternative polynomial representation is meaningful and useful. The design is identified with the set of lines through the origin of the x-space and a point in $\mathcal{D}$, to indicate that in a pure mixture experiment the relative proportions of the component are of interest irrelevant of the total mixture amount. The set of all such lines is called the *design cone* and indicated as $\mathcal{C}_\mathcal{D}$. The set of all polynomials vanishing on all points of $\mathcal{C}_\mathcal{D}$ is $\mathrm{Ideal}(\mathcal{C}_\mathcal{D})$.

Example 2. The simplex centroid design, $\mathcal{D}$ in $\mathbb{R}^q$ (see Scheffé (1963)) is the projection, on the simplex in $\mathbb{R}^q$ with respect to the origin, of the full factorial design with levels 0 and 1. This shows that it has $2^q - 1$ points, that the coordinates of each point are either zero or equal to each other, and, moreover, it holds

$$\mathrm{Ideal}(\mathcal{C}_\mathcal{D}) = \langle x_i^2 x_j - x_i x_j^2 : i, j = 1, \ldots, q; i \neq j \rangle$$

In particular the design point with projective coordinates $(1 : \ldots : 1) \in \mathbb{R}^q$ is the barycenter point of the simplex in $\mathbb{R}^q$. See (Maruri-Aguilar et al, 2006, §4.2).

In mathematical terms, a mixture experiment is thus to be interpreted as a projective variety. For the consequence of this interpretation we refer to Maruri-Aguilar et al (2006). Note that $\mathrm{Ideal}(\mathcal{C}_\mathcal{D})$ is a homogeneous ideal, that is an ideal generated by homogeneous polynomials. A polynomial is homogeneous of degree $s \in \mathbb{Z}_{\geq 0}$ if each one of its terms has the sum of its exponents equal to s. By convention $a \in \mathbb{R}$ has degree zero. Indeed if f is a homogeneous polynomial of degree s and $f(d) = 0$ for all $d \in \mathcal{D}_x$, then for $\lambda \in \mathbb{R}$ $f(\lambda d) = \lambda^s f(d)$. The converse is true; see Theorem 1 in Maruri-Aguilar et al (2006).

2.1 Homogeneous models for pure mixture experiments

Cone design ideals lead naturally to consideration of homogeneous polynomial regression models. We need to recall the basics of algebraic statistics for design of experiments (Pistone et al (2001)). Let $\mathbb{R}[x_1, \ldots, x_d]$ be the set of all polynomials in $x_1, \ldots, x_d$ with real coefficients. The set of real valued functions over $\mathcal{D}$ is isomorphic to the quotient space $\mathbb{R}[x_1, \ldots, x_d]/\mathrm{Ideal}(\mathcal{D})$ defined by the equivalence relationship stating that two polynomials f and g are equivalent if they take the same values over all the points of $\mathcal{D}$. The quotient space is a $\mathbb{R}$-vector space, of dimension equal to the number of points in $\mathcal{D}$ and it admits vector space bases formed by monomials.

In Lemma 3 and Theorem 4 of Maruri-Aguilar et al (2006) it is proved that if $\mathcal{D}$ is a mixture design then there are bases formed by monomials of the same total degree larger than a suitable integer and an algorithm to compute them is provided. Any such basis can be used to construct homogeneous polynomial regression models of the Kronecker type (Draper and Pukelsheim (1998)) and submodels.

Example 3. For the design in Example 2 the largest set of degree three, linearly independent monomials in the quotient space is

$$x_i^3, \text{ for all } i = 1, \ldots, q \qquad x_i^2 x_j, \text{ for all } i < j, i, j = 1, \ldots, q$$
$$x_i x_j x_k, \text{ for all } i < j < k, i, j, k = 1, \ldots, q$$

A full basis of a given degree, equivalently a saturated homogeneous model identified by $\mathcal{D}$, can be retrieved only for a degree larger than three.

3 A model selection algorithm

Consider a product design $\mathcal{D} = \mathcal{D}_x \times \mathcal{D}_z$ with no replicated runs. Let $E_x = \{x^\alpha : \alpha \in L_x\} \subset \mathbb{R}[x_1, \dots, x_q]$ be a set of linearly independent monomials in $\mathbb{R}[x_1, \dots, x_q]/\operatorname{Ideal}(\mathcal{D}_x)$ and $E_z = \{z^\alpha : \alpha \in L_z\} \subset \mathbb{R}[z_1, \dots, z_k]$ a set of linearly independent monomials in $\mathbb{R}[z_1, \dots, z_k]/\operatorname{Ideal}(\mathcal{D}_z)$. Let $E_x \otimes E_z$ be the Kronecker product of E_x and E_z. Then by the basic property of Kronecker and tensor products, $E_x \otimes E_z$ is a set of linearly independent monomials in $\mathbb{R}[x, z]/\operatorname{Ideal}(\mathcal{D})$. Moreover if also $\mathcal{D}_z$ and $\mathcal{D}_x$ have no replicated points, then it is a $\mathbb{R}$-vector space basis and it has dimension $n_x n_z$ where n_i is the number of points in $\mathcal{D}_i$, $i = z, x$. Typically E_x is a set of monomials of the same degree, two or three, and E_z is an order ideal. In Section §3.5 of Pistone et al (2001) and Section §3 of Maruri-Aguilar et al (2006) algorithms are provided to compute E_x and E_z for generic $\mathcal{D}_x$ and $\mathcal{D}_z$.

In practice and when q and k are not small, it has to be expected: a) that $E_x \otimes E_z$ is large and restriction to a subset has to be considered to generate response surface models for the problem at hand, and also b) that not all runs in $\mathcal{D}_x \times \mathcal{D}_z$ can be implemented. We suggest an algorithm for selection of a subset L of $E_x \otimes E_z$ to be used as support for a model identifiable by a given fraction $\mathcal{F} \subset \mathcal{D}$. The subset L is selected according to a "statistical" criterion.

The design/model matrix for $\mathcal{D}$ and a model supported over a subset of $E_x \otimes E_z$ must be of full rank, independently of the selected representatives of the homogeneous points in $\mathcal{C}_{\mathcal{D}_x}$. (This is an immediate corollary of Lemma 3 in Maruri-Aguilar et al (2006)). That is, identifiability does not depend on the homogeneous coordinates. However, other properties of the design/model matrix are strongly effected by the representatives used; for example the eigenvalues of the corresponding information matrix.

We choose to minimise the condition number λ of the information matrix. It is defined as $\lambda = \lambda_{max}/\lambda_{min}$ where λ_{max} and $\lambda_{min} \geq 0$ are the maximum and minimum eigenvalues of the information matrix, $X_L^t X_L$ where ^{t} indicates transpose. If $X_L^t X_L$ is close to singular then its columns are almost linearly dependent and this is signaled by a minimum eigenvalue close to zero. Thus a small condition number indicates more stability in the least square estimates and a smaller variance inflation factor than for larger condition numbers.

In Draper and Pukelsheim (1998) Kronecker type models are studied for pure mixture experiments and in Prescott et al (2002) quadratic Kronecker models are conjectured to be the most robust to miss-specification of the information matrix among second order models for experiments with mixtures. In particular in Corollary 1 the authors in Prescott et al (2002) show that any model in a K-chain has higher maximum eigenvalue of the information matrix than the Kronecker type model. A K-chain is a chain of mixture models all of the same size; a model in the chain differs from the next one by one term; and the final model is of the Kronecker type. Thus E_x is of Kronecker type.

Other statistical criteria can be considered. A referee, whom the authors thank, suggested coupling the condition number criterion with a criterion related to the goodness of fit of the model. The major change in the algorithm below is in the definition of λ_i which could become a vector or remain scalar and should now correspond to the new criterion or combinations of criteria. In Example 6 of Section 3.2 we simply checked that the R^2 values expressing the goodness of fit of the selected sub-model were not worse than those of previous analyses presented in the literature.

3.1 Selection based on term orderings

As mentioned, the number of columns in the design/model matrix, X, for $E_x \otimes E_z$ and $\mathcal{D}$ or $\mathcal{F}$ might render a full search prohibitive. Nevertheless term orderings can be used to guide this search. The search might start with a Kronecker type model E_x (likely to have a low condition number). Terms are substituted one at a time according to a term ordering which favours the x-indeterminates. We suggest and sketch a variation of the algorithms in Babson et al (2003); Maruri-Aguilar et al (2006) to scan the class of models obtained with term orderings. This class of models is typically smaller than the class of sub-matrices of X which are of full rank and of size n. The search we suggest is based on the algorithms for exchange of bases introduced in Faugère et al (1993), developed in Babson et al (2003), and described for designs in Section §3.5 of Pistone et al (2001) and Section §3 of Maruri-Aguilar et al (2006). In Bates et al (2003) an algorithm is given for listing all saturated models which are order ideal and have the same support size.

The idea is to order the finite set of monomials $E_x \otimes E_z$, equivalently the columns of X they label, in all possible ways that can be extended to a full term ordering. We do so by using vectors of weights, i.e. $w \in \mathbb{Z}_{\geq 0}^{q+k}$. It turns out that a finite set of weighing vectors is sufficient to describe all such possible ways. The set of all weighing vectors, W, depends only on the exponents of monomials in the candidate set $E_x \otimes E_z$. Thus W can be computed once for each set $E_x \otimes E_z$ independently of the design and become part of a library. The computation of W is straightforward for models in two dimensions, but for models in higher dimensions there is still need for efficient algorithms. This is largely investigated in Maruri-Aguilar (2007) to which we refer for discussion on the computation of W. See also Example 6 for a method to compute W approximately.

Example 4. There are only two ways of ordering the three monomials $x_1^2, x_1 x_2,$ $x_2^2 \in \mathbb{R}[x_1, x_2]$ which are extendable to term ordering. They are $x_1^2 < x_1 x_2 < x_2^2$ and $x_2^2 < x_1 x_2 < x_1^2$. The first one corresponds to the weighing vector $w = (1, 2)$, indeed $((1, 2) \cdot (2, 0)) = 2 < ((1, 2) \cdot (1, 1)) = 3 < ((1, 2) \cdot (0, 2)) = 4$. Many other weighing vectors can be equivalently considered.

Call X_w the matrix X whose columns are reordered according to w. Then the first n linearly independent columns of X_w can be used as support for regression models. We select the model with n terms and the smallest condition number across $w \in W$. The algorithm can be outlined as follows.

Input: $\mathcal{D}_x$ and $\mathcal{D}_z$, a fraction $\mathcal{F} \subseteq \mathcal{D}_x \times \mathcal{D}_z$ and the number of terms in the final submodel n and the sets of monomials E_x and E_z, which are determined following the guidelines at the beginning of Section 3. The final submodel size n cannot be greater than the number of points in $\mathcal{F}$ to ensure identifiability.

Output: a submodel L_0 with a minimal condition number λ_0. The final submodel is formed by the smallest terms of $E_x \otimes E_z$ with respect to a term ordering.

Technique: the search space of candidate submodels is generated by ordering $E_x \otimes E_z$ with different weight vectors, and within this search space we look for the submodel with the smallest condition number.

Step 1: Compute the design-model matrix X using the points in $\mathcal{F}$ and the terms in $E_x \otimes E_z$. Compute (see also Example 6), the set of weight vectors, $W := \{w_1, \ldots, w_s\}$. Initialize $i := 1$, $\lambda_0 := \infty$ and $L_0 := []$.

Step 2: Order $E_x \otimes E_z$ and the corresponding columns of X using the weight vector w_i. Let L be the first n monomials of $E_x \otimes E_z$ such that the rank of X_L is n. Let λ_i be the condition number of $X_L^t X_L$.

Step 3: If $\lambda_i < \lambda_0$ then $\lambda_0 \leftarrow \lambda_i$ and $L_0 \leftarrow L$.

Step 4: Update $i \leftarrow i + 1$. If $i \leq s$ then repeat from Step 3.1, otherwise L_0 is the set of terms of the wanted model.

The algorithm clearly ends as W is a finite set (Babson et al (2003)). Moreover, any weight vector identifies a model of size n and thus the algorithm gives an answer. The algorithm is of order $O((n_x n_z)^{2(qk-1)} n^2)$ and, as the dimensions q and k are fixed, the algorithm is of polynomial order in $(n_x n_z)$ (see Babson et al (2003). This argument is detailed in Chapter 4 of Maruri-Aguilar (2007)). The search space is certainly more restricted than the full combinatorial search of exponential order $\binom{n_x n_z}{n} = O((n_x n_z)^{n_x n_z})$. However the final model respects a hierarchical structure, unlike many of the models in a combinatorial search and the search is clearly much faster.

3.2 Examples

Example 5. A mixture-amount design $\mathcal{D}$ is given in the left-hand side of Table 1 in affine coordinates. Here x_1 and x_2 are proportions of a total amount m. The ideals of interest are $\text{Ideal}(\mathcal{D}) = \langle x_1 + x_2 - m, (m-1)(m-2), (x_2 - 1)(m-2), (x_2 - 1)(x_2 - 2) \rangle$, $\text{Ideal}(\mathcal{C}_{\mathcal{D}_x}) = \langle x_1 x_2 (x_1 - x_2) \rangle$ and $\text{Ideal}(\mathcal{D}_m) = \langle (m-1)(m-2) \rangle$, from which we have that $E_x = \{x_1^2, x_1 x_2, x_2^2\}$ and $E_z = \{1, m\}$. The corresponding X matrix is shown in the right side of Table 1. The algorithm of Section 3 returns $L_0 = \{x_1^2, x_2^2, x_1 x_2, m x_2^2\}$ for the weighing vector $w = (1, 2, 3)$. In this simple case we can additionally perform a full combinatorial search, which returns the same result.

x_1	x_2	m	x_1^2	$x_1 x_2$	x_2^2	$m x_1^2$	$m x_1 x_2$	$m x_2^2$
0	1	1	0	0	1	0	0	1
0	2	2	0	0	4	0	0	8
1	1	2	1	1	1	2	2	2
2	0	2	4	0	0	8	0	0

Table 1. Mixture-amount design and matrix X for Example 5.

Example 6. We consider the well-known bread experiment introduced in Næs et al (1998), for which $\mathcal{D}_x$ is a simplex lattice with three factors and 10 runs and $\mathcal{D}_z$ is a factorial 3^2 design. The analysis in Prescott (2004) returns a final model of $n = 15$ terms and with condition number 86.83; see Equation (11) in Prescott (2004).

For the natural sets $E_x = \{x_1, x_2, x_3, x_1^2, x_2^2, x_3^2, x_1 x_2, x_1 x_3, x_2 x_3\}$ and $E_z = \{1, z_1, z_2, z_1^2, z_1 z_2, z_2^2\}$, the set $\mathrm{E}_x \otimes \mathrm{E}_z$ has 54 monomials, and the number of submodels with fifteen terms is $\binom{54}{15} \approx 8.6 \times 10^{12}$. A full search on this space is impossible but the algorithm of Section 3 can be applied to select a candidate model with a small condition number.

Instead of computing the full set of weighing vectors W, which, as mentioned, can be computationally expensive, we mimic it as follows. The $(q + k - 1)$-dimensional simplex intersects all "cones of equivalence classes" of the weighing vectors. In this sense every point on the simplex is equivalent to an element of W. We apply our algorithm with a sample of random vectors uniformly distributed over the simplex. If the sample is large enough, there is a high chance of picking at least one w from each equivalence class. This alternative is properly quantified in Chapter 2 of Maruri-Aguilar (2007).

The variables are listed as $(x_1, x_2, x_3, z_1, z_2)$. The algorithm returns the submodel $L_0 = (\{x_1, x_2, x_3\} \otimes \{1, z_1, z_2\} \cup \{x_2, x_3\} \otimes \{z_1^2, z_1 z_2, z_2^2\})$ for $w = (17, 12, 10, 3, 2)$. The model L_0 traded the monomials $x_1 z_1^2, x_1 z_2^2$ in Equation(11) of Prescott (2004) for $x_2 z_1 z_2$ and $x_3 z_1 z_2$, and this slight asymmetry allows for the reduction of the condition number to 47.47. With respect to the model in Prescott (2004), there is a slight increase in the root-mean-squared error, while R^2 is practically the same.

4 Final comments

In this note we considered a design $\mathcal{F} \subseteq \mathcal{D}_x \times \mathcal{D}_z$ and a set of linearly independent monomial functions over the vector space of real functions defined over $\mathcal{D}_x \times \mathcal{D}_z$.

An algorithm for selecting a model identified by $\mathcal{F}$, with a given number of terms and of minimal condition number is described. It has polynomial complexity in the number of design points and model size. Its search space is smaller than the one of a full search. In the authors' experience, (see also Babson et al (2003)), not only it is fast (especially when coupled with a search of the W vectors over a grid as in Example 6) but also it performs well in returning good models. One possible drawback is that it might exclude models which are symmetric in the factors. This is inherent in the use of term orderings and thus w vectors. Indeed there is no term ordering for which $x_2^2 < x_1^2 < x_1 x_2$. Symmetric models might be added to the search space or one can use only partially weighing vectors w. Methods for working with monomial bases of the quotient space which are free of term-ordering computations and are based on multiplication tables are being studied in the algebraic community; see Rouillier (1999) for a first example. Other criteria can be substituted for the minimal condition number criterion and general designs, even with replicated runs, can be considered. We focused on mixture designs with process variables or mixture amount experiments. The final model we obtain is usually not one suggested in the literature; it differs usually from the model obtained by running the standard algorithm in §3.5 of Pistone et al (2001), and when comparable it performs statistically at least as well as other models suggested in the literature for the examples we tried.

References

Babson E, Onn S, Thomas R (2003) The Hilbert zonotope and a polynomial time algorithm for universal Gröbner bases. Adv Appl Math 30(3):529–544

Bates RA, Giglio B, Wynn HP (2003) A global selection procedure for polynomial interpolators. Technometrics 45(3):246–255

Cornell JA (2002) Experiments with mixtures, 3rd edn. Wiley Series in Probability and Statistics, Wiley-Interscience [John Wiley & Sons], New York

Draper NR, Pukelsheim F (1998) Mixture models based on homogeneous polynomials. J Statist Plann Inference 71(1-2):303–311

Faugère JC, Gianni P, Lazard D, Mora T (1993) Efficient computation of zero-dimensional Gröbner bases by change of ordering. J Symb Comp 16(4):329–344

Maruri-Aguilar H (2007) Algebraic statistics in experimental design. Ph.D. thesis (submitted), Department of Statistics, University of Warwick

Maruri-Aguilar H, Notari R, Riccomagno E (2006) On the description and identifiability analysis of mixture designs. Accepted for publication in Statistica Sinica

Næs T, Færgestad EM, Cornell J (1998) A comparison of methods for analysing data from a three component mixture experiment in the presence of variation created by two process variables. Chem Int Lab Syst 41:221–235

Pistone G, Riccomagno E, Wynn HP (2001) Algebraic Statistics, Monographs on Statistics and Applied Probability, vol 89. Chapman & Hall/CRC, Boca Raton

Prescott P (2004) Modelling in mixture experiments including interactions with process variables. Qual Tech & Qual Manag 1(1):87–103

Prescott P, Draper NR, Dean AM, Lewis SM (2002) Mixture experiments: ILL-conditioning and quadratic model specification. Technometrics 44(3):260–268

Rouillier F (1999) Solving zero-dimensional systems through the rational univariate representation. Appl Algebra Engrg Comm Comput 9(5):433–461

Scheffé H (1958) Experiments with mixtures. J Roy Statist Soc Ser B 20:344–360

Scheffé H (1963) The simplex-centroid design for experiments with mixtures. J Roy Statist Soc Ser B 25:235–263

D-optimal Designs for Nonlinear Models Possessing a Chebyshev Property

Viatcheslav B. Melas

Faculty of Mathematics and Mechanics, St.Petersburg State University, University avenue 28, Petrodvoretz, 198504 St. Petersburg, Russia
v.melas@pobox.spbu.ru

Summary. The paper is devoted to experimental design for nonlinear regression models, whose derivatives with respect to parameters generate a generalized Chebyshev system. Most models of practical importance possess this property. In particular it is seen in exponential, rational and logistic models as well as splines with free knots. It is proved that support points of saturated locally D-optimal designs are monotonic and real analytic functions of initial values for those parameters on which models depend nonlinearly. This allows one to represent the functions by Taylor series. Similar properties of saturated maximin efficient designs are also investigated.

Key words: nonlinear regression models, exponential, rational and logistic models, locally D-optimal designs, maximin efficient D-optimal designs, functional approach

1 Introduction

Assume experimental results $y_1, \ldots, y_N \in \mathbf{R}$ arise under the standard nonlinear regression model

$$y_j = \eta(x_j, \Theta) + \varepsilon_j, \quad j = 1, \ldots, N, \tag{1}$$

where $x_j \in X$, $j = 1, \ldots, N$ are observation points, X is a set of possible values for these points, $\Theta = (\theta_1, \ldots, \theta_m)^T$ is the vector of unknown parameters, ε_i, $i = 1, \ldots, N$ are independent identically distributed random values such, that $E\varepsilon_j = 0, D\varepsilon_j = \sigma^2 > 0$, σ^2 is unknown constant.

Let the regression function be of the form

$$\eta(x, \Theta) = \sum_{i=1}^{k} a_i \eta_i(x, \Lambda), \quad \Lambda = (\lambda_1, \ldots, \lambda_{m-k})^T, \tag{2}$$

where $a_i \neq 0$, $i = 1, \ldots, k$, $\Lambda \in \Omega$, Ω is an open set in $\mathbf{R}^{m-k}$ such that $\min_{i \neq j} |\lambda_i - \lambda_j| \geqslant \triangle > 0$, X is a finite or semi-infinite interval in $\mathbf{R}$, $\triangle$ is a given small value, $\Theta = (a_1, \ldots, a_k, \lambda_1, \ldots, \lambda_{m-k})^T$.

116   Viatcheslav B. Melas

Consider the following assumptions.

A1 Assume that the functions $\eta_i(x, \Lambda)$, $i = 1, \ldots, k$ are real analytic in $x \in X$ for any fixed $\Lambda \in \Omega$.

A2 Assume that for all fixed $\Lambda \in \Omega$ the functions

$$f_j(x) = \eta_j(x, \Lambda), j = 1, \ldots, k, \tag{3}$$

$$f_{k+j}(x) = \frac{\partial}{\partial \lambda_j} \sum_{i=1}^{k} a_i \eta_i(x, \Lambda), \ j = 1, \ldots, m - k \tag{4}$$

are linearly independent and such that any linear combination

$$\sum_{i=1}^{m} \alpha_i f_i(x),$$

where $\{\alpha_i\}$ are arbitrary numbers from $\mathbf{R}$, possesses no more than $m - 1$ isolated roots allowing for their multiplicity.

We will say that model (1)–(2) possesses a generalized Chebyshev property if it satisfies A2. Thus we impose on the function system $f_1(x), \ldots, f_m(x)$ the requirement which is more strong than weak Chebyshev property and is more weak than the extended Chebyshev property in the sense of (Karlin and Studden (1966), Ch.1).

It is easy to check that many nonlinear models used in practice satisfy A1 and A2. Consider a few examples.

Example 1 (General exponential model). Let

$$\eta(x, \theta) = \sum_{i=1}^{s} a_{0i} x^{i-1} + \sum_{j=1}^{n-k} \sum_{\nu=1}^{l_j} a_{j\nu} x^{\nu-1} e^{-\lambda_j x}, \tag{5}$$

$$a = (a_1, \ldots, a_k)^T = (a_{01}, \ldots, a_{0s}, a_{11}, \ldots, a_{m-k,l_{m-k}})^T;$$

$$x \in [d_1, d_2]; \ \lambda_1, \ldots, \lambda_{m-k} > 0; \ k = s + \sum_{j=1}^{m-k} l_j.$$

The corresponding functions $f_i(x)$, $i = 1, \ldots, m$ generate an extended Chebyshev system (see Karlin and Studden (1966), Ch. 1) and consequently the model possesses property A2. Since exponential functions are real and analytic we have also property A1. Functions of the form (5) generate an important class of solutions of linear differential equation systems. For this reason they are widely used in practice. Frequently occurring particular cases are:

$$a_1 e^{-\lambda_1 x}, \ a_1 + a_2 e^{-\lambda_1 x}, \ a_1(e^{-\lambda_1 x} - e^{-\lambda_2 x}), \ a_1 e^{-\lambda_1 x} + a_2 e^{-\lambda_2 x}$$

(see, e.g., Han and Chaloner (2003) and Becka and Urfer (1996) among many others).

Example 2 (Rational models). The regression functions

$$\frac{a_1 x}{x + \lambda_1}, \quad \frac{a_1 x + a_2}{x^2 + \lambda_1 x + \lambda_2}$$

are widely used in many fields (see, e.g. , Ratkowsky (1990)).

It is easy to check that these functions with $\lambda_1, \lambda_2 > 0$, $x \in [d_1, d_2]$, $d_1 \geq 0$ satisfy A1 and A2. Under similar restrictions the general rational model

$$\eta(x, \theta) = \sum_{i=1}^{k} a_i x^{i-1} \left/ \left(\sum_{i=1}^{m-k} \lambda_i x^{i-1} + x^{n-k-1} \right) \right.$$

also possesses properties A1 and A2 (see, e.g., Melas (2006), Ch.5).

Example 3 (Splines with free knots). Let

$$\eta(x, \theta) = \sum_{i=1}^{k} a_i x^{i-1} + \sum_{i=1}^{m-k} (x - \lambda_i)_+^{k-1},$$

where $(a)_+ = \max\{0, a\}$; $d_1 \leq \lambda_1 < \ldots < \lambda_{m-k} \leq d_2$; $x \in [d_1, d_2]$.

This function satisfies A2 (see, e.g., Dette et al (2006)). Note that the property A1 is satisfied only for $x \in (d_1, \lambda_1) \cup (\lambda_1, \lambda_2) \cup \ldots \cup (\lambda_{n-k}, d_2)$.

The Chebyshev property of basic regression functions has been mainly exploited for constructing c- and E-optimal designs for linear models (see Studden (1968) and Imhof and Studden (2001)). For nonlinear (in parameters) rational models as in Example 2 similar results were obtained for locally c- and E-optimal designs Dette et al (2004a). Here we will concentrate on the D-criterion. For nonlinear models (1)–(2) D-optimal designs usually depend on the true values of parameters, here $\lambda_1, \ldots, \lambda_{m-k}$. We will use two approaches in order to overcome this difficulty: locally optimal and maximin efficient designs. The locally optimal approach consists of replacing true values by initial ones and was first implemented in Chernoff (1953).

The maximin approach was considered in Melas (1978). It consists of maximizing the determinant (or other functional) of the information matrix for the least favorable values of nonlinear parameters. Müller (1995), Dette (1997) and others consider the problem of maximizing the ratio of two determinants. One of these is the determinant of the information matrix of a given design and the other is a similar determinant for a locally D-optimal design. The corresponding designs are usually called maximin efficient D-optimal designs.

Locally D-optimal designs for regression functions from Example 1 are studied in Melas (1978) on the base of a functional approach. This approach consists of considering support points of optimal designs as implicit functions of the initial values of the parameters. This enables a proof that support points of locally D-optimal designs are monotonic and it also allows representation of the functions by Taylor series. In Melas (2005), Melas (2006) this approach was entended to models possessing properties A1 and A2. And in Dette et al (2006) it was implemented for splines with free knots.

Here we present similar results in a more simple and general form for locally D-optimal as well as for maximin efficient D-optimal designs. We will consider designs with minimal support, that is designs with the number of support points equal to the number of unknown parameters. Such designs will be called saturated. For many models it has been proved theoretically or empirically that optimal designs are saturated (see, e.g., He et al (1996) and Melas (2006)).

2 Locally D-optimal designs

Let experimental results satisfy relations (1)–(2) and let assumptions A1 and A2 be fulfilled. A discrete probability measure given by the table

$$\xi = \begin{pmatrix} x_1 \ldots x_n \\ \omega_1 \ldots \omega_n \end{pmatrix},$$

where $x_i \neq x_j$ $(i \neq j)$; $\omega_i > 0$; $\sum \omega_i = 1$; $x_i \in X; i = 1, \ldots, n$ will be called as usual an (approximate) experimental design. Consider the value

$$\det M(\xi, \Theta), \tag{6}$$

where

$$M(\xi, \Theta) = \sum_{i=1}^{n} f(x_i) f^T(x_i) \omega_i.$$

The functions $f(x)$, generally speaking, depend on Θ and are determined by (3)–(4). It is easy to check that the design maximizing (6) does not depend on the vector a, but can depend on Λ.

A design maximizing (6) over the class of all designs with $n = m$ support points under a fixed Λ will be called saturated locally D-optimal design (SLD-design). In many cases (see, e.g., He et al (1996) and Melas (2006)) such a design is unique and maximizes (6) among all (approximate) designs. It is easy to check (see, e.g., Fedorov (1972)) that in such a design all weight coefficients are the same: $\omega_i = 1/m$, $i = 1, \ldots, m$.

Without loss of generality assume that

$$d_1 \leq x_1 < \ldots < x_m \leq d_2.$$

Let $\tau = (\tau_1, \ldots, \tau_s)$, $s = m$, $s = m - 1$ or $s = m - 2$ be the vector of design points not coinciding with the bounds of the interval. Then the design $\xi = \xi_\tau$ depends only on the vector τ. Assume that

A3 All SLD-designs with $\Lambda \in \Omega$ have the same type.

A4 For a point $\Lambda = \Lambda^0 \in \Omega$ there exists a unique SLD-design, denoted it by $\xi_0 = \xi_{\tau_{(s)}}$.

As was verified in papers Melas (2005), Melas (2006), and Dette et al (2006) many nonlinear models possess properties A1–A4. The following result holds.

Theorem 1. *For regression model (1)–(2) let assumptions A1–A4 be fulfilled. Then for any $\Lambda \in \Omega$ there exists a unique SLD-design $\xi = \xi_\tau, \tau = \widehat{\tau}(\Lambda)$. The vector function*

$$\widehat{\tau}(\Lambda) : \Lambda \in \Omega \to \widehat{\tau}(\Lambda) \in [d_1, d_2]^s$$

is real analytic and all its components monotonically depend on each of the parameters $\lambda_1, \ldots, \lambda_{m-k}$.

A proof of the theorem is based on the Implicit Function Theorem of Gunning and Rossi (1965) and a representation for the Jacobian matrix. It is similar to arguments in (Melas (2006), Ch. 2). Here we give a more simple and more universal formulation of condition A2.

Remark 1. If the design $\xi_{\tau_{(0)}}$, $\tau_{(0)} = \hat{\tau}(\Lambda^0)$ is found (analytically or numerically) the vector function $\tau(\Lambda)$ can be expanded into a Taylor series using recurrent formulae given in Melas (2005) and Melas (2006).

3 Maximin efficient designs

Since a locally D-optimal design does not depend on a we can set, without loss of generality, $a_1 = a_2 = \ldots = a_k = 1$. Denote

$$M(\xi, \Lambda) := M(\xi, \Theta),$$

where $\Theta = (a_1, \ldots, a_k, \lambda_1, \ldots \lambda_{m-k})^T$ and $a_1 = a_2 = \ldots = a_k = 1$.
    Let

$$\Omega = \overline{\Omega} = \{\Lambda : (1-z)c_i \le \lambda_i \le (1+z)c_i, i = 1, \ldots, m-k\},$$

where $c = (c_1, \ldots, c_{m-k})^T$ is an initial value for Λ and $z > 0$ is the value of the relative error of this approximation. Denote by

$$\Omega_c = \{(1-z)c, (1+z)c\},$$

the set of the two extreme points of $\overline{\Omega}$. A design ξ will be called a saturated maximin efficient (D-optimal) design for Ω or, briefly, $SMME$ design if

$$\min_{\Lambda \in \Omega} \frac{\det M(\xi, \Lambda)}{\det M(\xi^*(\Lambda), \Lambda)} = \max_{\xi} \min_{\Lambda \in \Omega} \frac{\det M(\xi, \Lambda)}{\det M(\xi^*(\Lambda), \Lambda)}, \tag{7}$$

where $\xi^*(\Lambda)$ is an SLD-design and maximin is taken over all saturated designs. The value of the left side taken to degree $1/m$ will be called the minimal efficiency of the design ξ. Note that the ratio

$$\det M(\xi, \Lambda)/\det M(\xi^*(\Lambda), \Lambda)$$

can be redefined while preserving continuity if $\min_{i \neq j} |\lambda_i - \lambda_j| = 0$ (see Melas and Pepelyshev (2005)). A design ξ will be called a maximin efficient (D-optimal) design with a minimal structure if it is an $SMME$ design for $\Omega = \Omega_c$.

The following result allows substantial simplification of the problem.

Theorem 2. *Let regression model (1)–(2) satisfy assumptions A1–A4. Then there exists a positive value z^* such that for any $z \leq z^*$ there exists a unique $MMEMS$ design and it is a unique $SMME$ design for $\Omega = \overline{\Omega}$. The weights of this design are uniform and its nontrivial points are real analytic functions of z with $0 < z < z^*$.*

A proof of this theorem is based on the Taylor expansion of the criterion value. It is obtained by the same arguments as in the proof of Theorem 4.2 in Dette et al (2006). In that paper the recurrent formulae for expanding these functions into Taylor a series can also be found.

Let us demonstrate the approach by the following example. Let

$$\eta(x, \Theta) = a_1 e^{-\lambda_1 x} + a_2 e^{-\lambda_2 x}; \ x \in [0, \infty); \ a_1, a_2 \neq 0, \ \Lambda = (\lambda_1, \lambda_2)^T,$$

$$\Lambda \in \overline{\Omega} = \{\Lambda : (1 - z)c_i \leq \lambda_i \leq (1 + z)c_i, i = 1, 2\}.$$

In Dette et al (2004b) it is proved that the locally D-optimal design for this model is unique and is of the form

$$\xi = \begin{pmatrix} 0 & \tau_1 & \tau_2 & \tau_3 \\ 1/4 & 1/4 & 1/4 & 1/4 \end{pmatrix}. \tag{8}$$

Thus, properties A3 and A4 hold and A1 and A2 are in evidence for the general exponential model as was discussed above. The locally D-optimal designs for this model were constructed in Melas (2005). Let us consider $SMME$ designs for $\overline{\Omega}$. Without loss of generality we can assume that $c_1 \leq c_2$. It can be checked that the $SMME$ design has the form (8) with $\tau_1 = \widehat{\tau}_i(c_1, c_2, z), i = 1, 2, 3$ and $\widehat{\tau}_i(c_1, c_2, z) = \widehat{\tau}_i(1, c_2/c_1, z)/c_1, i = 1, 2, 3$.

Thus it will do to consider the case $c_1 = 1$. Take $c_2 = 5$ (in other cases we obtained similar results). The Taylor expansions obtained by formulae from Dette et al (2006) are the following

$$\widehat{\tau}_1(z) = 0.161 + 0.038z^2 + 0.021z^4 + \ldots,$$
$$\widehat{\tau}_2(z) = 0.626 + 0.232z^2 + 0.163z^4 + \ldots,$$
$$\widehat{\tau}_3(z) = 1.819 + 0.813z^2 + 0.513z^4 + \ldots.$$

Note that it is difficult to study the radius of convergence of these series analytically. However, we found numerically that they converge with arbitrary $z < 1$, and with $z \leq 0.5$ the first three terms are sufficient to obtain design points with precision 10^{-3}. Also we found that the corresponding designs practically coincide with $SMME$ designs for $z \leq 0.28$. (This can be checked numerically, using an equivalence theorem.)

The minimal efficiency of the $MMEMS$ design and the locally D-optimal design for the point $\Lambda = c$ is shown in Figure 1.

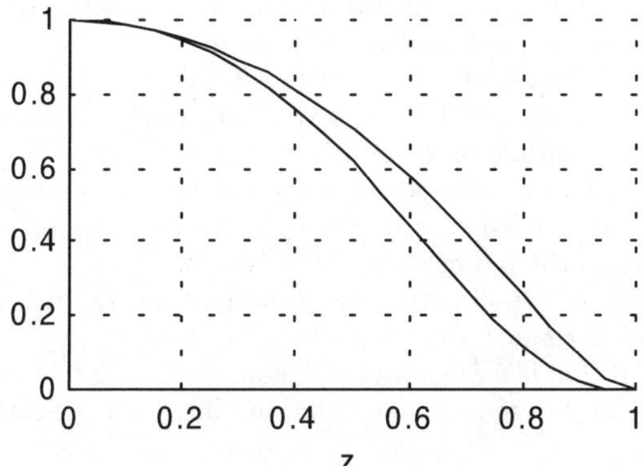

Fig. 1. The efficiency of the $MMEMS$ design (higher curve) and the locally optimal design in c (lower curve).

We can see that $MMEMS$ designs have considerable advantage when $z > 0.2$. However, for $z > 0.28$ it is possible to construct more efficient designs. For example, for $z = 0.5$ we constructed numerically a design, which is MME among all approximate designs. This design has six support points and is approximately equal to

$$\begin{pmatrix} 0 & 0.140 & 0.440 & 1.048 & 1.75 & 3.25 \\ 0.24 & 0.18 & 0.19 & 0.16 & 0.13 & 0.10 \end{pmatrix}.$$

The minimal efficiency of this design is equal to 0.8431, whereas for the MMEMS design this efficiency is equal to 0.7045. For the equidistant design in 10 points on the interval $[0, 2]$, which is the best among equidistant designs, this efficiency equals 0.5904.

Acknowledgement. The work was partly supported by Russian Foundation of Basic Research (project N 07-01-00519). The author is indebted to two anonymous referees for helpful comments for an earlier version of the paper.

References

Becka M, Urfer W (1996) Statistical aspects of inhalation toxicokinetics. Environmental Ecolog Statist 3:51–64

Chernoff H (1953) Locally optimal designs for estimating parameters. Ann Math Statistics 24:586–602

Dette H (1997) Designing experiments with respect to "standardized" optimality criteria. J Roy Statist Soc Ser B 59(1):97–110

Dette H, Melas V, Pepelyshev A (2004a) Optimal designs for a class of nonlinear regression models. Ann Statist 32(5):2142–2167

Dette H, Melas VB, Wong W (2004b) Locally d-optimal designs for exponential regression. Preprint Ruhr-Universität Bochum

Dette H, Melas VB, Pepelyshev A (2006) Optimal designs for free knot least squares splines. Preprint Ruhr-Universität Bochum Http://www.ruhr-uni-bochum.de/mathematik3/preprint.htm

Fedorov VV (1972) Theory of optimal experiments. Academic Press, New York, translated from the Russian and edited by W. J. Studden and E. M. Klimko, Probability and Mathematical Statistics, No. 12

Gunning RC, Rossi H (1965) Analytic functions of several complex variables. Prentice-Hall Inc., Englewood Cliffs, N.J.

Han C, Chaloner K (2003) D- and c-optimal designs for exponential regression models used in viral dynamics and other applications. J Statist Plann Inference 115(2):585–601

He Z, Studden WJ, Sun D (1996) Optimal designs for rational models. Ann Statist 24(5):2128–2147

Imhof LA, Studden WJ (2001) E-optimal designs for rational models. Ann Statist 29(3):763–783

Karlin S, Studden WJ (1966) Tchebycheff systems: With applications in analysis and statistics. Pure and Applied Mathematics, Vol. XV, Interscience Publ. John Wiley & Sons, New York-London-Sydney

Melas V (1978) Optimal designs for exponential regression. Mathematische Operationsforshung und Statistik, Ser Statistics 9:45–59

Melas V, Pepelyshev A (2005) On representing maximin efficient designs by taylor series. Proceedings of the 5th StPetersburg Workshop on Simulation pp 469–474

Melas VB (2005) On the functional approach to optimal designs for nonlinear models. J Statist Plann Inference 132(1-2):93–116

Melas VB (2006) Functional approach to optimal experimental design, Lecture Notes in Statistics, vol 184. Springer, New York

Müller CH (1995) Maximin efficient designs for estimating nonlinear aspects in linear models. J Statist Plann Inference 44(1):117–132

Ratkowsky D (1990) Handbook of Nonlinear Regression Models. Dekker, New York

Studden WJ (1968) Optimal designs on Tchebycheff points. Ann Math Statist 39:1435–1447

A New Tool for Comparing Adaptive Designs; a Posteriori Efficiency

José A. Moler[1] and Nancy Flournoy[2]

[1] Departamento de Estadística e Investigación Operativa. Universidad Pública de Navarra. Campus de Arrosadia s/n. 31006-Pamplona, Spain
jmoler@unavarra.es
[2] Department of Statistics. University of Missouri - Columbia. 146 Middlebush Hall Columbia, MO 65211-6100, USA flournoyn@missouri.edu

Summary. In this work, we consider an adaptive linear regression model designed to explain the patient's response in a clinical trial. Patients are assumed to arrive sequentially. The adaptive nature of this statistical model allows the error terms to depend on the past which has not been permitted in other adaptive models in the literature.

Some techniques of the theory of optimal designs are used in this framework to define new concepts: *a-posteriori efficiency* and *mean a-posteriori efficiency*. We then explicitly relate the variance of the allocation rule to the mean a-posteriori efficiency. These measures are useful for studying the comparative performance of adaptive designs. As an example, a comparative study is made among several design-adaptive designs to establish their properties with respect to a criterion of interest.

Key words: optimal designs, adaptive allocation, adaptive regression

1 Introduction

In this work we relate two topics: optimal designs and adaptive designs for clinical trials.

Adaptive designs constitute a class of sequential designs in which the probability distribution of the nth allocation depends on the accrued information obtained from the previous $n-1$ treament allocations and/or responses to treatment. The use of adaptive allocation for randomization can reduce sources of bias and permit the inclusion of ethical issues in the assignment of treatments; see Rosenberger and Lachin (2002).

The theory of optimal designs deals with the construction of an experimental design which optimizes some criterion given a statistical model; see, for instance, Atkinson and Donev (1992).

The general framework is a clinical trial for comparing the performance of L different treatments and a linear regression model which relates the response of the patient to the treatment allocation. Thus, an experimental design ξ indicates the proportion of patients assigned to each treatment i, $i = 1, \ldots, L$. Some recent papers deal with this problem; see, for instance, Zhang and Rosenberger (2006), Rabie and Flournoy (2004) and Dragalin et al (2006).

We are interested in using the classical techniques of the theory of linear optimal designs to measure the performance of an adaptive design with respect to a criterion. However, when an adaptive design is used to allocate patients to treatments, a complicated correlation structure is generated among the errors of the linear model. In the following, we introduce an adaptive linear model which allows us to deal with this situation.

Let $\{\boldsymbol{\delta}_j\}_{j \geq 1}$ be a sequence of L-dimensional random vectors such that, if the j-th patient is assigned to treatment i, then $\delta_{ji} = 1$ and $\delta_{jr} = 0$ for $r \neq i$. Given n sequential treatment assignments, the L-random vector $\mathbf{N}_n/n = (N_{n1}/n, \ldots, N_{nL}/n)$ is generated, where $N_{ni} := \sum_{j=1}^{n} \delta_{ji}$, $i = 1, \ldots, L$. Then, the resulting design is denoted

$$\xi_n := \left\{ \begin{array}{ccc} 1 & \cdots & L \\ N_{n1}/n & \cdots & N_{nL}/n \end{array} \right\}.$$

Consider the sequence of random vectors $\{\mathbf{Z}_n\} = \{(Z_{n1}, Z_{n2}, \ldots, Z_{nL})\}$, where Z_{ni} represents the response of the nth patient to the ith treatment, $i = 1, \ldots, L$. We assume throughout the paper that

[**A1**] for each treatment i , $\{Z_{ni}\}_{n \geq 1}$ is a sequence of identically distributed random variables, such that $\mu_i = E[Z_{ni}]$, $\sigma_i^2 = Var[Z_{ni}] > 0$ and Z_{ni} is independent of the past history of the trial $\sigma(\boldsymbol{\delta}_1, \ldots, \boldsymbol{\delta}_{n-1}, \mathbf{Z}_1, \ldots, \mathbf{Z}_{n-1})$ and of the current allocation rule $\boldsymbol{\delta}_n$.

Let Y_n be the observed response of the n-th patient. As we allocate patients following an adaptive design, consider the adaptive regression model

$$\frac{Y_n}{\sigma_n} = \sum_{i=1}^{L} \mu_i \frac{\delta_{ni}}{\sigma_n} + \frac{\varepsilon_n}{\sigma_n}, \quad \sigma_n^2 = \sum_{i=1}^{L} \delta_{ni}\sigma_i^2. \tag{1}$$

Then, from [**A1**], $\{\varepsilon_n\}$ is a sequence of martingale differences with respect to the increasing sequence of σ-algebras $\{\sigma(\boldsymbol{\delta}_1, \ldots, \boldsymbol{\delta}_j)\}_{j \geq 1}$ such that, for each n, $\varepsilon_n := Y_n - E[Y_n|\boldsymbol{\delta}_1, \ldots, \boldsymbol{\delta}_n]$. This model permits errors to depend on the past history of the trial, which is necessary for many adaptive designs. No assumptions are made on the response distribution and we use ordinary least squares (OLS) estimation. This setup contrasts with that of the study of optimal adaptive designs by Baldi Antognini and Giovagnoli (2005) in that they rely on maximum likelihood estimators

Let $\hat{\boldsymbol{\mu}}_n = (\hat{\mu}_{n1}, \ldots, \hat{\mu}_{nL})$ denote the OLS estimator when n patients have been treated. Lai and Wei (1982) show that the conditional variance of the OLS estimator is the diagonal matrix $diag(\sigma_1^2/N_{n1}, \ldots, \sigma_L^2/N_{nL})$ and provide sufficient conditions for the strong consistency of, and the existence of a central limit theorem for, the sequence $\{\hat{\boldsymbol{\mu}}_n\}_{n \geq 1}$, namely: [**C1**] for each $i = 1, \ldots, L$,

$N_{ni} = O(n)$, that is, there exist constants c_1, $c_2 > 0$ such that for each treatment i

$$c_1 < \lim_{n \to \infty} N_{ni}/n < c_2, \quad a.s.,$$

and **[C2]** $E[\|\varepsilon_n\|^\alpha | \delta_1, \ldots, \delta_n] < \infty$, for $\alpha > 2$.

Some adaptive designs satisfy the following condition

$$\textbf{[A2]} \ \mathbf{N}_n/n \to \boldsymbol{\pi} \ a.s., \qquad \sqrt{n}(\mathbf{N}_n/n - \boldsymbol{\pi}) \to N(\mathbf{0}, \Sigma), \ [D]$$

where $\boldsymbol{\pi} = (\pi_1, \ldots, \pi_L)$ is the asymptotic allocation, $[D]$ denotes convergence in distribution and $\Sigma = (\Sigma_{ij})$ is the covariance matrix of the asymptotic distribution. Observe that if $\pi_i > 0$, $i = 1, \ldots, L$, **[C1]** holds.

The paper is organized as follows. In section 2, we introduce the necessary tools related to the concept of design efficiency with practical interpretations. In section 3, we describe how to use these new measures of efficiency for several optimality criteria. Finally, in section 4, we illustrate how to use these measures in a comparative study of several design-adaptive designs. These measures can be used for response-adaptive designs as well.

2 A-posteriori and mean a-posteriori efficiencies

In the theory of optimal designs for linear models, an optimality criterion is a real convex (or concave) function ϕ which takes values in the space of information matrices. For a design ξ, it takes the value $\phi(\xi)$. Let $\xi^* = min_\xi \phi(\xi)$ be the design for which the information matrix minimizes (maximizes) the criterion function.

Because $\mathbf{N}_n$ is a random vector which takes values depending on the n-length realization of the adaptive design, conditional on the adaptive treatment allocation, we define **a-posteriori efficiency** as

$$\mathcal{PE}_n := \frac{\phi(\xi^*)}{\phi(\xi_n | \mathbf{N}_n)}. \tag{2}$$

Using the random value $\mathcal{PE}_n$, we avoid evaluating exact moments of the adaptive design over time, which is a difficult task, but is required for the traditional definition of efficiency: $\phi(\xi^*)/E[\phi(\xi_n)| \mathbf{N}_n]$. The behaviour of the stochastic process $\{\mathcal{PE}_n\}$ is useful for comparing different adaptive designs because it is closely related to that of $\mathbf{N}_n$, as we will see in section 3 for selected criteria. We also define **mean a-posteriori efficiency** as

$$\mathcal{ME}_n := E_{N_n}[\mathcal{PE}_n] = \phi(\xi^*) E_{N_n}\left[\frac{1}{\phi(\xi_n)}\right]. \tag{3}$$

Observe that the sequence $\{\mathcal{ME}_n\}$ is not random and, under some conditions, an explicit relation with the two first moments of the adaptive allocation can be obtained. For this reason, it appears to be an appropriate tool to compare the performance of adaptive designs.

Both expressions (2) and (3) take values in the interval $[0, 1]$ when the criterion functions are convex and positive, a usual situation among the classical optimality criteria. This allows interesting interpretations in practical problems. When $\mathcal{PE}_n = 1$, the design applied is optimal. The a-posteriori loss with the design applied is then $\mathcal{PL}_n = n(1 - \mathcal{PE}_n)$. Similar expressions can be obtained with the efficiency (3).

3 Efficiencies

In this section, we use the efficiencies introduced in the previous section to compare the behavior of several adaptive designs for a clinical trial with two treatments and the linear model (1). We explicitly relate the variance of the allocation rule given by a design to the mean a-posteriori efficiency. This is useful for a comparative study. In the following subsections, we illustrate the procedure with two criteria. The first one focusses on optimizing a statistical property of the estimators, whereas the second one focusses on optimizing the ethics of the design.

3.1 Criterion: minimize variance of mean treatment differences

The 2 group D_A-optimal design minimizes $Var[\hat{\mu}_{1n} - \hat{\mu}_{2n}|\boldsymbol{\delta}_1, \ldots, \boldsymbol{\delta}_n]$. Given a design ξ_n, the a-posteriori efficiency reaches the value zero when all patients are allocated to only one treatment and the value one when the optimal allocation ξ^* is applied.

Assuming equal variance, σ^2, for the patient responses, given a design ξ_n, it follows that $Var[\hat{\boldsymbol{\mu}}_n|\boldsymbol{\delta}_1, \ldots, \boldsymbol{\delta}_n] = \sigma^2 diag(1/N_{n1}, 1/N_{n2})$. Then, $Var[\hat{\mu}_{1n} - \hat{\mu}_{2n}|\boldsymbol{\delta}_1, \ldots, \boldsymbol{\delta}_n] = \sigma^2(1/N_{n1} + 1/N_{n2})$. The minimum value of this expression, $4\sigma^2/n$, occurs under equal allocation. So we can write

$$\mathcal{PE}_n = \frac{4\sigma^2/n}{\sigma^2(1/N_{n1} + 1/N_{n2})}$$
$$= 4\left[\pi_1(1 - \pi_1) + N_{n1}/n - \pi_1 - ((N_{n1}/n)^2 - \pi_1^2)\right] \tag{4}$$

When [**A2**] holds and $\pi_1 \neq \pi_2$, from Slutsky's theorem we have that $\sqrt{n}(\mathcal{PE}_n - 4\pi_1\pi_2) \to N(0, 16(\pi_2 - \pi_1)^2 \Sigma_{11})$; otherwise, it converges to zero. This result shows that high variability in the allocation rule, Σ_{11}, implies that the adaptive design can yield allocations with poor a-posteriori efficiencies. This relationship appears explicitly if we have that $E[N_{n1}/n] = \pi_1$, which maybe an appropriate approximation for moderate to large samples; taking expectations in (4), we obtain

$$\mathcal{ME}_n = 4\pi_1(1 - \pi_1) - 4Var[\frac{N_{n1}}{n}]. \tag{5}$$

Observe that the higher the variance of the design, the smaller is the mean a-posteriori efficiency of the design, even if the design is balanced asymptotically (which is the optimal situation with this criterion).

When the responses do not have equal variances, given a design ξ_n, $Var[\hat{\mu}_{1n} - \hat{\mu}_{2n}|\boldsymbol{\delta}_1, \ldots, \boldsymbol{\delta}_n] = \sigma_1^2/N_{n1} + \sigma_2^2/N_{n2}$ reaches the minimum value, $(\sigma_1 + \sigma_2)^2/n$, with Neyman allocation, which is $\pi_1 = \sigma_1/(\sigma_1 + \sigma_2)$. Let $r := \sigma_2/\sigma_1$ and $b_n := N_{n1}/n - \pi_1$. Then, we can write

$$\begin{aligned}
\mathcal{PE}_n &= \frac{(\sigma_1 + \sigma_2)^2}{\sigma_1^2/(N_{n1}/n) + \sigma_2^2/(N_{n2}/n)} \\
&= (1 + r)^2 \frac{N_{n1}/n(1 - N_{n1}/n)}{1 - (1 - r^2)N_{n1}/n} \\
&= \frac{(1 + r)^2}{1 - \pi_1(1 - r^2)}(\pi_1 + b_n)(1 - \pi_1 - b_n)\frac{1}{1 - \dfrac{1 - r^2}{1 - \pi_1(1 - r^2)}b_n}.
\end{aligned} \tag{6}$$

Consider $f(r) := (1 - r^2)/(1 - \pi_1(1 - r^2))$ and $g(r) := f(r)(1 + r)/(1 - r)$. Now, under the assumption that $|f(r)| < 1$, the Taylor expansion of (6) is

$$\begin{aligned}
\mathcal{PE}_n &= g(r)[\pi_1(1 - \pi_1) + (1 - 2\pi_1)b_n - b_n^2][1 + f(r)b_n + f(r)^2 b_n^2 + o(b_n^2)] \\
&= g(r)\pi_1(1 - \pi_1) + g(r)\left(1 - 2\pi_1 + f(r)\pi_1 - f(r)\pi_1^2\right) b_n \\
&\quad -g(r)\left(1 - f(r)(1 - 2\pi_1) - f(r)^2\pi_1(1 - \pi_1)\right) b_n^2 + o(b_n^2) \\
&:= a(\pi_1) + b(\pi_1)b_n - c(\pi_1)b_n^2 + o(b_n^2).
\end{aligned} \tag{7}$$

The function $a(\pi_1)$ is a function of π_1 whose maximum is the Neyman allocation. When [A2] holds, reasoning as before, a central limit theorem holds for $\mathcal{PE}_n$ in which the variance of the asymptotic distribution is $b(\pi_1)^2\Sigma_{11}$.

Taking expectations of (7) under the assumption that $E[N_{n1}/n] = \pi_1$ yields

$$\mathcal{ME}_n \sim a(\pi_1) - c(\pi_1)Var[N_{n1}/n], \tag{8}$$

where $f(n) \sim g(n)$ means $f(n)/g(n) \to 1$ as $n \to \infty$. As $c'(\pi_1) = 3f(r)^2c(\pi_1)$, $c(\pi_1)$ takes only positive values in the interval $[0, 1]$ and thus, the higher the variability of the adaptive allocation, the smaller will be the efficiency of the resulting designs.

3.2 Criterion: minimize the mean response

The ethical criterion we study is to minimize the total expected number of failures. So $\xi^* = argmin_\xi\{n_1q_1 + n_2q_2\}$, where q_1 and q_2 are the probability of failure under treatments 1 and 2, respectively. A generalization of this criterion, to include continuous responses and, assuming less response is better for the patient, is $\{n_1\mu_1 + n_2\mu_2\}$, where μ_1 and μ_2 are expected responses under treatments 1 and 2, respectively. In Zhang and Rosenberger (2006) this criterion is considered, but subject to a known variance.

Assume positive mean responses in this section, i.e., $\mu_1 > 0$ and $\mu_2 > 0$. Without loss of generality, let $\mu_1 < \mu_2$. Then $min_{\xi_n}\{n_1\mu_1 + n_2\mu_2\} =$

$min_{\xi_n}\{n\mu_2 + (\mu_1 - \mu_2)n_1\} = n\mu_1$, which corresponds to the design which allocates all patients to the best treatment. The a-posteriori efficiency of an adaptive design relative to one that minimizes the total number of failures is

$$\mathcal{PE}_n = \frac{\mu_1}{\mu_2 + (\mu_1 - \mu_2)N_{n1}/n}$$
$$= \frac{\mu_1}{\mu_2} \frac{1}{\left[1 - \pi_1(1 - \frac{\mu_1}{\mu_2}) - (1 - \frac{\mu_1}{\mu_2})(N_{n1}/n - \pi_1)\right]}. \tag{9}$$

Define $c := \mu_1/\mu_2$, $\bar{c} := 1 - \mu_1/\mu_2$ and $b_n := N_{n1}/n - \pi_1$. Then, when $|\frac{\bar{c}}{1 - \bar{c}\pi_1}| < 1$, we have

$$\mathcal{PE}_n = \frac{c}{1 - \bar{c}\pi_1} \frac{1}{1 - \frac{\bar{c}}{1 - \bar{c}\pi_1}b_n}$$
$$= \frac{c}{1 - \bar{c}\pi_1}[1 + \frac{\bar{c}}{1 - \bar{c}\pi_1}b_n + (\frac{\bar{c}}{1 - \bar{c}\pi_1})^2 b_n^2 + o(b_n^2)]$$
$$:= a(\pi_1) + b(\pi_1)b_n + c(\pi_1)b_n^2 + o(b_n^2). \tag{10}$$

As $c < 1$, $a(\pi_1)$ is an increasing function when $\pi_1 \in [0, 1]$. When most patients are allocated to treatment 1, $a(\pi_1)$ is close to one, yielding the maximum efficiency.

Observe that $\sqrt{n}(\mathcal{PE}_n - a(\pi_1)) \to N(0, b(\pi_1)^2\Sigma_{11})$. Since b_n^2 is the squared deviation of N_{n1}/n from its limit, it is a measure of the variability in the design at stage n. As $0 < c < 1$ we have that, $(1 - c)\pi < 1$, and so, $c(\pi_1)$ is positive in $[0, 1]$. Thus, $\mathcal{PE}_n$ increases with b_n^2. This trade-off between efficiency and variability of the allocation rule is reflected more clearly when we take expectations in (10) under the assumption $E[N_{n1}/n] = \pi_1$, because then we obtain

$$\mathcal{ME}_n \sim a(\pi_1) + c(\pi_1)Var\left[\frac{N_{n1}}{n}\right]. \tag{11}$$

So, a high variance of the allocation rule results in an increment of the efficiency for the ethical criterion. This agrees with previous results in the literature. For instance, in Hu and Rosenberger (2003) several adaptive designs are compared for dichotomous responses. They conclude that the Play the Winner rule has high variance and, then, the estimators obtained with this design have high variability. However, from the ethical point of view, it provides the best performance among the designs considered in their paper.

Minimizing expected failures is a criterion function that reaches a global minimal value in the allowable set; that is, $[0, 1] \times [0, 1]$. The transformation $(\mathcal{PE}_n - \mu_1/\mu_2)(\mu_2/(\mu_2 - \mu_1))$ takes values in $[0, 1]$ and allows us to interpret the a-posteriori efficiency in the usual terms where 0 and 1 correspond with the worst and best efficiency, respectively. Slight modifications in the a posteriori efficiency function allow us to deal with negative values for the mean response.

4 Example

In this section we illustrate how to use $\mathcal{PE}_n$ and $\mathcal{ME}_n$ as a comparative tool among adaptive designs. We use the mean efficiency given by (5) for comparing the difference between mean responses with equal variance, as in Atkinson (2002); but it can be performed with any other appropriate criterion function.

A simulation study was carried out with 5000 replications of each design. For a deeper description of each design shown in Table 1, see chapter 3 in Rosenberger and Lachin (2002) for [A], [C], [F] and [G], Chen (2000) and Baldi Antognini (2005) for [B] and [D] and Atkinson (2002) for [E]. With the notation used in [A2], we have $\pi_1 = 1/2$ for all these designs but a central limit theorem does not exist for all of them. From (5), we conclude that the higher the value of the design variance, the smaller will be the mean a-posteriori efficiency for comparing mean treatment differences. In Table 1, the designs are ranked according to the increasing value of the estimated design variance when 50 patients have participated in the design.

When the total average loss is plotted for each design as a function of n, the graph looks virtually the same as Figure 1 in Atkinson (2002). However. our figure is generated by model (1), which includes dependence among the errors. The graph also agrees with the Table 1 ranking. Complete randomization is the least efficient and Efron's design is the most efficient due to its small variability.

Table 1. For $n = 10, 25, 50$ patients: average of allocations to Treatment 1 (n times sample variance of allocations to Treatment 1).

	$n = 10$	$n = 25$	$n = 50$
[A] Efron's design	0.46 (0.037)	0.49 (0.025)	0.49 (0.010)
[B] Ehrenfest model* ($w = 10$)	0.50 (0.062)	0.50 (0.025)	0.50 (0.012)
[C] Smith's design	0.55 (0.034)	0.52 (0.027)	0.51 (0.025)
[D] General Efron	0.54 (0.045)	0.52 (0.047)	0.52 (0.037)
[E] Atkinson design	0.55 (0.067)	0.52 (0.055)	0.51 (0.052)
[F] Wei's urn (1, 3)	0.55 (0.096)	0.52 (0.090)	0.51 (0.090)
[G] Complete Randomization	0.50 (0.248)	0.50 (0.253)	0.50 (0.247)

* Exact values of $nVar[N_{n1}/n]$ following Baldi Antognini (2005)

5 Conclusion

In this paper, we have obtained a relationship between a measure of efficiency and the variance of the design. This provides the basis for using the variance of the design to compare the efficiency of adaptive designs. We consider an adaptive linear model to relate the patient response with the treatment assigned. This model accommodates dependence among errors resulting from the use of adaptive designs to allocate patients.

For the sake of brevity, the possibilities of the proposed tools have not been developed to their full extent. First, model (1) can be extended to the case when covariates are included; see Moler et al (2007). Second, the examplary comparisons are presented for illustrative purposes, but a similar study could be carried out for any other criterion and for response-adaptive designs. Finally, composition of criteria could be used to make comparisons among designs with respect to several targets at the same time.

References

Atkinson A (2002) The comparison of designs for sequential clinical trials with covariate information. Journal of the Royal Statistical Society, A 162:349–373

Atkinson A, Donev AN (1992) Optimum Experimental Designs. Oxford, Oxford

Baldi Antognini A (2005) On the speed of convergence of some urn designs for the balanced allocation of two treatments. Metrika 62:309–322

Baldi Antognini A, Giovagnoli A (2005) On the large sample optimality of sequential designs for comparing two or more treatments. Sequential Designs 24:205–217

Chen Y (2000) Which design is better? ehrenfest urn versus biased coin. Advances in Applied Probability 32:738–749

Dragalin V, Fedorov VV, Yuehui W (2006) Optimal designs for bivariate probit model. GSK BDS Technical Report 01

Hu F, Rosenberger W (2003) Optimality, variability, power: evaluating response-adaptive randomization procedures for treatment comparisons. Journal of the American Statistical Association 463:671–678

Lai T, Wei C (1982) Least square estimates in stochastic regression models with applications to identification and control of dynamic systems. The Annals of Statistics 10:154–166

Moler JA, Plo F, San Miguel M (2007) A sequential design for a clinical trial with a linear prognostic factor. To appear in Journal of Statistical Planning and Inference

Rabie H, Flournoy N (2004) Advances in Model-Oriented Design and Analysis mODa 7., Physica-Verlag, Berlin Heidelberg New York, chap Optimal Designs for the Contingent Response Model

Rosenberger WF, Lachin JM (2002) Randomization in Clinical Trials. Theory and practice. Wiley, New York

Zhang L, Rosenberger W (2006) Response-adaptive randomization for clinical trials with continuous outcomes. Biometrics 62:562–569

Optimal Cutpoint Determination: The Case of One Point Design

The Nguyen and Ben Torsney

Department of Statistics, University of Glasgow, 15 University Gardens, Glasgow G12 8QW, U.K.

the@stats.gla.ac.uk   bent@stats.gla.ac.uk

Summary. The paper briefly describes results on determining optimal cutpoints in a survey question. We focus on the case when we offer all respondents a set of cutpoints: a one point design. Applications in the social sciences will be cited, including contingent valuation studies, which aim to assess a population's willingness to pay for some service or amenity, and in market research studies. The problem will be formulated as a generalized linear model. The formula for the Fisher information matrix is constructed. Search methods are used to find optimal solutions. Results are reported and illustrated pictorially.

Key words: information matrix, design objectives, D-, A-, E-, e_1-, e_2-optimality, design points, cutpoints, categories

1 Introduction

Suppose that we are concerned about a characteristic of a population such as income or expenditure on a particular product and a survey is conducted. Categorical information is to be recommended if respondents are likely to be reluctant to be very specific or to have poor memory recall. In this case, the best way to get information from respondents is to offer them consecutive ranges of values of the response variable with these ranges chosen in advance. So, the problem arises of how to choose such ranges optimally. This kind of design could also be applied in surveying general practitioners in respect of what percentage of patients they assign to a specific drug, or to a new market expansion in which a company wants to investigate the population's expenditure potential for a new product or in a new market. In contingent valuation studies the primary aim is to assess a population's willingness to pay for some ecosystem, environmental services, non-market goods or towards an increase in charges for some public services. Since respondents may never have considered such questions it is unrealistic to expect them to state a specific 'willingness to pay value'. In a simple dichotomous choice question they are offered a single 'bid' question; e.g. 'are you willing to pay \$20.?' In a double bounded approach they would then be offered a second bid, lower, e.g. \$10, if their response to the first 'bid' is NO and higher, e.g. \$30, otherwise. We would then know into which of the four ranges, below \$10, between \$10 and \$20, between \$20 and \$30, above \$30, a respondent's willingness to pay falls. See Alberini (1995), Kanninen (1996) and Torsney and Gunduz (1999). Our problem extends further when we consider offering respondents a range of bid values. A fundamental question is: what bid values should be offered to respondents?

Put more technically, we denote X, on a continuous scale, as the variable of interest. In practice, however, we can not measure this variable precisely on the sample members. An alternative is that we record only to which of a finite number of categories they belong, possibly determining this by a process of elimination. Our main task is how to determine these categories optimally.

2 The formal problem

Suppose that we know that $X \in \mathcal{X} = [C, D]$, so that this is a sample space (which in theory, but not in practice, could be the real line). Suppose that we wish to place responses into one of k categories determined by cutpoints $x_1, x_2, \ldots, x_{k-1}$, chosen in advance, satisfying $C = x_0 < x_1 < x_2 < \ldots < x_{k-1} < x_k = D$. Thus we have partitioned $\mathcal{X}$ into k sub-intervals.

What sets of values should be chosen for these cut-points? This defines a non-linear regression design problem, in which the design variable is the vector $\underline{x} = (x_1, x_2, \ldots, x_{k-1})$. The solution should depend on the underlying distribution of X in the population of interest.

We make the simple but widely used assumption that X (or it could be some function $h(X)$, e.g. $ln(X)$ when X is positive, as in the case of 'Willingness to Pay') has distribution function:

$$P(X \leq x) = F((x - \mu)/\sigma), \quad x \in \mathcal{X} \tag{1}$$

where μ and σ are unknown location and scale parameters respectively, and $F(z)$ is a standardised distribution function. Equivalently:

$$P(X \leq x) = F(\alpha + \beta x), \quad x \in \mathcal{X} \tag{2}$$

where $\alpha = -(\mu/\sigma), \quad \beta = 1/\sigma$.

This is a Generalised linear model in the parameters α, β. Let $\gamma = (\alpha, \beta)^T$. We have a two parameter model and our objective is good estimation of some aspects of these parameters. Often μ is of particular interest.

3 Some design objectives

We wish to choose a design which will ensure good estimation of some aspects of our model. We could be interested in efficient estimation of either both

parameters, or , in this context, possibly only of μ. For the latter we then
wish to minimise $Var(\hat{\mu})$.

Since $\mu = -\alpha/\beta$, $\hat{\mu} = -\hat{\alpha}/\hat{\beta}$; $Var(\hat{\mu}) \cong Var(\underline{c}^T \hat{\underline{\lambda}})$, $\underline{c} = \partial\mu/\partial\underline{\lambda} \propto -(1, \mu)^T/\beta$
This is an example of the c-optimal criterion. Alternatively, minimizing
$Var(\hat{\sigma})$, $(\sigma = 1/\beta)$ corresponds to this criterion with $\underline{c} = -(0, 1)^T/\beta^2$.

If we want good estimation of both parameters then we wish to make
$C = Cov(\hat{\lambda})$ 'small'. Possible targets are to minimise either: $det(C)$ (D-
optimality); $tr(C)$ (A-optimality); or the maximum eigenvalue of C (E-
optimality).

For the moment we note that for non-linear models optimal designs typically depend on the unknown parameters of such models. They are locally
optimal designs. Provisional estimates of parameters are needed for these to
be of practical value. We will focus on the construction of such designs. We
can characterise this parameter dependence through a parameter dependent
transformation to a standardised problem.

Let: $Z = (X - \mu)/\sigma = \alpha + \beta X$, $z = (x - \mu)/\sigma = \alpha + \beta x$, $A = (C - \mu)/\sigma = \alpha + \beta C$, $B = (D - \mu)/\sigma = \alpha + \beta D$ where $\alpha = -(\mu/\sigma)$, $\beta = 1/\sigma$.

Then

$$P(X \leq x) = P(Z \leq z) = F(z), \quad z \in \mathcal{Z} = [A, B] \tag{3}$$

We have in Z a transformed standardised version of X. We can focus on deter-
mining cutpoints $z_1, z_2, \ldots, z_{k-1}$ satisfying $A = z_0 < z_1 < z_2 < \ldots < z_{k-1} < z_k = B$. We have a design problem with design vector $z = (z_1, z_2, \ldots, z_{k-1})$.
Of course $z_j = (x_j - \mu)/\sigma = \alpha + \beta x_j$, $j = 0, 1, 2, \ldots, k$. Ford et al (1992) used
this approach for the two-category case.

4 Two category case

We briefly review this simplest case where the vector $\underline{x} = x_1$ is scalar, which
means that only one cut-point is offered to each respondent. Consequently we
have two categories. Let $x_1 = x \in \mathcal{X} = [C, D]$, so that x is a single design
variable.

We focus on construction of design measures ξ, because if both parameters
need to be estimated, at least two support points are needed. That is, we
seek a distribution ξ_x on $\mathcal{X}$ which will identify the optimal proportions of
observations to take at each point in $\mathcal{X}$. This means that respondents will be
split into groups according to these optimal proportions and different groups
will be offered different single cutpoints.

Note that we are assuming that we are free to take x to be any value in
$\mathcal{X} = [C, D]$, even if $\mathcal{X} = \mathbb{R}$. This can be permissible. However, we could
be restricted to a subset of $\mathcal{X}$, say $[c, d]$. We denote by $M(\xi_x)$ the expected
information matrix per observation. We have:

$$Cov(\hat{\underline{\gamma}}) \propto M^{-1}(\xi_x), \tag{4}$$

where $\hat{\gamma} = (\hat{\alpha}, \hat{\beta})$ as denoted above.

If the distribution ξ_x assigns weight ξ_i to a discrete set of values $x_1, x_2, \ldots$ and $\xi_i \geq 0$, $\sum \xi_i = 1$, then:

$$M(\xi_x) = E_{\xi_x}(I_x) = \sum \xi_i I_{x_i}, \tag{5}$$

where I_x is the expected information matrix of a single observation at x or a one point design at x. Here:

$$I_x = w(z) \begin{pmatrix} 1 \\ x \end{pmatrix} (1 \ x) \tag{6}$$

Clearly the function $w(.)$ is playing the role of a weight function. We assume it is measurable. It has the form:

$$w(z) = \frac{\{f(z)\}^2}{\{F(z)[1 - F(z)]\}}, \quad f(z) = F'(z) \text{ and } z = \alpha + \beta x$$

We are now considering a standardized problem under the parameter dependent transformation:

$$\begin{pmatrix} 1 \\ z \end{pmatrix} = B \begin{pmatrix} 1 \\ x \end{pmatrix}; \quad B = \begin{pmatrix} 1 & 0 \\ \alpha & \beta \end{pmatrix} \tag{7}$$

So, we have:

$$I_x = w(z) (B^{-1}) \begin{pmatrix} 1 \\ z \end{pmatrix} (1 \ z)(B^{-1})^T \tag{8}$$

Hence:

$$I_x = B^{-1} I_z (B^{-1})^T, \tag{9}$$

where

$$I_z = w(z) (1, z)^T (1, z) \tag{10}$$

Extending these results to the expected per observation information matrix, we have:

$$M(\xi_X) = B^{-1} M(\xi_Z)(B^{-1})^T, \tag{11}$$

where ξ_Z is the distribution induced on $Z = [A, B]$ by ξ_X on $X = [C, D]$. Hence we have:

$$M(\xi_Z) = E_{\xi_z}\{I_z\} = \sum \xi_i I_{z_i} \quad \text{and} \quad det\{M(\xi_X)\} \propto det\{M(\xi_Z)\}$$

$$\underline{c}^T M(\xi_x)\underline{c} = \underline{c}_B^T M(\xi_z)\underline{c}_B; \quad \underline{c}_B = B\underline{c}.$$

Thus D-optimal and c-optimal criteria, as functions of ξ_X, transform respectively to the D-optimal and other c-optimal criteria as functions of ξ_Z.

Thus, we focus on finding the design ξ_Z which either

$$\text{maximizes } det[M(\xi_Z)] \Rightarrow D - \text{optimality}$$

or

$$\text{minimizes } \underline{c}_B^T M^{-1}(\xi_Z)\underline{c}_B \Rightarrow c - \text{optimality}.$$

We consider two cases related to the previous section:

- If $\underline{c} = -(1, \mu)^T / \beta \Rightarrow \underline{c}_B = B\underline{c} = (-1/\beta, 0)^T$

So, we minimize $\underline{c}_B^T M^{-1}(\xi_Z)\underline{c}_B$ which is equivalent to minimizing $(1,0)^T M^{-1}(\xi_z)(1,0)$ i.e. $\underline{c}_B \propto \underline{e}_1 = (1,0)^T$.

- If $\underline{c} = -(0, 1)^T / \beta^2 \Rightarrow \underline{c}_B \propto \underline{e}_2 = (0,1)^T$

So we are interested in the criteria:

e_i-**optimality:** A design is called e_i-optimal if it maximizes the value of the function:

$$-\underline{e}_i^T M^{-1}(\xi_Z)\underline{e}_i; \quad i = 1, 2$$

where $\underline{e}_1 = (1,0)^T$; $\underline{e}_2 = (0,1)^T$.

5 One point designs: k categories

In the two-category case (one cutpoint), to ensure estimation of both parameters in the model, we need at least two support points. This is why we can not use the same cutpoint for all respondents. We have to use at least two distinct values for a single cutpoint.

However if we offer each respondent at least two cutpoints, we are free to use the same set for all respondents since all parameters can be estimated; i.e. we can settle for a one point design. We now consider this case.

Suppose there are k categories and hence $k - 1$ cutpoints. Let the original cutpoints be $x_1, x_2, \ldots, x_{k-1}$ and $x_0 = C$, $x_k = D$. The vector $\underline{x} = (x_1, x_2, \ldots, x_{k-1})$ represents our single design point. Let:

$$\theta_1 = P(X \le x_1) = F(\alpha + \beta x_1) = F(z_1), \quad \theta_k = 1 - F(z_{k-1}),$$
$$\theta_i = P(x_{i-1} \le X \le x_i) = F(z_i) - F(z_{i-1}), \quad i = 2, 3, \ldots, k - 1.$$

Then the Fisher Information matrices at the vector (design point) $\underline{x} = x_1, x_2, \ldots, x_{k-1}$ or at $\underline{z} = z_1, z_2, \ldots, z_{k-1}$, $z_i = \alpha + \beta x_i$ are:

$I_X = XQX^T$, $I_Z = ZQZ^T$ (non-singular for $k \ge 3$),

where:

$X^T = (\underline{1}_{k-1}|\underline{x})$, $Z^T = (\underline{1}_{k-1}|\underline{z})$, $\underline{1}_n = (1, 1, \ldots, 1) \in \mathbb{R}^n$

$Q = D_f H D_\theta^{-1} H^T D_f;$ $H = (I_{k-1}|\underline{0}_{k-1}) - \underline{0}_{k-1}|I_{k-1});$

$D_f = \mathrm{diag}\{f(z_1), f(z_2), \ldots, f(z_{k-1}\}, \quad f(z) = F'(z);$

$D_\theta = \mathrm{diag}(\theta_1, \theta 2, \ldots, \theta_k), (\theta_i :\text{Cell probabilities}).$

$\underline{0}_n = (0, 0, \ldots, 0)^T \in \mathbb{R}^n; \quad I_n :$ Identity matrix of order n.

Then:

$Z = BX, \quad I_X = B^{-1} I_Z (B^{-1})^T.$

We can focus on determining an optimal z^*. We first assume that the distribution of Z is symmetric; for example, logistic, normal, double reciprocal and double exponential. It is intuitive that any set of cutpoints must be symmetrical too (about zero). We consider the following cases:

$k = 3 : \underline{z}^* = (-z^*, z^*); \quad k = 4 : \underline{z}^* = (-z^*, 0, z^*);$
$k = 5 : \underline{z}^* = (-z_2^*, -z_1^*, z_1^*, z_2^*); \quad k = 6 : \underline{z}^* = (-z_2^*, -z_1^*, 0, z_1^*, z_2^*).$
The criteria considered are :

- D-Optimality: Maximise $\{log\,det(I_z)\}$.
- A-Optimality: Maximise $\{-tr(I_z^{-1})\}$.
- e_i-Optimality: Maximise $\{-e_i^T I_z^{-1} e_i\}$ $i = 1, 2$.
- E-optimality: Maximize$\{-\lambda_{max}\}$; λ_{max} is maximum eigenvalue of I_z^{-1}

We wish to choose $\underline{z}^*$ to maximise $\phi(z) = \psi(I_z)$ for $Z^T = (\underline{1}_{k-1}|\underline{z})$, $\psi(.)$ being one of the above criteria.

6 Results

By simple searching through z^* or (z_1^*, z_2^*) values, we have the following results for the logistic distribution.

Table 1. Numerical results for logistic distribution, k=3 and k=4

		k=3			k=4	
Criterion	z^*	$F(z^*)$	$\phi(z^*)$	z^*	$F(z^*)$	$\phi(z^*)$
D-optimality	1.4700	0.8131	-1.5567	1.9800	0.8787	-1.2483
A-optimality	1.1600	0.7613	-5.0182	1.7100	0.8468	-4.3789
e_1-optimality	0.6900	0.6660	-3.3750	1.1000	0.7503	-3.2000
e_2-optimality	2.1700	0.8975	-1.0226	2.1700	0.8975	-1.0226
E-optimality	0.6900	0.6660	-3.3750	1.1000	0.7503	-3.2000

Table 2. Numerical results for logistic distribution, k=5

Criterion	z_1^*	z_2^*	$F(z_1^*)$	$F(z_2^*)$	$\phi(z_1^*, z_2^*)$
D-optimality	0.8500	2.5100	0.7006	0.9248	-1.0709
A-optimality	0.6100	2.1600	0.6479	0.8966	-4.1245
e_1-optimality	0.4100	1.3900	0.6011	0.8006	-3.1251
e_2-optimality	1.5900	3.1700	0.8306	0.9597	-0.8284
E-optimality	0.4100	1.3900	0.6011	0.8006	-3.1251

Table 3. Numerical results for logistic distribution, k=6

Criterion	z_1^*	z_2^*	$F(z_1^*)$	$F(z_2^*)$	$\phi(z_1^*, z_2^*)$
D-optimality	1.3300	2.9100	0.7908	0.9483	-0.9788
A-optimality	1.0500	2.5400	0.7408	0.9269	-3.9942
e_1-optimality	0.6900	1.6100	0.6660	0.8334	-3.0857
e_2-optimality	1.5900	3.1700	0.8306	0.9597	-0.8284
E-optimality	0.6900	1.6100	0.6660	0.8334	-3.0857

We note the not surprising conclusion that all criteria increase or do not change with k the number of categories with virtually no change from $k= 5$ to 6. This suggests that five categories suffices. The e_2-criteria does not change from $k = 3$ to $k = 4$ or from $k = 5$ to $k = 6$, i.e. when zero is inserted as an extra cutpoint into a symmetric set of non-zero cutpoints. The explanation of this is that the difference in the Fisher information matrix (between after and before zero is inserted) is a diagonal matrix with a zero second diagonal entry. We also investigated the change in the Fisher information matrix when an arbitrary cutpoint is inserted between two other cutpoints and find that the difference (with-without extra cutpoint) in the Fisher information matrix is non-negative definite. This confirms that the criteria increase or do not change with k. We have similar results for other distributions.

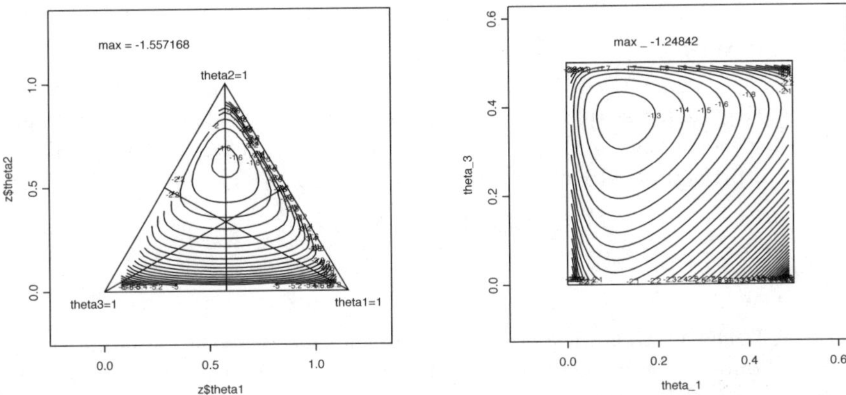

Fig. 1. Contour plots in 3 and 4 category cases, D-optimality and logistic distribution

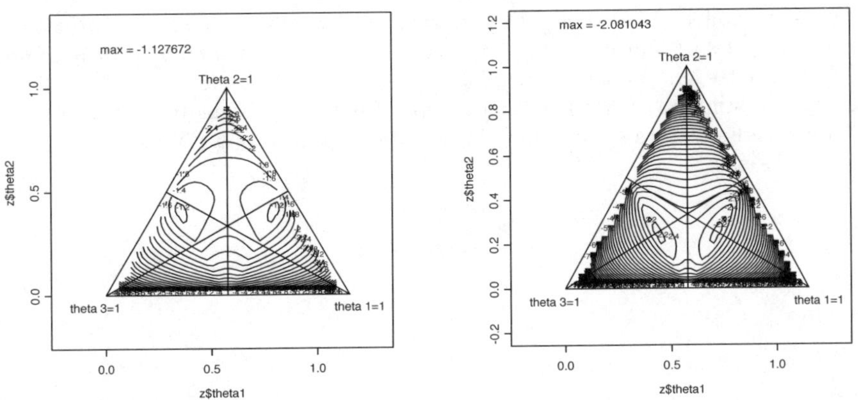

Fig. 2. Contour plots in 3 category cases, D-optimality, double exponential and double reciprocal distribution

The above results are partially confirmed for D-optimality in the case of $k=3$, 4 by the contour plots in Figures 1 and 2. The triangle simplex is the plot of the criterion against the three cell probabilities $\theta_1, \theta_2, \theta_3$ of the three category case. The rectangle is a plot of the criterion against θ_1 and θ_3 in the four category case but subject to $\theta_1 + \theta_2 = \theta_3 + \theta_4$. Similar results are found for the other symmetric distributions. There are some unusual results for the three category case when the distribution is double exponential and double reciprocal. Looking at Figure 2, we can see that there are two optimal design points in two different positions. These two points are symmetrical with respect to the perpendicular from the top vertex of the triangular.

7 Future work

Future work will focus on asymmetric distributions, multiple design points, multivariate responses and the use of a multiplicative algorithm; also the bivariate approach of Alberini (1995) and Gunduz (1999).

References

Alberini A (1995) Optimal designs for discrete choice contingent valuation surveys: single bound, double bound, and bivariate model. Journal of Environmental Economics and Management 28:287–306

Ford I, Torsney B, Wu C (1992) The use of a canonical form in the construction of locally optimal designs for non-linear problems. JR Statist Soc SerB 54(2):569–583

Gunduz N (1999) D-optimal designs for weighted linear regression and binary regression models. PhD thesis, Department of Statistics, University of Glasgow

Kanninen B (1996) Optimal experimental design for double-bounded dichotomous choice contingent valuation. Land Economics 69(2):138–146

Torsney B, Gunduz N (1999) A brief review of optimal designs in contingent valuation studies. Tatra Mountains Mathematics Publications 17:185–195

D–Optimal Designs for Regression Models with Length-Biased Poisson Response

Isabel Ortiz, Carmelo Rodríguez, and Ignacio Martínez

Dpto. Estadística y Mat. Aplicada, Universidad de Almería, Edificio CITE-III,
Cra. Sacramento s/n, La Cañada de San Urbano, 04120 Almería, Spain
iortiz@ual.es   crt@ual.es   ijmartin@ual.es

Summary. This paper is concerned with the search for locally optimal designs when the observations of the response variable arise from a weighted distribution in an exponential family. The expression for the information matrices for length-biased distributions from an exponential family are obtained. Locally D–optimal designs are derived for regression models whose response variable follows a weighted Poisson distribution. Two link functions are considered for these models.

Key words: optimal designs, generalized linear model, length-biased response, exponential family

1 Introduction

In this paper, attention is confined to regression models with a biased response, i.e. models where the response variable is a weighted distribution. The concept of weighted distributions can be traced back to the study of the effect of methods of ascertainment upon estimation of frequencies by Fisher (1934), although these models were first formulated in an unified way by Rao (1965). From an original density function $f(y)$ of a variable Y, the weighted density function $f^*(y)$ with respect to a non-negative real weight function $w(.)$ is defined as

$$f^*(y) = \frac{w(y)f(y)}{E_f(w(y))} \tag{1}$$

where $0 < E_f(w(y)) < \infty$. The subindices f and f^* in the expected values and variances are used to distinguish between the original distribution and its weighted counterpart. A special case of interest arises when the weight function is of the form $w(y) = y^k$ with $k > 0$. Such distributions are known as size-biased distributions of order k and the most common cases, called length-biased, occur when $k = 1$. Biased random variables frequently arise in biological and epidemiological studies, economics, survey sampling and many other fields (see Patil (1981), Nanda and Jain (1999) and Navarro et al (2001)) to model unequal sampling probabilities. Some concrete examples are the time period between two blood donations which contains a seroconversion (development of detectable antibodies to a virus) (see Satten et al (2004)); or to explain the famous waiting time paradox, namely that the chance that an interval between two buses brackets the arrival of an individual at a bus stop is proportional to the length of the interval, see Feller (1971).

The general form of a linear regression model is

$$E(y|x) = \beta^T \eta(x),$$

where the random variable Y is the response variable, x is the explanatory variable, chosen in a design space $\mathcal{X} \subset I\!R$, β is a $k \times 1$ vector of unknown parameters and $\eta(x)$ is the vector of regression functions.

Optimal design theory has been extensively developed for linear models where the response variable Y is normally distributed. In this paper, optimal designs for Poisson regression models with length-biased responses are discussed. In Section 2 the model under consideration and the notation related to optimal designs are introduced. Expressions for the information matrices for weighted distributions are studied in Section 3. Some results for locally D–optimal designs for Poisson distributions are included in Section 4. Finally, in Section 5, some concluding remarks are given.

2 Model and notation

The Generalized Linear Model (GLM) is a generalization of the normal linear model and assumes that

$$E(y|x) = \mu(x, \beta),$$

where $\mu(x, \beta)$ is related to a linear predictor $\alpha = \beta^T \eta(x)$ by means of a link function. The exponential family of distributions are particularly interesting cases of the GLM. The density function of the distributions in the one-parameter exponential family considered in Rohatgi (1988), takes the form

$$f(y|\alpha) = \exp\{yq(\alpha) - b(\alpha) + c(y)\} \tag{2}$$

for some specific functions $q(.)$, $b(.)$ and $c(.)$. Distributions such as the Normal, Gamma, Inverse Gaussian, Poisson and Binomial with only one unknown parameter are included in this family. The characteristics of a variable Y with a distribution in the exponential family are

$$E(y) = b'(\alpha)/q'(\alpha),$$
$$Var(y) = \frac{-q''(\alpha)b'(\alpha) + q'(\alpha)b''(\alpha)}{q'(\alpha)^3},$$

where primes denote differentiation with respect to α. In what follows, the function (2) is denoted as $f(y|x,\beta)$ since $\alpha = \beta^T \eta(x)$.

In this paper, we will assume that Y has a weighted distribution or is a biased response. More specifically, the distribution of Y is obtained as a weighting distribution in the exponential family (original distribution) as in (1), from a weight function $w(y)$. Length-biased responses are obtained when $w(y) = y$.

For a model from the exponential family the regression model is $E_f(y|x) = b'(\alpha)/q'(\alpha)$ with $\alpha = \beta^T \eta(x)$, and for the weighted distribution with $w(y) = y$, the regression model will be

$$E_{f^*}(y|x) = \frac{b'(\alpha)}{q'(\alpha)} + \frac{b''(\alpha)}{q'(\alpha)b'(\alpha)} - \frac{q''(\alpha)}{q'(\alpha)}. \tag{3}$$

In the context of regression models, the optimal design of an experiment tries to determine an optimal collection of N values $\{x_1, x_2, \ldots, x_N\}$ of a variable x at which we will observe the response variable Y. It can be re-written in terms of the n distinct points $\{x_1, x_2, \ldots, x_n\}$ at which observations are taken (called the support points) and their relative frequencies (masses) $\{p_1, p_2, \ldots, p_n\}$ with $\sum_{i=1}^n p_i = 1$. An approximate design ξ can be defined as any probability measure in $\mathcal{X}$. Optimal designs for GLM's have been studied in several papers (see, for instance, Ford et al (1992) or Burridge and Sebastiani (1992)).

The Fisher information matrix for β, given an observation at design point x, is

$$I(x,\beta) = E_f \left(-\frac{\partial^2 \log f(y|x,\beta)}{\partial \beta^2} \right) \tag{4}$$

and the Fisher information matrix for a design ξ is $M(\xi,\beta) = \int_{\mathcal{X}} I(x,\beta)\xi(dx)$. The information matrix becomes the main tool when we look for optimal designs for god estimation of β. Its inverse matrix is proportional to the covariance matrix of the estimators of the parameters in the model. The D–optimal design minimizes the generalized variance of the parameter estimates or equivalently, maximizes the determinant of the information matrix. When the information matrix depends on unknown parameters some additional information to obtain optimal designs is needed. This information can consist of initial values for the parameters, these designs are called locally optimal designs, (see Chernoff (1953)). Interest in obtaining locally optimal designs is summarized in Ford et al (1992).

3 Information matrices for biased response

The Fisher information matrix (4) for β at point x for a density function $f(y|x,\beta)$ in the exponential family (2) is

$$I(x,\beta) = \left(b''(\alpha) - \frac{q''(\alpha)b'(\alpha)}{q'(\alpha)} \right) \eta(x)\eta(x)^T.$$

For a biased response with weight function $w(y)$ which does not depend on the parameter vector β, the Fisher information matrix $I^*(x,\beta)$ from (1) and (4) is

$$I^*(x,\beta) = E_{f^*}\left(-\frac{\partial^2 \log f(y|x,\beta)}{\partial \beta^2} \right) + \frac{g_{\beta^2}(x,\beta)}{g(x,\beta)} - \frac{g_\beta(x,\beta)g_\beta^T(x,\beta)}{g^2(x,\beta)},$$

where $g(x,\beta)$ denotes the mean $E_f(w(y))$ which can depend on β; also $g_\beta(x,\beta) = \partial g(x,\beta)/\partial \beta$ and $g_{\beta^2}(x,\beta) = \partial^2 g(x,\beta)/\partial \beta^2$. If the density function $f(y|x,\beta)$ is in the exponential family (2), then

$$E_{f^*}\left(-\frac{\partial^2 \log f(y|x,\beta)}{\partial \beta^2} \right) = E_{f^*}\left(b''(\alpha) - yq''(\alpha) \right)\eta(x)\eta^T(x) =$$

$$= \left(b''(\alpha) - q''(\alpha)E_{f^*}(y) \right)\eta(x)\eta^T(x)$$

and therefore

$$I^*(x,\beta) = I(x,\beta) + q''(\alpha)\left(\frac{b'(\alpha)}{q'(\alpha)} - E_{f^*}(y) \right)\eta(x)\eta^T(x) +$$

$$+ \frac{g_{\beta^2}(x,\beta)}{g(x,\beta)} - \frac{g_\beta(x,\beta)g_\beta^T(x,\beta)}{g^2(x,\beta)}.$$

For a design ξ in $\mathcal{X}$, the information matrix $M^*(\xi,\beta)$ for the parameter vector β when the response is biased can be written as

$$M^*(\xi,\beta) = M(\xi,\beta) + \int_{\mathcal{X}}\left[q''(\alpha)\left(\frac{b'(\alpha)}{q'(\alpha)} - E_{f^*}(y) \right)\eta(x)\eta^T(x) + \right.$$

$$\left. + \frac{g_{\beta^2}(x,\beta)}{g(x,\beta)} - \frac{g_\beta(x,\beta)g_\beta^T(x,\beta)}{g^2(x,\beta)} \right]\xi(dx), \tag{5}$$

where $M(\xi,\beta)$ is the information matrix of the design ξ for the original distribution (non-biased distribution).

Lemma 1. *The information matrix for a model from the exponential family with length-biased response is*

$$M^*(\xi,\beta) = M(\xi,\beta) + \int_{\mathcal{X}} \frac{b'''(\alpha)b'(\alpha) - b''(\alpha)^2}{b'(\alpha)^2}\eta(x)\eta(x)^T\xi(dx) +$$

$$+ \int_{\mathcal{X}}\left(\frac{q''(\alpha)b''(\alpha)}{q'(\alpha)b'(\alpha)} - \frac{Q'''(\alpha)}{q'(\alpha)} \right)\eta(x)\eta(x)^T\xi(dx).$$

Proof. For length-biased responses $w(y) = y$ and we have

$$g(x,\beta) = b'(\alpha)/q'(\alpha),$$
$$g_\beta(x,\beta) = (b''(\alpha)q'(\alpha) - b'(\alpha)q''(\alpha))\eta(x)/q'(\alpha)^2,$$
$$g_{\beta^2}(x,\beta) = (b'''(\alpha)q'(\alpha)^2 - 2b''(\alpha)q'(\alpha)q''(\alpha) - b'(\alpha)Q'''(\alpha)q'(\alpha)$$
$$+ 2b'(\alpha)q''(\alpha)^2)\eta(x)\eta(x)^T/q'(\alpha)^3.$$

From (3) and substituting in (5) we obtain the asserted expression of the information matrix.

4 Locally D-optimal designs for length-biased Poisson response

A locally D–optimal design ξ^* maximizes the logarithm of the determinant of the information matrix for some best guess for the unknown parameters in the regression model. It is equivalent to minimizing the volume of the confidence ellipsoid of the estimators of the parameters in the model.

We consider the following case: the regression function vector $\eta(x) = (1,x)^T$, the vector of unknown parameters $\beta = (\beta_0,\beta_1)^T$ and the design space $\mathcal{X} = [x_{min},x_{max}]$. Since there are two parameters in the regression model, it follows from Carathéodory's Theorem that the locally D–optimal design is supported at two or three different points. By a standard argument (see Silvey (1980)) the two–point D–optimal designs put equal masses at both points. The notation $\beta^0 = (\beta_0^0,\beta_1^0)^T$ is used as the best guess for the parameters.

Following Ford et al (1992), a canonical form is used to find D–optimal designs. The local D–optimal criterion is invariant under transformations of the regression function $\eta(x)$ of the form $B\eta(x)$ where B is a non-singular 2×2 matrix, depending on β^0. If $\eta(x) = (1,x)^T$ and $B = \begin{pmatrix} 1 & 0 \\ \beta_0^0 & \beta_1^0 \end{pmatrix}$ then the variable x is mapped to $z = \beta_0^0 + \beta_1^0 x$, resulting in an induced design space Z which varies with β^0. This leads to a canonical version of the design problem which can be solved independently of β^0. Hence, solving the transformed problem for any Z yields the solution in $\mathcal{X}$ for the best guess β^0.

The information matrix for the model with length-biased response can be written in the form

$$M^*(\xi,\beta) = \int_{\mathcal{X}} h(\alpha)\eta(x)\eta(x)^T \xi(dx),$$

for $\alpha = \beta^T\eta(x)$ and a given $h(.)$. Then the equivalent problem in the space Z will consist of finding the D–optimal design given the information matrix

$$M_z(\xi_z) = \int_Z h(z) \begin{pmatrix} 1 \\ z \end{pmatrix} \begin{pmatrix} 1 & z \end{pmatrix} \xi_z(dz). \tag{6}$$

A geometrical method based on the set

$$S = \left\{ s = (s_1, s_2)^T / s_1 = \sqrt{h(z)}, \ s_2 = z s_1 \text{ and } z \in Z = [a, b] \right\}, \qquad (7)$$

can be used to construct D–optimal designs in the space Z. The information matrix (6) will be

$$M_z(\xi_z) = \int_Z ss^T \xi_z(dz). \qquad (8)$$

and the points of Z in the support of the D–optimal design for the transformed model can be determined geometrically: at these points, the set S is in contact with the smallest ellipsoid centered on the origin and containing S.

Hence D–optimal designs will be determined in two steps. Firstly we look for the best two–point equally supported design in Z. If this design is not the D–optimal design, we will find the points and the masses of the three–point D–optimal design. Kiefer and Wolfowitz's Equivalence Theorem will be used to check the optimality of a design. In the next step, the optimal support points from Z are transformed to the design space $\mathcal{X}$.

Poisson distribution.

The above method is applied to Poisson distribution for two different link functions, as a particular case in the exponential family. Let Y be a Poisson distribution with mean λ. Its probability function is $f(y) = \frac{e^{-\lambda}\lambda^y}{y!} = exp\{y \log \lambda - \lambda - \log y!\}$. It is a model of the exponential family (2).

Two link functions are considered. A first case is $\lambda = e^\alpha$, then $\log(E_f(y)) = \alpha = \beta^T \eta(x)$. This is the usual link function. And a second case is when $\lambda = \alpha^r$, $0 < |r| < \infty$, $\alpha > 0$ with $\lambda^{1/r} = (E_f(y))^{1/r} = \alpha = \beta^T \eta(x)$.

For $\lambda = e^\alpha$, the probability function is $f(y|\alpha) = \exp\{y\alpha - e^\alpha - \log(y!)\}$ with $q(\alpha) = \alpha$, $b(\alpha) = e^\alpha$ and $c(y) = -\log y!$. In this case, the Fisher information matrix for β at point x for the original distribution and for the length-biased response, from Lemma 1, are the same

$$I(x, \beta) = I^*(x, \beta) = e^\alpha \eta(x)\eta(x)^T.$$

So for all design ξ we have that $M(\xi, \beta) = M^*(\xi, \beta)$ and the optimal designs are the same for both models. For $\eta(x) = (1, x)^T$, $z = \alpha = \beta_0 + \beta_1 x$ the D–optimal design in $Z = [a, b]$ is equally supported at $\max\{a, b - 2\}$ and b Ford et al (1992).

For the second link function $\lambda = \alpha^r$, the probability function is $f(y|\alpha) = \exp\{yr \log \alpha - e^\alpha - \log y!\}$ with $q(\alpha) = r \log \alpha$, $b(\alpha) = \alpha^r$ and $c(y) = -\log y!$. The Fisher information matrix for β at point x is

$$I(x, \beta) = r^2 \alpha^{r-2} \eta(x)\eta^T(x).$$

The D–optimal design for $\eta(x) = (1, x)^T$, $z = \alpha = \beta_0 + \beta_1 x$ and $z \in [a, b]$ is equally supported at a and $\min\{b, ra/(r + 2)\}$ if $r < -2$; at a and b if $-2 \leq r \leq 0$ and at $\max\{a, rb/(r + 2)\}$ and b if $r > 0$ Ford et al (1992).

The information matrix at point x when the response is length-biased is

$$I^*(x,\beta) = I(x,\beta) - 2r\alpha^{-2}\eta(x)\eta^T(x).$$

Particular cases are:

$$r = 1 \quad I(x,\beta) = \alpha^{-1}\eta(x)\eta^T(x) \quad I^*(x,\beta) = I(x,\beta) - 2\alpha^{-2}\eta(x)\eta^T(x);$$
$$r = 2 \quad I(x,\beta) = 4\eta(x)\eta^T(x) \quad I^*(x,\beta) = I(x,\beta) - 4\alpha^{-2}\eta(x)\eta^T(x).$$

In this case, the D–optimal design for $\eta(x) = (1,x)^T$, with $z = \alpha = \beta_0 + \beta_1 x$ in the design space $[a,b]$ is equally supported at two points, which are, the endpoints of the design interval when $r < 0$ and $\max\{a, z_1\}$ and b when $r \geq 1$, z_1 being a solution of the equation

$$r^2 z_1^{r+1} - br(r-2)z_1^r - 4b = 0. \tag{9}$$

For example, if $r = 1$ the solution is $z_1 = \frac{1}{2}(-b + \sqrt{b(16+b)})$, for $r = 2$ the solution is $z_1 = b^{1/3}$.

When $0 < r \leq 1$ the D-optimal support points are the endpoints of the design interval, provided the lower endpoint of the interval (a) is greater that $(2/r)^{1/r}$. Otherwise the support points are the upper endpoint (b) and the solution z_1 to equation (9). For example, suppose the design space is $[20, 110]$. Then if $r = 0.8$ the support points of the D-optimal design are the endpoints, 20 and 110. In contrast, if $r = 0.5$, the points are the upper endpoint, 110, and the solution of equation (9), namely $z_1 = 24.6305$. The critical limit for r is $r = 0.4779$.

The efficiency study allows us to evaluate the performance of a D–optimal design for the model with Poisson response when it is used to fit the model with length-biased Poisson response. The usual definition of D–efficiency for any design ξ and the D–optimal design ξ^* is

$$\mathrm{eff}_D(\xi,\beta) = \left(\frac{\det M(\xi,\beta)}{\det M(\xi^*,\beta)} \right)^{\frac{1}{7}k},$$

where k is the number of unknown parameters in the model.

For $r = 0.8$ the D–optimal design for the original model is equally supported at 31.4286 and 110. Its D–efficiency is 0.6948. For $r = 0.5$ the D–optimal design support points are 22 and 110, with a higher efficiency of 0.9859.

5 Concluding remarks

The results of this paper extend optimal design theory to Generalized Linear models from the exponential family in which there is a biased structure in the response. The search for optimal designs is more difficult when a model with a weighted distribution for the response is considered. In this paper, expressions for the information matrices have been derived assuming length-biased distributions for the response of the regression model. These expressions are used to search for optimal designs.

Finally, we add to previous results for the Poisson distribution. The local D–optimal designs for this distribution are characterized for two link functions. When $\lambda = e^\alpha$ the optimal designs for any criterion for the length-biased response are the same as those for a 'standard' regression model with a basic Poisson distribution for the response. In these cases both information matrices coincide. If $\lambda = \alpha^r$, the information matrices for the basic and length-biased models are different, and then local D-optimal designs can be very different, as has been shown in several examples.

Acknowledgement. The support of the Spanish Ministerio de Educación y Ciencia (Grant MTM2004-06641-C02-2) is gratefully acknowledged. The authors are also grateful to two referees for their constructive comments, which led to a substantial improvement of an earlier version of this paper.

References

Burridge J, Sebastiani P (1992) Optimal designs for generalized linear models. J Ital Statist Soc 2:183–202

Chernoff H (1953) Locally optimal designs for estimating parameters. Ann Math Statist 24:586–602

Feller W (1971) An Introduction to Probability Theory and its Applications. John Wiley & Sons, New York

Fisher R (1934) The effect of methods of ascertainment upon the estimation of frequencies. Annals of Eugenics 6:13–25

Ford I, Torsney B, Wu C (1992) The use of a canonical form in the construction of locally optimal designs for non-linear problems. J Roy Statist Soc Ser B 54:569–583

Nanda A, Jain K (1999) Some weighted distributions results on univariate and bivariate cases. J Statist Plann Inference 77:169–180

Navarro J, del Águila Y, Ruiz J (2001) Characterizations through reliability measures from weighted distributions. Statist Papers 42:395–402

Patil G (1981) Studies in statistical ecology involving weighted distributions. Indian Statistical Institute Jubilee International Conference on Statistics: Applications and New Directions 1:478–503

Rao C (1965) On discrete distributions arising out of methods of ascertainment. Sankhyā Series A 27:311–324

Rohatgi V (1988) An Introduction to Probability Theory and Mathematical Statistics. John Wiley & Sons, New York

Satten G, Kong F, Wright D, Glynn S, Schreiber G (2004) How special is a special interval: modeling departure from length-biased sampling in renewal processes. Biostatistics 5:145–151

Silvey S (1980) Optimum design. Chapman and Hall, London

Efficient Sampling Windows for Parameter Estimation in Mixed Effects Models

Maciej Patan[1] and Barbara Bogacka[2]

[1] Institute of Control and Computation Engineering, University of Zielona Góra, ul. Podgorna 50, 65-246 Zielona Gora, Poland   M.Patan@issi.uz.zgora.pl
[2] School of Mathematical Sciences, Queen Mary, University of London, Mile End Road, London E1 4NS, U.K.   B.Bogacka@qmul.ac.uk

Summary. In the paper we present a method of calculating an efficient window design for parameter estimation in a non-linear mixed effects model. We define a window population design on the basis of a continuous design for such a model. The support points of the design belong to intervals whose boundaries are determined in a way which ensures that the efficiency of the design is high; also the width of the intervals is related to the dynamic system's behaviour.

Key words: mixed effects non-linear model, population experimental design, equivalence theorem

1 Introduction

Optimum experimental design for parameter estimation in mixed effects pharmacokinetic models has gained considerable attention in the statistical literature (cf. Mentré et al (1997) and Mentré et al (2001)). The advantage of high precision of estimation of the population parameters is clear. However, in advanced phases of clinical trials when a drug is tested in a population of patients, it may be impossible to maintain accurate timing of blood sampling for every patient. This may discourage a practitioner from applying a suggested optimum sampling schedule and may result in an inefficient experiment and so loss of resources. Sampling windows, that is time intervals assuring some minimum required efficiency, are a good solution to this problem. Several authors have proposed various methods for deriving such windows, based on a design efficiency factor, see Green and Duffull (2003); Pronzato (2002); Graham and Aarons (2006).

The main objective of this paper is to give a method of calculating sampling windows for mixed effects non-linear models which would not only assure a required minimum efficiency of population parameter estimation but would also give a window size reflecting parameter sensitivity, as in Bogacka et al

(2006) who derived such a method for calculating sampling windows for a fixed non-linear model. The method is based on a condition of the Equivalence Theorem for D-optimality. It gives less flexibility (narrower windows) when it is important to obtain an observation at a time close to the optimum schedule and more flexibility (wider windows) when it is less important. In Section 4 we show how the method can be applied to our considered class of non-linear models. First, however, in Section 2 we introduce the notion of a population design, as given in Patan and Bogacka (2006), and in Section 3 we briefly present how to calculate such designs. Section 5 explains the theory behind the method through an example of a mixed effects PK model. Some concluding remarks are given in Section 6.

2 Population experimental design

2.1 Class of considered models

In what follows, we suppose that there is a population of N individuals (patients, units, systems etc.) for each of which n_k measurements are gathered, possibly according to different time schedules, that is, the model for each observation can be written as

$$y_i^k = \eta(t_i^k; \theta_i^k) + \varepsilon_i^k, \quad i = 1, \ldots, n_k, \quad k = 1, \ldots, N, \tag{1}$$

where y_i^k is an observation at time $t_i^k \in T = [0, t_{max}]$, ε_i^k are i.i.d. random measurement errors with a known (including the error variance) density g, $y_i^k | \theta_i^k \sim g(y_i^k | \theta_i^k, t_i^k)$, and η is a known possibly nonlinear function.

The p-dimensional vectors $\theta_i^k \in \Theta$ are assumed to be independent realizations of a random vector θ with probability density $h(\theta_i^k; \psi)$. The function h depends on the $2p$-dimensional population parameter vector $\psi = (\mathrm{E}(\theta), var(\theta))^{\mathrm{T}} = (\psi_1, \ldots, \psi_{\widetilde{p}})^{\mathrm{T}}$, where $\widetilde{p} = 2p$, $\mathrm{E}(\theta)$ and $var(\theta)$ denote the vectors of expectations and variances of θ, respectively. We also assume that the elements of θ are uncorrelated. Accurate estimation of ψ, the vector of constant parameters ψ_i, is of primary interest.

Note the difference between model (1) and a typical population model, where it is assumed that each subject is represented by a single parameter value, say θ^k. Here, θ_i^k varies with each observation and it allows us to assume independence of all the observations, which is very useful in deriving the information matrix for the population design.

2.2 Experimental design

We assume that a population of N patients consists of G groups of sizes $N_j, j = 1, 2, ..., G$, and the individuals in the same group follow the same schedule of measurements (design). We define the population experimental design ζ as in Patan and Bogacka (2006):

$$\zeta = \left\{ \begin{array}{ccc} (\xi_1, n_1) & \cdots & (\xi_G, n_G) \\ \alpha_1 & \cdots & \alpha_G \end{array} \right\}; \quad \sum_{j=1}^{G} \alpha_j = 1, \tag{2}$$

where $\alpha_j \in (0, 1]$ represents the proportion of N subjects in group j,

$$\xi_j = \left\{ \begin{array}{ccc} t_1^j & \cdots & t_{s_j}^j \\ w_1^j & \cdots & w_{s_j}^j \end{array} \right\}; \quad w_i^j \in (0, 1], \quad \sum_{i=1}^{s_j} w_i^j = 1. \tag{3}$$

The design $\xi_j \in \Xi$, where Ξ denotes a set of admissible designs defined by (3), is a continuous measure on a set of s_j distinct (support) points in a design region T. Note that the individual design ξ_j does not preserve the information about the number of measurements n_j. Hence, the whole experimental system per individual is described by the pair (ξ_j, n_j).

3 Optimum population design

The purpose of the considered optimal design problem is to determine the population sampling scheme which guarantees accurate estimation. As a quantitative measure of the precision of estimation we use a function defined on the Fisher Information Matrix (FIM) as is commonly done in the optimum experimental design theory (cf. Atkinson and Donev (1992)). We denote by

$$M(t_i^j) = \mathrm{E}\left\{ -\frac{\partial^2 \ell(\psi | y_i^j, t_i^j)}{\partial \psi \partial \psi^{\mathrm{T}}} \right\} \tag{4}$$

the elementary FIM for the observation made at time instant t_i^j, where

$$\ell(\psi | y_i^j) = \log \int_{\Theta} g(y_i^j | \theta, t_i^j) h(\theta; \psi) \, d\theta \tag{5}$$

is the loglikelihood function for ψ. The randomness of θ is accounted for in the form of ℓ.

The assumption of independent observations allows us to sum the FIMs for all single observations. Normalizing the sum by the limit, in practice, on the total number of measurements N_0, we obtain the average FIM for the population design ζ

$$M(\zeta, N) = \frac{N}{N_0} \sum_{j=1}^{G} \alpha_j n_j \sum_{i=1}^{s_j} w_i^j M(t_i^j). \tag{6}$$

Since for nonlinear response models, in general, integral (5) is analytically intractable, in order to evaluate the FIM, some approximation procedures are required. In the statistical literature there exist a variety of methods such as numerical integration, stochastic approximation (Retout and Mentré (2003)) or linearisation of the model around the expected value of the random-effect (Retout et al (2001, 2002)). In the context of this work the first mentioned techniques was exploited in our simulations.

In this paper we further consider the most commonly used D-optimal criterion, $\Psi(M) = -\log \det M$; that is the log-determinant of FIM, which in a linear case and under normality of random effects, minimizes the volume of the confidence ellipsoid for the parameter vector. In non-linear models this property holds asymptotically.

We formulate the optimum observation strategy in terms of the following optimization problem:

$$\Psi\big[M(\zeta, N)\big] \rightarrow \min \quad \text{subject to} \quad N \sum_{j=1}^{G} \alpha_j n_j \leq N_0. \tag{7}$$

If the number of all individuals in the population is not predetermined *a priori* and has to be estimated, it is convenient to relax the restriction of N being an integer and allow it to take any positive real value. Then, as we show in Patan and Bogacka (2006), the optimal solution to (7) is on the boundary of the constraint and the inequality becomes an equality. Therefore, in the following, N is allowed to take positive real values.

Although the formulation of this problem seems quite simple, the necessity of simultaneous calculation of the numerous coefficients of a two-level population design quickly leads to a very cumbersome task. ζ 'carries' a large number of unknowns and there is no unique parsimonious solution to (7). Here we suggest a solution based on 'folding' ζ into a simple form design $\widetilde{\omega}$ belonging to Ξ, and then, having found an optimum $\widetilde{\omega}$, 'unfolding' it to the original form of ζ. Note that the FIM does not change if we replace ζ defined in (2) by a simpler notation

$$v = \left\{ \begin{array}{ccc} \xi_1 & \cdots & \xi_G \\ \beta_1 & \cdots & \beta_G \end{array} \right\} ; \quad \beta_j = \frac{N}{N_0} \alpha_j n_j ; \quad \sum_{j=1}^{G} \beta_j = 1. \tag{8}$$

That is $M(\zeta, N) = M(v)$. Furthermore, introducing

$$\omega = \left\{ \begin{array}{cccccc} t_1^1 & \cdots & t_{s_j}^1 & \cdots & t_1^G & \cdots & t_{s_G}^G \\ q_1^1 & \cdots & q_{s_j}^1 & \cdots & q_1^G & \cdots & q_{s_G}^G \end{array} \right\} ; \quad q_i^j = \beta_j w_i^j ; \quad \sum_{j=1}^{G} \sum_{i=1}^{s_j} q_i^j = 1 \tag{9}$$

we have $M(v) = M(\omega)$. Different groups do not have to have completely different sets of support points; that is, some points t_i^j may be the same for different j's. Consequently, it is sensible to further introduce weights $q_1, \ldots, q_s$, which are the sums of q_i^j's for the repeated time instants. This allows us to rewrite ω in the more compact form

$$\widetilde{\omega} = \left\{ \begin{array}{ccc} t_1 & \cdots & t_s \\ q_1 & \cdots & q_s \end{array} \right\} ; \quad \sum_{k=1}^{s} q_k = 1. \tag{10}$$

Such reformulation makes it possible to solve the problem of finding the two level hierarchical optimal population design in terms of finding the equivalent one level design, since instead of (7) we have the minimization of $\Psi(M(\widetilde{\omega}))$ subject to $\sum_{k=1}^{s} q_k = 1$, a classical design problem studied thoroughly in the literature. Note, that $\widetilde{\omega} \in \Xi$ and we call it a *global design*. Obviously,

$$M(\zeta, N) = M(\widetilde{\omega}). \tag{11}$$

The information about groups is included in q_i^j and so in q_k. This information is later recovered after an optimum design $\widetilde{\omega}^{\star}$ has been found. Then, the weights β_j are determined as a solution of the system of nonlinear equations in (9), and finally the parameters α_j, n_j and N are recovered via solution of the system of the equations defined in (8). Since such a solution is not necessarily unique we need to assume that we know some of these values. For example, if we assume a priori the number of groups G and the numbers of observations per individual n_j's, then an optimum population design $\zeta^{\star}$ consists of G subsets of the support points of $\widetilde{\omega}^{\star}$ with the optimally recovered values of all weights $\alpha_j^{\star}$ and an optimum number $N^{\star}$ of all individuals in the experiment.

4 Efficient sampling windows

As was mentioned before, specific optimum time points for taking measurements are not always feasible in practice. For example, in pharmacokinetic studies there are possible delays in seeing patients by medical personnel. Additionally, individuals are often non-compliant with respect to taking the prescribed dose of a drug at a specified time. Then the sampling times lead to a suboptimal design and there is no information on the loss of efficiency. To control the loss it may be better to design sampling times within some intervals (windows), i.e. $t_i^j \in [a_i^j, b_i^j]$. The question is how to determine the boundaries a_i^j and b_i^j to obtain the required accuracy of parameter estimation. Here we follow the idea of Bogacka et al (2006), where they use the Equivalence Theorem for the choice of the windows in a way which ensures some minimum efficiency. We apply the efficiency factor as defined in Atkinson and Donev (1992):

$$\mathrm{Eff}_D(\zeta, N; \psi) = \left(\frac{\det(M(\zeta, N; \psi))}{\det(M(\zeta^{\star}, N^{\star}; \psi))} \right)^{(1/\widetilde{p})}, \tag{12}$$

and we define an *efficient sampling window population design* as a design ζ^{W} which assures some minimal level of efficiency (12) while keeping fixed the D-optimal individual design weights w_i^j and the numbers of observations n_i^j as well as the D-optimum group proportions α_j .

Due to (11), we have $\mathrm{Eff}_D(\zeta, N; \psi) = \mathrm{Eff}_D(\widetilde{\omega}; \psi)$. This allows us to apply the windows calculated for the optimum global design $\widetilde{\omega}^{\star}$ to an optimum design $\zeta^{\star}$. According to the Equivalence Theorem (in fixed models) the minimization of the function Ψ is equivalent to the minimization of the maximum variance of the response prediction, which is bounded from above by the number of the model parameters and it achieves this bound at the optimum points. A corresponding interpretation in our case is not so clear and requires further investigation. Cutting the function along a constant slightly smaller than $\widetilde{p}$ will give intervals $[a_i^j, b_i^j]$ including the optimum points, as shown in Figure 1. Based on this observation we can derive a general scheme of calculating the sampling windows:

Step 1. Calculate a locally D-optimum global population design $\widetilde{\omega}^\star$.

Step 2. Choose the minimum efficiency $c_{\min}$ of the window design and a small $\lambda \in (0,1)$.

Step 3. Calculate time windows $T_k = [a_k, b_k]$ solving the equation $d(t, \widetilde{\omega}^\star, \psi^0) = \lambda\widetilde{p}$, where $d(t, \widetilde{\omega}^\star, \psi^0) = \text{trace}\,[M(\widetilde{\omega}^\star, \psi^0)^{-1}M(t)]$ and ψ^0 is some initial estimate of the population parameters.

Step. 4. If minimum efficiency is assured (i.e. $\min_{\widetilde{\omega}} \text{Eff}_D(\widetilde{\omega}) \geq c_{\min}$, where $\widetilde{\omega}$ denotes any design with support points $t_k \in \widetilde{T}_k$, $k = 1, \ldots, s$, and weights equal to the weights of $\widetilde{\omega}^\star$) then STOP, else increase λ and repeat Step 3.

Then, having found the efficient sampling windows for the support points of $\widetilde{\omega}^\star$ we can directly apply them to obtain an efficient ζ^W. The most cumbersome part of the proposed scheme is the execution of the global optimization problem over a hypercube present in Step 4. In order to solve this task, the stochastic procedure based on the Adaptive Random Search strategy (Walter and Pronzato (1997)) was successfully applied.

5 Example

As an illustrative example we use the one-compartment model with first-order drug absorption (Jonsson et al (1996)):

$$y = \frac{Dk_a}{V(k_a - k_e)}\left(e^{-k_e t} - e^{-k_a t}\right) + \varepsilon, \tag{13}$$

where k_a and k_e are the first-order absorption and elimination rates, respectively, V is the apparent volume of distribution, D is a known dose and ε is an additive zero-mean uncorrelated Gaussian measurement noise with a constant variance. (It is assumed that such additive noise will be a good approximation of the real random process.) The regression parameters $\theta = (V, k_a, k_e)^{\mathrm{T}}$ are independent and normally distributed. The prior values of the population parameters are:

$$\psi^0 = \left(\mathrm{E}(\theta), var(\theta)\right)^{\mathrm{T}} = (100, 2.08, 0.1155, 0.3, 0.3, 0.03)^{\mathrm{T}} \quad \text{and} \quad var(\varepsilon) = 0.15.$$

We are looking for a D-optimum population design to estimate the population parameters as precisely as possible. We assume that the concentration of the drug can be measured within the design space $T = [0.25, 12]$ scaled in hours after administration and the total number of measurements is assumed to be $N_0 = 900$. The global D-optimum design obtained for these priors is a set of three equally distributed points:

$$\widetilde{\omega}^\star = \begin{Bmatrix} 0.45 & 1.86 & 9.90 \\ 0.33 & 0.33 & 0.33 \end{Bmatrix}.$$

Unfolding it back to an optimum population design $\zeta^\star$ gives non-unique solutions which depend on what is assumed to be known. Three possible optimum population designs are given below.

Identical design (one group design); $G = 1, n_1 = 9$ is assumed:

$$\zeta^\star = \left\{ \left(\begin{Bmatrix} 0.45 & 1.86 & 9.90 \\ 0.33 & 0.33 & 0.33 \\ 1 \end{Bmatrix}, 9 \right) \right\}; \quad N^\star = 100.$$

This means that for each patient, out of the optimum number of a hundred patients, we have to conduct exactly three measurements at each time instant.

One-point population design; $G = 3$, $n_1 = n_2 = n_3 = 10$ is assumed:

$$\zeta^\star = \left\{ \left(\{ {}^{0.45}_{\ \ 1} \}, 10 \right) \left(\{ {}^{1.86}_{\ \ 1} \}, 10 \right) \left(\{ {}^{9.90}_{\ \ 1} \}, 10 \right) \right\}; \quad N^\star = 90.$$
$$\qquad\quad 0.33 \qquad\qquad 0.33 \qquad\qquad 0.33$$

Here, each group consist of 30 patients; each patient in a group should have 10 replications at the same time point.

Arbitrarily structured design; $G = 3$, $n_1 = n_2 = n_3 = 10$ is assumed:

$$\zeta^\star = \left\{ \left(\{ {}^{0.45}_{0.57} {}^{9.90}_{0.43} \}, 10 \right) \left(\{ {}^{0.45}_{0.26} {}^{1.86}_{0.52} {}^{9.90}_{0.22} \}, 10 \right) \left(\{ {}^{1.86}_{0.62} {}^{9.90}_{0.38} \}, 10 \right) \right\}; \quad N^\star = 90.$$
$$\qquad\quad 0.40 \qquad\qquad\qquad 0.40 \qquad\qquad\qquad 0.20$$

After rounding of the weights there are 36 patients in groups 1 and 2 and 18 patients in group 3. Patients in group 1 have two sampling times replicated 6 and 4 times respectively; in group 2 three sampling times are replicated 3, 5 and 2 times, and in group 3 two sampling times are replicated 6 and 4 times, respectively. Rounding will lower the efficiency, but with $N_0 = 900$ this is negligible. (The efficiency of the rounded design is equal 0.999.) In Fig. 1 we see the way of generating sampling windows for various efficiency levels and also for various minimal window lengths. The latter are calculated with a different objective, namely minimising the length of a sampling window while keeping efficiency as high as possible. This may be important when, for practical reasons, the former gives intervals which are too narrow. The function $d(t, \omega^\star, \psi^0)$ reflects the behaviour of the model function: it has sharp peaks when the concentration changes fast and a flat peak at the area of slow drug elimination. It gives narrower windows when it is important to take measurements close to the optimum times and wider windows when it is less important. Table 1 shows the windows for the three optimum sampling times, both for some chosen minimum levels of efficiency and for some chosen minimal window lengths.

An efficient window population design ζ^W follows the form of the population continuous design with its time points belonging to the respective windows.

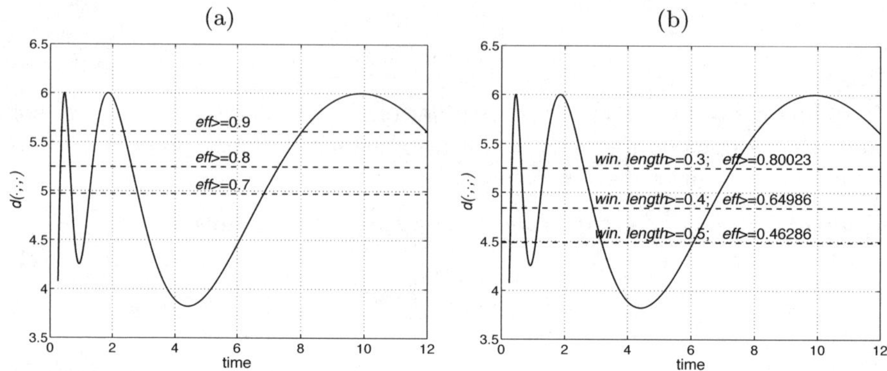

Fig. 1. The Equivalence Theorem condition and time windows generated by different thresholds with guaranteed efficiency (a) and with guaranteed minimal length (b).

Table 1. Sampling windows assuring minimal efficiency or minimal window length

Min. efficiency	Windows	$\lambda\widetilde{p}$
0.9	$[0.3545, 0.5635]$, $[1.4755, 2.3505]$, $[8.0275, 11.9915]$	5.61
0.8	$[0.3215, 0.6235]$, $[1.3325, 2.6085]$, $[7.2965, 12.0000]$	5.25
0.7	$[0.3015, 0.6705]$, $[1.2395, 2.8055]$, $[6.8225, 12.0000]$	4.97

Min. length	Windows	$\lambda\widetilde{p}$
0.5	$[0.2725, 0.7735]$, $[1.0765, 3.1805]$, $[6.0625, 12.0000]$	4.49
0.4	$[0.2925, 0.6945]$, $[1.1975, 2.8995]$, $[6.6145, 12.0000]$	4.84
0.3	$[0.3215, 0.6235]$, $[1.3325, 2.6085]$, $[7.2965, 12.0000]$	5.25

6 Conclusions

In the paper we present the definition of a population design and briefly discuss a way of dealing with the optimization problem of a large dimension which the definition creates. We then follow the idea of Bogacka et al (2006) to calculate efficient window designs. The way we define the population design allows us to use the Equivalence Theorem not only for finding the optimum design but also for the purpose of determining efficient windows. The technique for generating efficient sampling windows for population designs is relatively simple, assures satisfactory efficiency and indicates the importance of accurate timing of the sampling. However, the results are based on the assumption of independent observations and so further work is needed to solve this problem when this assumption can be relaxed.

Acknowledgement. We are grateful to the referees for their very helpful comments which improved the paper. This research was funded by the UK Engineering and Physical Science Research Council under grant EP/C54171/1.

References

Atkinson AC, Donev AN (1992) Optimum Experimental Design. Oxford University Press, Oxford

Bogacka B, Johnson P, Jones B, Volkov O (2006) D-efficient window experimental designs. Accepted to JSPI

Graham G, Aarons L (2006) Optimum blood sampling time windows for parameter estimation in population pharmacokinetic experiments. Statistics in Medicine In press

Green B, Duffull S (2003) Prospective evaluation of a D-optimal designed population pharmacokinetic study. J of Pharmacokinetics and Pharmacodynamics 30:145–161

Jonsson EN, Wade JR, Karlsson MO (1996) Comparison of some practical sampling strategies for population pharmacokinetic studies. J of Pharmacokinetics and Pharmacodynamics 24(2):245–263

Mentré F, Mallet A, Baccar D (1997) Optimal design in random-effects regression models. Biometrika 84(2):429–442

Mentré F, Dubruc C, Thénot JP (2001) Population pharmacokinetic analysis and optimization of the experimental design for mizolastine solution in children. J of Pharmacokinetics and Pharmacodynamics 28(3):229–319

Patan M, Bogacka B (2006) Optimum experimental design for uncorrelated mixed effects models. In preparation (http://wwwissiuzzgorapl/~mpatan/m220906pdf)

Pronzato L (2002) Information matrices with random regressors. Application to experimental design. J of Statistical Planning and Inference 108:189–200

Retout S, Mentré F (2003) Further developments of the Fisher information matrix in nonlinear mixed effects models with evaluation in population pharmacokinetics. Journal of Biopharmaceutical Statistics 13(2):209–227

Retout S, Duffull S, Mentré F (2001) Development and implementation of the population Fisher information matrix for the evaluation of population pharmacokinetic designs. Computer methods and Programs in Biomedicine 65:141–151

Retout S, Mentré F, Bruno R (2002) Fisher information matrix for non-linear mixed-effect models: evaluation and application for optimal design of enoxaparin population pharmacokinetics. Statistics in Medicine 21:2623–2639

Walter É, Pronzato L (1997) Identification of Parametric Models from Experimental Data. Communications and Control Engineering, Springer-Verlag, Berlin

Quantile and Probability-level Criteria for Nonlinear Experimental Design*

Andrej Pázman[1] and Luc Pronzato[2]

[1] Department of Applied Mathematics and Statistics, Faculty of Mathematics, Physics and Informatics, Comenius University, 84248 Bratislava, Slovakia
`pazman@center.fmph.uniba.sk`

[2] Laboratoire I3S, Les Algorithmes, Bâtiment Euclide, 2000 route des lucioles, BP 121, 06903 Sophia Antopolis cedex, France
`pronzato@i3s.unice.fr`

Summary. We consider optimal experimental design for parameter estimation in nonlinear situations where the optimal experiment depends on the value of the parameters to be estimated. Setting a prior distribution for these parameters, we construct criteria based on quantiles and probability levels of classical design criteria and show how their derivatives can easily be approximated, so that classical algorithms for local optimal design can be used for their optimisation.

Key words: robust design, minimax optimal design, average optimal design, quantiles, nonlinear models

1 Introduction

Classical criteria for optimum experimental design for parameter estimation are functions of a normalized information matrix, which generally takes the form

$$\mathbf{M}(\xi, \theta) = \int_{\mathcal{X}} \mathcal{M}(x, \theta)\, \xi(dx) \tag{1}$$

with $\theta \in \mathbb{R}^p$ the parameters of interest and ξ the design, that is, a probability measure on some given region $\mathcal{X}$ of $\mathbb{R}^q$. Typically, in nonlinear situations, the $p \times p$ matrix $\mathcal{M}(x, \theta)$ depends on θ, the parameters to be estimated. For instance, a design ξ_D is D-optimal for LS estimation in the nonlinear regression

* The research of the first author has been supported by the VEGA grant No. 1/3016/06. The work of the second author was partially supported by the IST Programme of the European Community, under the PASCAL Network of Excellence, IST-2002-506778. This publication only reflects the authors' view.

model with scalar observations $Y_k = \eta(x_k, \theta) + \varepsilon_k$ and i.i.d. errors ε_k with zero mean and finite variance ($k = 1, 2 \ldots$) when it maximizes $\log \det \mathbf{M}(\xi, \theta)$ with $\mathcal{M}(x, \theta)$ the rank-one matrix

$$\mathcal{M}(x, \theta) = \frac{\partial \eta(x, \theta)}{\partial \theta} \frac{\partial \eta(x, \theta)}{\partial \theta^\top} \,. \tag{2}$$

Classically, a prior guess $\hat{\theta}^0$ for θ is used to design the experiment, with the hope that the local optimal design for $\hat{\theta}^0$ will be close to the optimal one for the unknown θ. When the alternation of estimation and design phases is possible, sequential design permits progressively adapting the experiment to an estimated value of θ that (hopefully) converges to its unknown true value, see e.g. Wu (1985); Chaudhuri and Mykland (1993) for maximum-likelihood and Spokoinyi (1992) for Bayesian estimation. In many circumstances, however, the repetition of experimentation phases is impossible, and a single design ξ^* must be determined, based on the prior information available. Two types of approaches have been suggested to achieve some robustness with respect to a misspecification of θ. Let $\Phi(\xi, \theta)$ denote the criterion to be maximized with respect to ξ, for instance, $\Phi(\xi, \theta) = \Psi[\mathbf{M}(\xi, \theta)]$ with $\mathbf{M}(\xi, \theta)$ the information matrix (1) and $\Psi(\cdot)$ a concave function on the space of non-negative definite $p \times p$ matrices, with $p = \dim(\theta)$. *An average optimal design* puts a prior probability measure π on θ and maximizes

$$\Phi_A(\xi) = \mathbb{E}_\pi \{\Phi(\xi, \theta)\} = \int_\Theta \Phi(\xi, \theta) \, \pi(d\theta)$$

with $\Theta \subset \mathbb{R}^p$ the support of π, see, e.g., Fedorov (1980); Chaloner and Larntz (1989); Chaloner and Verdinelli (1995). In *maximin-optimal design* $\Phi(\xi, \theta)$ is replaced by its worst possible value for θ in Θ and the criterion to be maximized is

$$\Phi_M(\xi) = \min_{\theta \in \Theta} \Phi(\xi, \theta) \,,$$

see, e.g., Melas (1978); Fedorov (1980); Müller and Pázman (1998). Compared to local designs, average and maximim optimal designs do not create any special difficulties (other than heavier computations) for discrete designs of the form $\xi = (1/n) \sum_{i=1}^n \delta_{x_i}$, with δ_x the delta measure which puts mass 1 at x and with n fixed (usually, algorithms for discrete design do not exploit any special property of the design criterion, but only yield local optima). For computational reasons, the situation is simpler when π is a discrete measure and Θ is a finite set (however, a relaxation algorithm is suggested in (Pronzato and Walter, 1988) for maximin-optimal designs when Θ is a compact set, and stochastic approximation can be used for average-optimal designs in general situations, see, e.g., Pronzato and Walter (1985)). When optimizing a design measure (approximate design theory), the concavity of Φ is preserved, which yields Equivalence Theorems, and globally convergent algorithms can be constructed, see, e.g., Fedorov and Hackl (1997). Although attractive, average and maximim optimal designs nevertheless raise several important difficulties among which are the following:

(i) A design ξ_A^* optimal for Φ_A can perform poorly for "many" values of θ, in the sense that $\pi\{\Phi(\xi_A^*, \theta) < u\}$ may be larger than α for some unacceptably low value for u and high level α.

(ii) For $g(\cdot)$ an increasing real function, the maximization of $g[\Phi(\xi, \theta)]$ is equivalent to that of $\Phi(\xi, \theta)$, but maximizing $\mathbb{E}_\pi\{g[\Phi(\xi, \theta)]\}$ is not equivalent to maximizing $\mathbb{E}_\pi\{\Phi(\xi, \theta)\}$ in general, so that a single design criterion for local optimality yields infinitely many criteria for average optimality.

(iii) Quite often an optimal design ξ_M^* for Φ_M is such that $\min_{\theta \in \Theta} \Phi(\xi_M^*, \theta)$ is reached for θ on the boundary of Θ, which makes ξ_M^* very sensitive to the choice of Θ. Also, if Θ is taken too large, it may contain values of θ such that $\mathbf{M}(\xi, \theta)$ is singular for all ξ and an optimal design may not exist.

(iv) The maximin criterion Φ_M is not differentiable everywhere, which induces some difficulties for its optimisation; in particular, the steepest-ascent direction does not necessarily correspond to a one-point delta measure.

This paper suggests new stochastic design criteria based on the distribution of $\Phi(\xi, \theta)$ when θ is distributed with some prior probability measure π on $\Theta \subset \mathbb{R}^p$. In particular, we shall consider the probability levels

$$P_u(\xi) = \pi\{\Phi(\xi, \theta) \geq u\} \tag{3}$$

and the quantiles

$$Q_\alpha(\xi) = \max\{u : P_u(\xi) \geq 1 - \alpha\}, \quad \alpha \in [0, 1], \tag{4}$$

with u and α considered as free parameters, to be chosen by the user. When the range of possible values for Φ is known (which is the case for instance when Φ is an efficiency criterion with values in $[0, 1]$), one can specify a target level u and then maximize the probability $P_u(\xi)$ that the target is reached (or equivalently minimize the risk $1 - P_u(\xi)$ that it is not). In other situations, one can specify a probability level α which defines an acceptable risk, and maximize the value of u such that the probability that $\Phi(\xi, \theta)$ is smaller than u is less than α, which corresponds to maximizing $Q_\alpha(\xi)$. We shall assume that $\Phi[(1 - \gamma)\mu + \gamma\nu, \theta]$ is continuously differentiable in $\gamma \in [0, 1)$ for any θ and any probability measures μ, ν on $\mathcal{X}$ such that $\mathbf{M}(\mu, \theta)$ is non degenerate. We also assume that $\Phi(\xi, \theta)$ is continuous in θ and that the measure π has a positive density on every open subset of Θ. This implies that $Q_\alpha(\xi)$ is defined as the solution in u of the equation $1 - P_u(\xi) = \alpha$, see Figure 1.

One may notice that the difficulties (i-iv) mentioned above for average and maximin optimal design are explicitly taken into account by the proposed approach: the probability indicated in (i) is precisely $1 - P_u(\xi)$ which is minimized; (ii) substituting $g[\Phi(\xi, \theta)]$ for $\Phi(\xi, \theta)$ with $g(\cdot)$ increasing leaves (3) and (4) unchanged; (iii) the role of the boundary of Θ is negligible when a small probability is attached to it (and for instance probability measures with infinite support are allowed); (iv) kernel smoothing makes P_u and Q_α

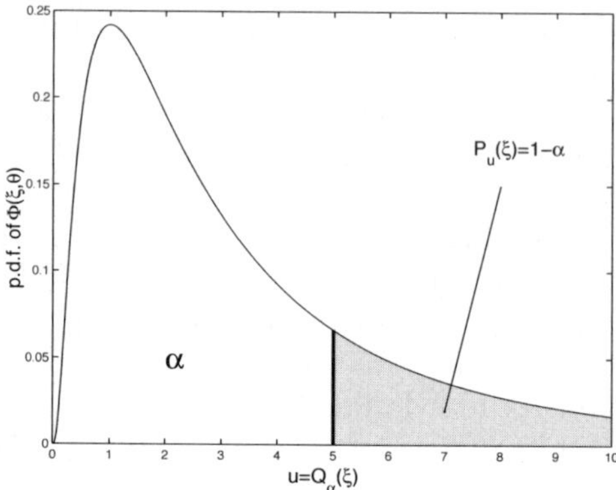

Fig. 1. Probability levels and quantiles for a design criterion $\Phi(\xi,\theta)$.

differentiable, see Sect. 2. When $\Phi(\xi,\theta)$ is concave in ξ for any θ, Φ_A and Φ_M are concave. Unfortunately, P_u and Q_α are generally not, which is probably the main drawback of the approach. However, Q_α obviously satisfies the following: let Θ denote the support of π, and suppose it is compact; then, $Q_\alpha(\xi) \to \Phi_M(\xi)$ when $\alpha \to 0$, and a design optimal for Q_α will tend to be optimal for Φ_M and *vice versa*. In the next section we show how the directional derivatives of $P_u(\xi)$ and $Q_\alpha(\xi)$ can be computed, to be used in steepest-ascent optimization algorithms which converge to a local optimum (at least). An illustrative example is presented in Sect. 3 and Sect. 4 gives some conclusions and perspectives.

2 Evaluations of criteria and their derivatives

Computation of derivatives. Let $\xi = (1-\gamma)\mu + \gamma\nu$ and consider the derivatives $\partial P_u(\xi)/\partial\gamma$ and $\partial Q_\alpha(\xi)/\partial\gamma$ at $\gamma = 0$. Since $Q_\alpha(\xi)$ satisfies the implicit equation $P_{Q_\alpha(\xi)}(\xi) = 1 - \alpha$, we can write

$$\{\partial P_u(\xi)/\partial\gamma + [\partial P_u(\xi)/\partial u][\partial Q_\alpha(\xi)/\partial\gamma]\}_{|u=Q_\alpha(\xi)} = 0\,,$$

which gives

$$\frac{\partial Q_\alpha(\xi)}{\partial\gamma} = -\left(\frac{\partial P_u(\xi)}{\partial\gamma}\bigg/\frac{\partial P_u(\xi)}{\partial u}\right)_{|u=Q_\alpha(\xi)}. \tag{5}$$

To compute the derivatives $\partial P_u(\xi)/\partial\gamma$ and $\partial P_u(\xi)/\partial u$ we write $P_u(\xi)$ as

$$P_u(\xi) = \int_\Theta I_{[u,\infty)}[\Phi(\xi,\theta)]\,\pi(d\theta) = \int_\Theta I_{(-\infty,\Phi(\xi,\theta)]}(u)\,\pi(d\theta)$$

with $I_A(\cdot)$ the indicator function of the set A. When approximating the indicator step-function by a normal distribution function with small variance σ^2, the two expressions above become respectively

$$P_u(\xi) \approx \int_\Theta \mathbb{F}_{u,\sigma^2}[\Phi(\xi,\theta)]\,\pi(d\theta) = \int_\Theta [1 - \mathbb{F}_{\Phi(\xi,\theta),\sigma^2}(u)]\,\pi(d\theta)$$

with $\mathbb{F}_{a,\sigma^2}$ the distribution function of the normal $\mathcal{N}(a,\sigma^2)$. Differentiating these approximations respectively with respect to γ and u, we get

$$\frac{\partial P_u(\xi)}{\partial \gamma}\bigg|_{\gamma=0} \approx \int_\Theta \varphi_{u,\sigma^2}[\Phi(\mu,\theta)]\,\frac{\partial \Phi(\xi,\theta)}{\partial \gamma}\bigg|_{\gamma=0}\,\pi(d\theta)\,, \tag{6}$$

$$\frac{\partial P_u(\xi)}{\partial u}\bigg|_{\gamma=0} \approx -\int_\Theta \varphi_{\Phi(\mu,\theta),\sigma^2}(u)\,\pi(d\theta)\,, \tag{7}$$

with φ_{a,σ^2} the density of $\mathbb{F}_{a,\sigma^2}$, which can be substituted in (5) to form an approximation of $\partial Q_\alpha(\xi)/\partial \gamma_{|\gamma=0}$. As shown below, this type of approximation can be related to another one, namely kernel smoothing.

Kernel smoothing. In order to estimate $P_u(\xi)$, $Q_\alpha(\xi)$ and their derivatives, one can also approximate the probability density function (p.d.f.) of $\Phi(\xi,\theta)$ by a standard kernel estimator $\phi_{n,\xi}(z) = 1/(nh_n)\sum_{i=1}^n K\left\{[z - \Phi(\xi,\hat\theta^i)]/h_n\right\}$. Here K is a symmetric kernel function (the p.d.f. of a probability measure on $\mathbb{R}$ with $K(z) = K(-z)$, e.g. $\varphi_{0,1}(\cdot)$) and $\hat\theta^i$ ($i = 1,\ldots,n$) is a sample of possible values for θ (e.g. independently randomly generated with the prior measure π). The bandwidth h_n tends to zero as $n \to \infty$. From this we obtain directly

$$P_u(\xi) \approx \hat P_u^n(\xi) = \int_{-\infty}^\infty \mathrm{I}_{[u,\infty)}(z)\,\phi_{n,\xi}(z)\,dz\,,$$

which is easily computed when $\int_u^\infty K(z)dz$ has a simple form. The value of $Q_\alpha(\xi)$ can then be estimated by $\hat Q_\alpha^n(\xi) = \{u : \hat P_u^n(\xi) = 1 - \alpha\}$, which is easily computed numerically. Consider now the computation of derivatives, with again $\xi = (1 - \gamma)\mu + \gamma\nu$. Direct calculations give

$$\frac{\partial \hat P_u^n(\xi)}{\partial \gamma}\bigg|_{\gamma=0} = \frac{1}{nh_n}\sum_{i=1}^n \frac{\partial \Phi(\xi,\hat\theta^i)}{\partial \gamma}\bigg|_{\gamma=0} K\left(\frac{u - \Phi(\mu,\hat\theta^i)}{h_n}\right)\,, \tag{8}$$

$$\frac{\partial \hat P_u^n(\xi)}{\partial u}\bigg|_{\gamma=0} = -\frac{1}{nh_n}\sum_{i=1}^n K\left(\frac{u - \Phi(\mu,\hat\theta^i)}{h_n}\right)\,. \tag{9}$$

Notice that taking $\sigma^2 = h_n$ and π the discrete measure with mass $1/n$ at each $\hat\theta^i$ in (6, 7) respectively gives (8) and (9) with $K = \varphi_{0,1}$, the density of the standard normal. Obviously, the accuracy of these kernel approximations improves as n increases (with the only limitation due to the computational cost that increases with n).

3 Example

To illustrate the feasibility of the approach we consider D-optimal designing for the nonlinear regression model $\eta(x,\theta) = \beta e^{-\lambda x}$, with $\theta = (\beta,\ \lambda)^\top$ the vector of parameters to be estimated. The information matrix $\mathbf{M}(\xi, \theta)$ for a design measure ξ then takes the form (1, 2). We suppose that $\beta > 0$ and take $\mathcal{X} = [0, \infty)$. The local D-optimal experiment $\xi_D(\theta)$ which maximizes $\det \mathbf{M}(\xi, \theta)$ puts mass $1/2$ at $x = 0$ and $x = 1/\lambda$, and the associated value of $\det \mathbf{M}(\xi, \theta)$ is $\det \mathbf{M}[\xi_D(\theta), \theta)] = \beta^2/(4e^2\lambda^2)$. We consider the D-efficiency criterion defined by $\Phi(\xi, \theta) = \{\det \mathbf{M}(\xi, \theta)/ \det \mathbf{M}[\xi_D(\theta), \theta]\}^{1/2}$, with $\Phi(\xi, \theta) \in [0, 1]$. Due to the linear dependency of $\eta(x, \theta)$ in β, $\xi_D(\theta)$ and $\Phi(\xi, \theta)$ only depend on λ and we shall simply write $\xi_D(\lambda)$, $\Phi(\xi, \lambda)$. Supposing that $\lambda = 2$ when designing the experiment, the efficiency $\Phi[\xi_D(2), \lambda]$ is the solid line depicted in Figure 2.

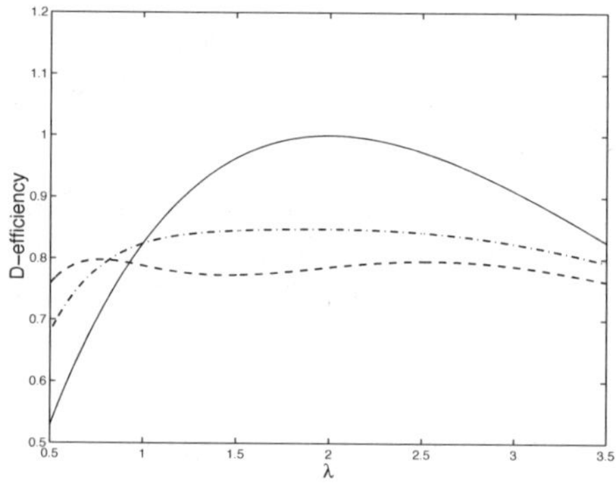

Fig. 2. D-efficiencies as function of λ for different designs; solid line: local D-optimal $\xi_D(2)$; dashed line: optimal for $P_{0.75}$; dash-dotted line optimal for $Q_{0.10}$.

Suppose now that we only know that $\lambda \in [1/2, 7/2]$ and we put a uniform prior for λ on that interval; $\xi_D(2)$ is then optimal for the midpoint, but its efficiency is less than 53% for the endpoint $\lambda = 1/2$. We approximate $P_u(\xi)$ and $Q_\alpha(\xi)$ by kernel smoothing with $K = \varphi_{0,1}$ for $n = 100$ values $\hat{\lambda}^i$ equally spaced in $[0.5, 3.5]$. No special care is taken for the choice of h_n, and we simply use the rule $h_n = \hat{\sigma}_n(\Phi)n^{-1/5}$ with $\hat{\sigma}_n(\Phi)$ the empirical standard deviation of the values $\Phi(\xi, \lambda^i)$, $i = 1, \ldots, n$. Figure 3 shows the estimated values $\hat{P}_u^n$ (left) and $\hat{Q}_\alpha^n$ (right), in dashed lines, as functions of u and α respectively, for $\xi = \xi_D(2)$. One can check the reasonably good agreement with the exact values of P_u and Q_α, plotted in solid lines (increasing n to $1\,000$ makes the curves almost indistinguishable).

The optimisation of $\hat{P}_{0.75}^n$ and $\hat{Q}_{0.10}^n$ with a vertex-direction (steepest-ascent) algorithm on the finite design space $\{0, 0.1, 0.2, \ldots 5\}$ respectively gives the four-point designs

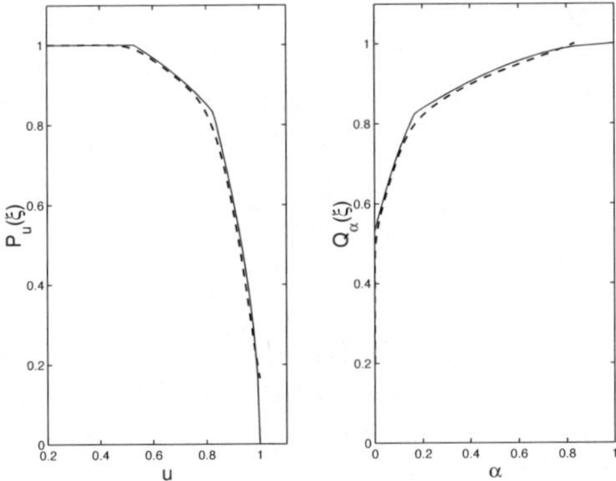

Fig. 3. Left: $\hat{P}_u^n(\xi)$ (dashed line) and $P_u(\xi)$ (solid line) as functions of u; right: $\hat{Q}_\alpha^n(\xi)$ (dashed line) and $Q_\alpha(\xi)$ (solid line) as functions of α; $\xi = \xi_D(2)$, $n = 100$.

$$\xi^*(P_{0.75}) \approx \left\{ \begin{array}{cccc} 0 & 0.3 & 0.4 & 1.7 \\ 0.4523 & 0.0977 & 0.2532 & 0.1968 \end{array} \right\},$$

$$\xi^*(Q_{0.10}) \approx \left\{ \begin{array}{cccc} 0 & 0.3 & 0.4 & 1.3 \\ 0.4688 & 0.1008 & 0.2634 & 0.1670 \end{array} \right\},$$

where the first row indicates the support points and the second one their respective weights. They satisfy $\hat{P}_{0.75}^n[\xi^*(P_{0.75})] \approx 0.9999$ and $\hat{Q}_\alpha^n[\xi^*(Q_{0.10})] \approx 0.783$. The efficiencies of these designs are plotted in Figure 2. The exact value $P_u[\xi^*(P_{0.75})]$ equals one, indicating that the efficiency is larger than 75% for all possible values of λ. The optimisation of $\hat{Q}_{0.01}^n$ gives a design very close to $\xi^*(P_{0.75})$ which, together with the shape of the dashed curve in Figure 2, suggests that $\xi^*(P_{0.75})$ is almost maximin optimal. Accepting a small loss of efficiency for about 10% of the values of λ produces a significant increase of efficiency on most of the interval, see the curve in dash-dotted line.

4 Conclusions and further developments

The paper shows the feasibility of optimal design based on quantiles and probability level criteria in the situations where the local optimal experiment depends on the unknown parameters to be estimated. In particular, kernel smoothing permits optimisation of design measures with classical algorithms borrowed from local optimal design theory. Adapting the sample size n, the kernel K and the bandwidth h_n to a particular problem, and maybe a particular algorithm, may deserve further studies. In particular, one might think of letting n grow with the number of iterations of the algorithm, as in stochastic approximation methods, see, e.g., Chapter 4 of (Kibzun and Kan, 1996).

Notice, finally that the ideas presented in this paper are very general and could also be applied to discrete designs based on more accurate descriptions of parameter uncertainty than functions of information matrices, such as the volumes of confidence regions (Hamilton and Watts, 1985), mean-squared error (Pázman and Pronzato, 1992; Gauchi and Pázman, 2006) or the entropy of the distribution of the LS estimator (Pronzato and Pázman, 1994).

References

Chaloner K, Larntz K (1989) Optimal Bayesian design applied to logistic regression experiments. Journal of Statistical Planning and Inference 21:191–208

Chaloner K, Verdinelli I (1995) Bayesian experimental design: a review. Statistical Science 10(3):273–304

Chaudhuri P, Mykland P (1993) Nonlinear experiments: optimal design and inference based likelihood. Journal of the American Statistical Association 88(422):538–546

Fedorov V (1980) Convex design theory. Math Operationsforsch Statist, Ser Statistics 11(3):403–413

Fedorov V, Hackl P (1997) Model-Oriented Design of Experiments. Springer, Berlin

Gauchi JP, Pázman A (2006) Designs in nonlinear regression by stochastic minimization of functionnals of the mean square error matrix. Journal of Statistical Planning and Inference 136:1135–1152

Hamilton D, Watts D (1985) A quadratic design criterion for precise estimation in nonlinear regression models. Technometrics 27:241–250

Kibzun A, Kan Y (1996) Stochastic Programming Problems. Wiley, New York

Melas V (1978) Optimal designs for exponential regressions. Math Operationsforsch und Statist, Ser Statistics 9:753–768

Müller C, Pázman A (1998) Applications of necessary and sufficient conditions for maximin efficient designs. Metrika 48:1–19

Pázman A, Pronzato L (1992) Nonlinear experimental design based on the distribution of estimators. Journal of Statistical Planning and Inference 33:385–402

Pronzato L, Pázman A (1994) Second-order approximation of the entropy in nonlinear least-squares estimation. Kybernetika 30(2):187–198, *Erratum* 32(1):104, 1996

Pronzato L, Walter E (1985) Robust experiment design via stochastic approximation. Mathematical Biosciences 75:103–120

Pronzato L, Walter E (1988) Robust experiment design via maximin optimization. Mathematical Biosciences 89:161–176

Spokoinyi V (1992) On asymptotically optimal sequential experimental design. Advances in Soviet Mathematics 12:135–150

Wu C (1985) Asymptotic inference from sequential design in a nonlinear situation. Biometrika 72(3):553–558

Optimal Designs for the Exponential Model with Correlated Observations

Andrey Pepelyshev

Faculty of Mathematics and Mechanics, St.Petersburg State University, University avenue 28, Petrodvoretz, 198504 St. Petersburg, Russia
`andrey@ap7236.spbu.ru`

Summary. In the exponential regression model with an autoregressive error structure exact D-optimal designs for weighted least squares analysis are investigated. It is shown that support points of a locally D-optimal design are discontinuous with respect to the correlation parameter. Also equidistant designs are proved to be considerably less efficient than maximin efficient D-optimal designs. A tool used in the study is the functional approach described in a recent book Melas (2006).

Key words: exponential regression, exact D-optimal designs, correlated observations, functional approach

1 Introduction

The present paper is devoted to constructing exact optimal designs for weighted least squares estimation in the exponential model with correlated observations. The exponential model is widely used in chemistry, pharmacokinetics, and microbiology. In microbiology, for example, this model is applied to describing growth and death of microorganisms, dose-response analysis and risk assessment; see Coleman and Marks (1998). Other applications can be found in Dette et al (2006b); Ucinski and Atkinson (2004). The exponential model with uncorrelated observations has been investigated in a number of papers Mukhopadhyay and Haines (1995); Han and Chaloner (2003); Dette and Neugebauer (1997), and Dette et al (2006b). In those papers locally, Bayesian and maximin D-, c- and e_k-optimal designs were constructed. However, constructing optimal designs for correlated observations is a more difficult problem. Hoel (1958) investigated the efficiency of equally spaced designs for the linear model. Abt et al. (1997,1998) studied optimal designs for linear and quadratic regression models with an autocorrelated error structure and a large number of support points. Dette et al (2006a) investigated exact D-optimal designs for the same models with a small number of points. Stehlík (2005) studied D-optimal designs for the linear model with several types of covariance function and a small number of points. A numerical algorithm for finding optimal designs is proposed in Fedorov and Hackl (1997); Müller and Pazman (2003); Ucinski and Atkinson (2004).

The present paper considers the problem of determining exact D-optimal designs for the exponential regression model and an autoregressive error structure and a small number of points.

In Section 2 the model is introduced. In Section 3 exact locally D-optimal designs are studied. It is shown that optimal points are discontinuous with respect to the level of correlation. This result is in agreement with the results obtained in Stehlík (2005); Dette et al (2006a). In Section 4 we study maximin D-optimal designs. We also investigate the efficiency of equally spaced designs.

2 Statement of problem

Consider the exponential regression model

$$Y_{t_i} = \eta(t_i) + \varepsilon_{t_i}, \ \eta(t) = ae^{-bt}, \ i = 1, \ldots, n,$$

where $t_i \in [c, \infty)$ are chosen by an experimenter, the parameters a and b are unknown and have to be estimated. Assume that the errors ε_{t_i} have zero expectation and the covariance of two measurements depends on the distance between the experimental conditions, that is

$$\mathrm{Cov}(Y_{t_i}, Y_{t_j}) = \sigma^2 e^{-\lambda|t_i - t_j|}, \ \lambda > 0,$$

where λ is some constant, which characterizes the level of correlation. We can assume without loss of generality that $\sigma = 1$ and $c = 0$.

An exact design $\xi = \{t_1, \ldots, t_n\}$ is a vector of n experimental conditions, $0 \le t_1 \le \ldots \le t_n$.

The weighted least squares estimate of $\beta = (a, b)^T$ is given by

$$\hat{\beta} = (X^T \Sigma^{-1} X)^{-1} X^T \Sigma^{-1} Y$$

with covariance matrix $\mathrm{Cov}(\hat{\beta}) = (X^T \Sigma^{-1} X)^{-1}$ (see Fedorov and Hackl (1997)), where

$$Y = (Y_{t_1}, \ldots, Y_{t_n})^T,$$
$$X^T = X_\xi^T = (f(t_1), \ldots, f(t_n)),$$
$$f(t) = (\partial\eta/\partial a, \partial\eta/\partial b)^T,$$
$$\Sigma = \Sigma_\xi = \left(e^{-\lambda|t_i - t_j|}\right)_{i,j=1,\ldots,n}.$$

An exact locally D-optimal design maximizes $\det M(\xi)$ where the information matrix is given by

$$M(\xi) = M(\xi, a, b, \lambda) = X_\xi^T \Sigma_\xi^{-1} X_\xi.$$

3 Locally D-optimal designs

As is shown in Dette et al (2006a), the information matrix can be presented in the form $M(\xi) = X_\xi^T V_\xi^T V_\xi X_\xi$ where V_ξ is a 2-diagonal matrix given by $V_\xi = (v_{i,j})$, $v_{i,i} = r_i$, $v_{i,i-1} = -s_i$, $r_1 = 1$, $s_1 = 0$,

$$r_i = 1\bigg/ \sqrt{1 - e^{-2\lambda(t_i - t_{i-1})}}, \;\; s_i = e^{-\lambda(t_i - t_{i-1})} r_i,$$

$i = 2, \ldots, n$. From the Cauchy-Binet formula we obtain the following expression for the determinant of the information matrix

$$\det M(\xi, a, b, \lambda) = a^2 \sum_{1 \le i < j \le n} e^{-2b(t_i + t_j)} \psi^2(t_i, t_{i-1}, t_j, t_{j-1}), \tag{1}$$

where

$$\psi(t_i, t_{i-1}, t_j, t_{j-1}) =$$
$$= \frac{(1 - e^{-(\lambda-b)d_i})(t_j - t_{j-1} e^{-(\lambda-b)d_j}) - (1 - e^{-(\lambda-b)d_j})(t_i - t_{i-1} e^{-(\lambda-b)d_i})}{(1 - e^{-2\lambda d_i})(1 - e^{-2\lambda d_j})},$$

$d_i = t_i - t_{i-1}$, $d_j = t_j - t_{j-1}$.

Analytical results on locally D-optimal designs are given in Lemma 1.

Lemma 1. *Let $\xi^* = \xi^*(a, b, \lambda) = \{t_1^*, \ldots, t_n^*\}$ be a locally D-optimal design for the exponential model with correlated observations. Then*
1) The design ξ^* does not depend on a.
2) The first point of the design is equal to zero, that is $t_1^* = 0$.
3) Points of the design ξ^* satisfy

$$t_i^*(\gamma b, \gamma \lambda) = \frac{1}{\gamma} t_i^*(b, \lambda)$$

for any $\gamma > 0$.

    Proof. The first statement simply follows from (1). Let $\xi_h = \{t_1 + h, \ldots, t_n + h\}$ where $h > 0$. Then

$$\det M(\xi_h) = e^{-4bh} \det M(\xi_0).$$

Thus, the second statement is proved. Observing (1) we note that

$$\det M(t_2, \ldots, t_n, \gamma b, \gamma \lambda) = \frac{1}{\gamma^2} \det M(\gamma t_2, \ldots, \gamma t_n, b, \lambda).$$

The last equality implies the third statement.

$\square$

    In view of Lemma 1, without loss of generality we can set $a = 1$.

    Since it is impossible to obtain the design explicitly further analysis will be based on the functional approach developed in Melas (2006). This approach allows one to construct a Taylor series for points of exact locally D-optimal designs as functions of some parameters.

3.1 The case $n = 2$

In this subsection we investigate 2-point locally D-optimal designs.

For the 2-point design $\{0, t_2\}$ the determinant of the information matrix has the form

$$\det M(t_2) = \det M(t_2, b, \lambda) = \frac{t_2^2 e^{-2bt_2}}{1 - e^{-2\lambda t_2}}.$$

It is easy to see that $\det M(t_2)$ has a unique maximum for all fixed b and λ. Let u^* be a unique solution of the equation

$$\frac{1}{1 - e^{-u}} = \frac{b}{\lambda} + \frac{2}{u}$$

on $u \in (-\infty, 0)$. A direct calculation shows that the second point of the D-optimal design equals

$$t^*(b, \lambda) = \frac{-u^*}{2\lambda}.$$

Now we will investigate the behavior of optimal designs for slightly and highly correlated observations.

As $\lambda \to \infty$, we obtain $u^* \to \infty$ and $2/u^* \to -b/\lambda$. Consequently, $t_2^*(b, \lambda) \to 1/b$ as $\lambda \to \infty$. This means that the locally D-optimal designs for correlated observations tend to the locally D-optimal design for independent observations since correlation is decreasing (as $\lambda \to \infty$) for $n = 2$. It is shown below, that this is true for $n > 2$.

For highly correlated observations we have $u^* \to 0$ as $\lambda \to 0$ and the equation for u^* can be rewritten in the following form $-\frac{1}{u} + o(\lambda) = \frac{b}{\lambda} + \frac{2}{u}$. Thus, $t_2^*(b, \lambda) \to 1/(2b)$ as $\lambda \to 0$.

3.2 The case $n = 3$

In this subsection we study locally 3-point D-optimal designs. Due to Theorem 1 it is sufficient to investigate locally optimal designs for fixed b. Let $b = 1$. For other values of b the optimal design can be obtained by rescaling the points.

Numerical calculations based on a routine `fminsearch` in MATLAB show that the function $\det M(t_2, t_3) = \det M(t_2, t_3, \lambda)$ has two local maxima for some values of λ. Let λ^* be the value of λ such that the function $\det M(t_2, t_3)$ has equal maxima. A direct computation shows that $\lambda^* \approx 0.22367$. Thus, points of a locally optimal design are discontinuous at $\lambda = \lambda^*$.

In order to study a locally optimal design for small λ, we note that

$$\det M(t_2, t_3, \lambda) = \sum_{j=-2}^{\infty} M_{(j)}(t_2, t_3)\lambda^j.$$

Table 1. Locally D-optimal design $\{0, t_2^*(\lambda), t_3^*(\lambda)\}$ for some values of λ with $b = 1$.

λ	0	$\lambda^* - 0$	$\lambda^* + 0$	∞
$t_2^*(\lambda)$	0.5395	0.5703	0.3401	1
$t_3^*(\lambda)$	3.3560	3.2386	0.8870	1

Thus, nonzero points of locally D-optimal designs tend to points which maximize $M_{(-2)}(t_2, t_3)$ as $\lambda \to 0$.

Table 1 contains locally D-optimal designs for some special values of λ with $b = 1$.

The implementation of the functional approach in Maple (see Melas (2006)) gives the following expansions. The expansions
$$t_2^*(\lambda) = 0.5395 + 0.1096\lambda + 0.1156\lambda^2 + 0.1077\lambda^3 + \ldots$$
$$t_3^*(\lambda) = 3.3560 - 0.6662\lambda + 1.8098\lambda^2 - 2.2812\lambda^3 + \ldots$$
converge for $\lambda \in (0, \lambda^*)$. Expansions
$$t_2^*(\lambda) = 0.5087 + 0.2687(\lambda - 1) - 0.0541(\lambda - 1)^2 - 0.0813(\lambda - 1)^3 + \ldots$$
$$t_3^*(\lambda) = 1.3056 - 0.4326(\lambda - 1) + 0.4930(\lambda - 1)^2 - 0.0819(\lambda - 1)^3 + \ldots$$
converge for $\lambda \in (\lambda^*, 2)$. Expansions
$$t_2^*(\lambda) = 0.6911 - 0.3836(\nu - 1/2) - 0.3161(\nu - 1/2)^2 + 1.14(\nu - 1/2)^3 + \ldots$$
$$t_3^*(\lambda) = 1.5177 - 0.0556(\nu - 1/2) - 1.4224(\nu - 1/2)^2 + 1.74(\nu - 1/2)^3 + \ldots,$$
where $\nu = 1/\lambda$, converge for $\lambda \in (1, \infty)$.

The 3-point D-optimal designs with $b = 1$ are depicted in Figure 1.

3.3 The case $n = 4$

Numerical calculations show that points of a locally D-optimal design are discontinuous at two points, say λ^* and λ^{**}. Table 2 contains locally D-optimal designs for some special values of λ with $b = 1$.

Table 2. Locally D-optimal designs $\{0, t_2^*(\lambda), t_3^*(\lambda), t_4^*(\lambda)\}$ for some values of λ with $b = 1$.

λ	0	$\lambda^* - 0$	$\lambda^* + 0$	$\lambda^{**} - 0$	$\lambda^{**} + 0$	∞
$t_2^*(\lambda)$	0.3288	0.3462	0.2491	0.6966	0.2030	0
$t_3^*(\lambda)$	0.8669	0.8919	0.5841	1.0074	0.9250	1
$t_4^*(\lambda)$	3.4058	3.3611	1.1180	1.3133	1.2490	1

The behavior of the 4-point D-optimal design with $b = 1$ is depicted in Figure 1.

The case $n = 5$ can be similarly studied. The behavior of the 5-point locally D-optimal design with $b = 1$ is depicted in Figure 1. Numerical calculations allow us to state the conjuncture that the number of points of discontinuity is increasing with n.

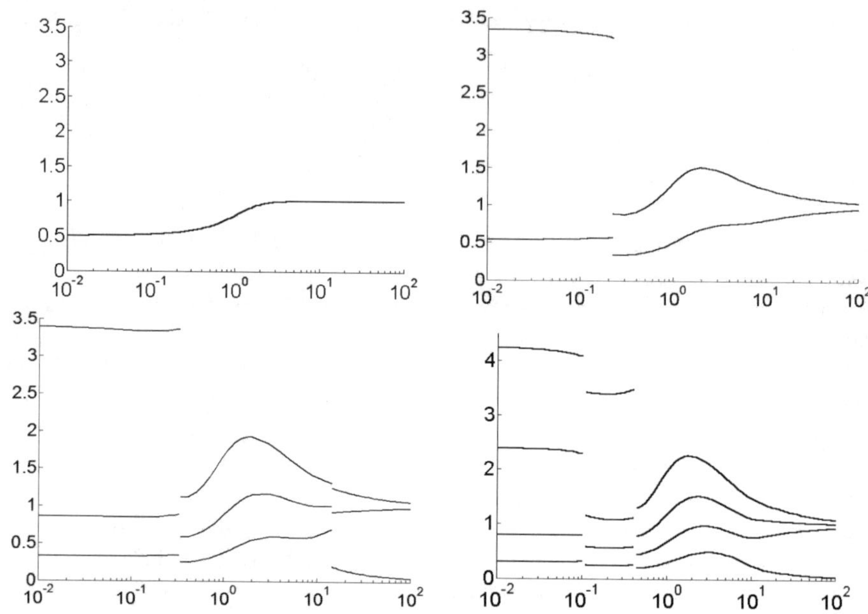

Fig. 1. Exact n-point locally D-optimal design $\{0, t_2^*(\lambda), \ldots, t_n^*(\lambda)\}$ with $b = 1$ for $n = 2$ (top left), $n = 3$ (top right), $n = 4$ (bottom left) and $n = 5$ (bottom right).

4 Maximin efficient D-optimal designs

Note that the implementation of locally optimal designs in practice requires a prior guess for the unknown parameters. This can raise confusion for an experimenter. The notion of maximin efficient designs seems to be more attractive and useful in practice; see Müller (1995).

The D-efficiency of a design ξ is given by

$$\mathrm{eff}_D(\xi) = \mathrm{eff}_D(\xi, a, b, \lambda) = \left[\frac{\det M(\xi, \bar{\beta})}{\det M(\xi_{loc}^*(\bar{\beta}), \bar{\beta})} \right]^{1/2}$$

where $\bar{\beta} = (a, b, \lambda)$ and ξ_{loc}^* is a locally D-optimal design. It is easy to see that the efficiency does not depend on a.

A design ξ^* is called a maximin (efficient) D-optimal design if it maximizes the worst D-efficiency over some set of the parameters Ω.

Analytical results about maximin D-optimal designs are given in Lemma 2.

Lemma 2. *Let* $\xi^* = \xi^*(\Omega) = \{t_1^*, \ldots, t_n^*\}$ *be a maximin D-optimal design for the exponential model with correlated observations. Then*
1) The first point of the design equals zero, that is $t_1^* = 0$.
2) The points of the design ξ^* *satisfy*

$$t_i^*(\gamma\Omega) = \frac{1}{\gamma} \, t_i^*(\Omega),$$

for any $\gamma > 0$.

The proof of Lemma 2 is similar to that of Lemma 1.

Consider a set Ω of the form

$$\Omega = \Omega(z) = \left\{ \bar{\beta} = (a, b, \lambda) : a = 1, \begin{array}{l} (1 - z)b_0 \leq b \leq (1 + z)b_0, \\ (1 - z)\lambda_0 \leq \lambda \leq (1 + z)\lambda_0 \end{array} \right\},$$

which seems appealing from a practical point of view. Values a_0 and λ are the initial guess and z can be interpreted as a relative error for the guess.

To study maximin designs $\xi^*(z)$ we implemented a special case of the functional approach introduced in Melas and Pepelyshev (2005). For example, suppose that $b_0 = 1$ and $\lambda_0 = 1$. Following this approach we obtain the expansions

$x_2(z) = 0.5088 - 0.1887z^2 - 0.0263z^4 - 0.4340z^6 + 12.7721z^8 + \dots,$

$x_3(z) = 1.3056 + 1.8505z^2 + 5.7658z^4 + 18.438z^6 + 89.7939z^8 + \dots$

which converge for $z \in [0, 0.4)$. These points are depicted in Figure 2, which also shows the dependence of the minimal efficiency of maximin designs and the equidistant design $\{0, 0.65, 1.3\}$ on z. We see that the maximin designs are more efficient than the equidistant design.

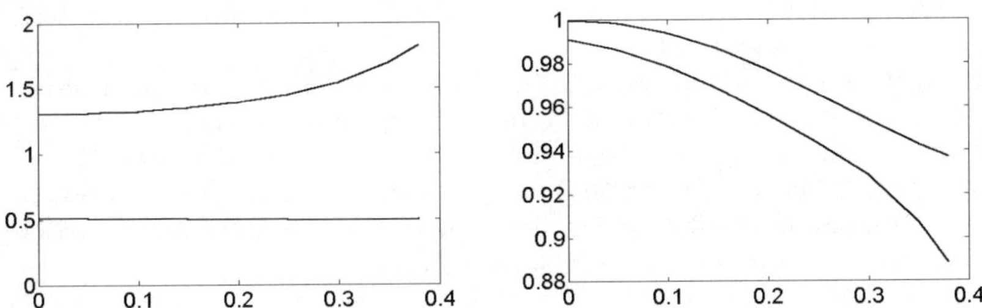

Fig. 2. Exact 3-point maximin D-optimal design $\xi^*(z) = \{0, t_2^*(z), t_3^*(z)\}$ with $b = 1$ for $n = 3$ (left), and the minimal efficiencies of the maximin design $\xi^*(z)$ and the equidistant design $\{0, 0.65, 1.3\}$ over $\Omega(z)$ (right).

Now we compare locally D-optimal designs supported on different numbers of points. Let $b = 1$, $\lambda = 1$. We use the efficiency defined by $\mathrm{eff}_n(\xi_p) = \sqrt{\det M(\xi_p)/\det M(\xi_n)}$. We obtain that $\mathrm{eff}_2(\xi_p)$ equals 1.125, 1.173, 1.196, 1.210, 1.218 for $p = 3, 4, 5, 6$ and 7. We see that the quantity of information is increasing very slowly with the number of design points.

References

Coleman M, Marks H (1998) Topics in dose-response modeling. J Food Protection 64(11):1550–1559

Dette H, Neugebauer HM (1997) Bayesian D-optimal designs for exponential regression models. J Statist Plann Inference 60(2):331–349

Dette H, Kunert J, Pepelyshev A (2006a) Exact optimal designs for weighted least squres analysis with correlated errors. Preprint RUB, Accepted Stat Sinica

Dette H, Martinez-Lopez I, Ortiz-Rodriguez I, Pepelyshev A (2006b) Maximin efficient design of experiment for exponential regression models. Preprint RUB, Accepted to JSPI

Fedorov VV, Hackl P (1997) Model-oriented design of experiments, Lecture Notes in Statistics, vol 125. Springer-Verlag, New York

Han C, Chaloner K (2003) D- and c-optimal designs for exponential regression models used in viral dynamics and other applications. J Statist Plann Inference 115(2):585–601

Hoel PG (1958) Efficiency problems in polynomial estimation. Ann Math Statist 29:1134–1145

Melas V, Pepelyshev A (2005) On representing maximin efficient designs by taylor series. Proceedings of the 5th StPetersburg Workshop on Simulation pp 479–484

Melas VB (2006) Functional approach to optimal experimental design, Lecture Notes in Statistics, vol 184. Springer, New York

Mukhopadhyay S, Haines LM (1995) Bayesian D-optimal designs for the exponential growth model. J Statist Plann Inference 44(3):385–397

Müller CH (1995) Maximin efficient designs for estimating nonlinear aspects in linear models. J Statist Plann Inference 44(1):117–132

Müller WG, Pazman A (2003) Measures for designs in experiments with correlated errors. Biometrika 90:423–434

Stehlík M (2005) Covariance related properties of d-optimal correlated designs. In: SM Ermakov VM, Pepelyshev A (eds) Proceedings of the 5th St.Petersburg Workshop on Simulations, St.Petersburg University Press, pp 645–652

Ucinski D, Atkinson AC (2004) Experimental design for processes over time. Studies in Nonlinear Dynamics and Econometrics 8(2), (Article 13). http://www.bepress.com/snde/vol8/iss2/art13

Determining the Size of Experiments for the One-way ANOVA Model I for Ordered Categorical Data

Dieter Rasch[1] and Marie Šimečková[2]

[1] University of Natural Resources and Applied Life Sciences, Vienna, Department of Landscape, Spatial and Infrastructure Sciences, IASC, Peter Jordanstr. 82, 1190 Wien, Austria
dieter.rasch@boku.ac.at
[2] Institute of Animal Science, Pratelstvi 815, 104 00 Prague Uhrineves, Czechia
simeckova.marie@vuzv.cz

Summary. The aim of the paper is to present a method of sample size determination for the Kruskal – Wallis test. The method is based on the concept of the relative effect between the two extreme distributions of those sampled and on the maxi-min size for the usual F-test.

Key words: size of experiment, ANOVA model I, ordered categorical data

1 Introduction

In this paper we combine ideas of Brunner and Munzel (2002) to describe the distance between distributions with results for determining the size of experiments as discussed in Herrendörfer et al (1997). We consider the one-way ANOVA model with a fixed factor. We assume that this fixed factor A has $a \geq 2$ levels. The null hypothesis to be tested is that the distribution of a random variable[3] $\boldsymbol{y}$ in all these levels is the same against the alternative that this is not the case.

Clearly the special case of the Wilcoxon two-sample test is covered also. Therefore our results could be compared with those from Chakraborti et al (2006) based on another approach.

In testing statistical hypotheses about a means of normal distributions the experimenter fixes a lower bound δ for the difference of practical interest (or importance; in application called the effect size). Further, the type-I-risk α of the test and its power $1 - \beta$ is fixed. If δ is given we also need prior information about the residual variance, σ^2, inside the groups. Even this is not needed, if

---

[3] Random variables are in bold print

the experimenter fixes the relative effect size[4] $\frac{\delta}{\sigma}$; in this paper this approach is used and w.l.o.g. put $\sigma = 1$.

2 The ANOVA model

2.1 Quantitative variables: the parametric F-test

The balanced model equation for a continuously distributed random variable $\boldsymbol{y}$ is written in the form:

$$\boldsymbol{y}_{ij} = E(\boldsymbol{y}_{ij}) + \boldsymbol{e}_{ij} = \mu + a_i + \boldsymbol{e}_{ij} \quad (i = 1, \ldots, a; \ j = 1, \ldots, n). \tag{1}$$

The main effects a_i of the factor A are real numbers, i.e. not random. The random errors $\boldsymbol{e}_{ij}$ are mutually independent with $E(\boldsymbol{e}_{ij}) = 0$, $var(\boldsymbol{e}_{ij}) = \sigma^2$ and $\sum_{i=1}^{a} a_i = 0$. Equal subclass numbers n are used because it is known that this is the optimal design for a given total sample size.

We want to design the experiment for testing the null hypothesis H_0: "All the a_i are equal" against the alternative hypothesis H_A: "At least two of the a_i's are different". If $\boldsymbol{y}$ has a normal distribution, the test statistic for testing the null hypothesis above is equal to

$$\boldsymbol{F} = \frac{\boldsymbol{MS}_A}{\boldsymbol{MS}_R}, \tag{2}$$

with $\boldsymbol{MS}_A$ and $\boldsymbol{MS}_R$ as the mean squares of factor A and residual, respectively. Under the null hypothesis F follows a (central) F-distribution with $f_1 = a - 1$ and $f_2 = a(n - 1)$ degrees of freedom. Otherwise it follows a non-central F-distribution with the same degrees of freedom but a non-centrality parameter λ. If the realization of $\boldsymbol{F}$ in (2) exceeds the $(1 - \alpha)$-quantile of the central F-distribution with f_1 and f_2 degrees of freedom the null hypothesis is rejected.

It was shown in Rasch and Guiard (2004) that the F-test is very robust against non-normality of continuous distributions and that we can use it for any continuous random variable. Therefore the following method for determining the minimal sample size can be used for any continuous random variable.

The power of the F-test depends on the non-centrality parameter λ of the F-distribution. It is proportional to the function of main effects

$$\sum_{i=1}^{a} (a_i - \bar{a})^2, \tag{3}$$

where $\bar{a} = \frac{1}{a} \sum_{i=1}^{a} a_i$. Of course if H_0 is true the expression (3) is equal to zero. If the a_i's are not all equal the value of λ depends on their values.

---

[4] In agricultural sciences δ/σ is often chosen from 1 up to 1.75 together with $\alpha = 0.05$ and $\beta = 0.2$.

The least favorable case from the point of view of the required sample size (leading to the maximal minimum number n for each factor level) is the case with the smallest possible value of λ if at least two of the a_i's are different. Let $a_{\max} = \max(a_i)$ be the largest and $a_{\min} = \min(a_i)$ be the smallest of the a effects a_i of the levels of A. Then (3) is minimized if the $a - 2$ remaining effects are equal to $(a_{\min} + a_{\max})/2$.

Using the triple $\{(a_{\max} - a_{\min})/\sigma; \alpha; \beta\}$ as the precision requirement for calculating of the minimal sample size we are always on the safe side. We call the corresponding minimal sample size the maxi-min size and denote it by $n_{\max}$.

2.2 Categorical variables: nonparametric Kruskal – Wallis test

The F-test for testing the hypothesis about the equality of means discussed in the previous section is based on the assumption that the observed variables are normally distributed and their distributions in different groups differ only in expected values. The Kruskal – Wallis test, which is considered here, can (but need not due to the above stated robustness) be used in cases when the normality assumption is questionable.

Let $\boldsymbol{y}_1, \ldots, \boldsymbol{y}_a$ be random variables with distribution functions $F_1, \ldots, F_a$; $\boldsymbol{y}_i$ corresponds to the observed variable in the i-th level of the factor A. We will test a more complex hypothesis "H_0: $F_1 = F_2 = \cdots = F_a$" against the alternative "H_0: $F_i \neq F_j$ for at least one pair of i, j".

The basic version of the Kruskal – Wallis test assumes that the distribution functions F_i's are continuous. In our case of categorical variables $\boldsymbol{y}_i$ this is not true and a corrected (for ties) Kruskal – Wallis tests is used.

More information about this test can be found e.g. in Lehmann (1975).

3 Ordered categorical variables and the relative effect

In the case of the ordered categorical variables the random variable $\boldsymbol{y}$ takes realizations belonging to r ordered categories $C_1 \prec C_2 \prec \cdots \prec C_r$ with $r > 1$. We used the symbol $\prec$ to denote the order relation. We need a measure for the distance between two distributions. For this we use the approach of BrunnerMunzel2002.

Definition 1. *For two random variables $\boldsymbol{y}_1$ and $\boldsymbol{y}_2$ with distribution functions $F_1(y)$ and $F_2(y)$ respectively, the probability*

$$p = P(\boldsymbol{y}_1 < \boldsymbol{y}_2) + \frac{1}{2}P(\boldsymbol{y}_1 = \boldsymbol{y}_2) = \int F_1 dF_2$$

is called the relative effect *of $\boldsymbol{y}_2$ with respect to $\boldsymbol{y}_1$. If $p = \frac{1}{2}$ we say that both distributions have equal tendency.*

The relative effect of $y_{\max} = \mu + a_{\max}$ with respect to $y_{\min} = \mu + a_{\min}$ is used to characterize the properties of the observed ordinal variables, together with the number of their categories r.

In our simulation experiment we generate ordered categorical variables by decomposition of the real line, on which a special continuous random variable takes its values, as described in the following definition.

Definition 2. *Assume a random variable x is continuously distributed. From it a new ordered categorical random variable y with r categories is derived, using a decomposition of the real line based on a set of values $\{\xi_1, \xi_2, \ldots, \xi_{r-1}\}$, $-\infty = \xi_0 < \xi_1 < \xi_2 < \cdots < \xi_{r-1} < \xi_r = +\infty$. Then $y = i$ when x lies in the interval $(\xi_{i-1}, \xi_i]$, $i = 1, \ldots, r$.*

Call the set $\{\xi_1, \xi_2, \ldots, \xi_{r-1}\}$ the support *of the decomposition.*

If F is the distribution function of a random variable x, then (for each decomposition $\{\xi_1, \xi_2, \ldots, \xi_{r-1}\}$) it is the case that $P(y = i) = F(\xi_i) - F(\xi_{i-1})$.

4 Relation between sample sizes for normally distributed variables and categorical variables

Let us assume a $(a \geq 2)$ continuously distributed random variables $x_1, \ldots, x_a$. We want to test whether their means are equal or there is at least one pair of these variables with different means.

Instead of these continuous variables, only the ordinal categorical variables $y_1, \ldots, y_a$ are observed. They are derived from the variables $x_1, \ldots, x_a$ using the decomposition based on the support $\{\xi_1, \xi_2, \ldots, \xi_{r-1}\}$, as is described in Definition 2 in Section 3.

The Kruskal – Wallis test is used to test the hypothesis of equal means. For assuring the appropriate type-II-risk β in the test, it is necessary to plan the experiment, i.e. determine the sample size (for the given significance level α). The explicit formula for the categorical variables is not known. In this section, the simulation to determine the type-II-risk β for a given sample size is described. Then some estimated formula is stated.

4.1 Discussed data

It is supposed that only the categorical variables are observed in the experiment. For the simulation experiment it is important to choose the mechanism of generating several distributions of the random variables of interest. We used six different distributions of the underlying continuous variable and, for each of them, two different supports of decompositions.

The variables of interest have in all a treatment groups the same type of distribution; they differ only in location (their expected values).

All the assumed underlying distributions are taken to have standard deviation equal to 1, this means that $\frac{\delta}{\sigma} = \delta$. They differ in the values of skewness and kurtosis. The first distribution is the normal distribution, i.e. both the skewness and the kurtosis are equal to 0. The second distribution is the uniform distribution in the interval $(-\sqrt{3}, \sqrt{3})$; its skewness is equal to 0, kurtosis to -1.2.

The other distributions arise from the Fleishman system, described in Rasch and Guiard (2004). This means that a random variable has the form $a + bx + cx^2 + dx^3$, where the x is a standard normally distributed random variable and a, b, c, d some given parameters. Information about the parameters and properties of the distribution used in our paper can be found in Table 1. For

Table 1. The parameters and properties of the used distributions.

No. of distr.	Skewness	Kurtosis	$c = -a$	b	d
1	0	3.75	0	0.748020807992	0.077872716101
2	0	7	0	0.630446727840	0.110696742040
3	1	1.5	0.163194276264	0.953076897706	0.006597369744
4	2	7	0.260022598940	0.761585274860	0.053072273491
5 (Normal)	0	0	0	1	0
6 (Uniform)	0	-1.2			

each of these distributions two different decompositions are explored. They are computed for distributions with zero expected values. First, the support points are equally distributed over the area in which 99 % of observations lie. Second, equal percentages of observations lie in all categories. For the uniform distribution, these two versions are identical.

4.2 Simulation

Consider one of the distributions described in the previous paragraph. Let us assume the expected value of the variable in the first group is $\mu_1 = -\delta/2$, in the second group is $\mu_2 = +\delta/2$, and in the (possible) remaining groups is zero.

The test was performed at the significance level $\alpha = 0.05$.

Let us assume a given underlying distribution, a given support of decomposition, a given number of groups a, and a given difference between the minimal and the maximal expected values of δ. Then the type-II-risk β was evaluated for each "reasonable" sample size n. "Reasonable" means that the largest assumed n is the maxi-min size for normally distributed variables and with β equal to 0.40; and the smallest is the first n for which the type-II-risk is smaller than 0.05. For subsequent analyses only β smaller than 0.40 was used.

For a fixed sample size n the first two steps of the simulation were:

1. The continuous random samples of size n were generated for each group with the appropriate expected value. Then they were transformed to the categorical variables, using the given support of decomposition.
2. The Kruskal – Wallis test was performed and the result was recorded.

These two steps were repeated $10,000$ times. The actual (estimated) type-II-risk β is for sample size n equal to the proportion of the non-significant tests in these repetitions. The simulation was performed using the environment R developed by R Development Core Team (2005).

The values for the number of groups a were chosen to be $2, 3, \ldots, 10$ for the normal distribution and $2, 3, 4, 6, 8, 10$ for the others. The difference between the minimal and the maximal expected values δ were chosen as $1.67, 1.25, 1.11$ and 1 (i.e. the standard deviation inside the groups σ was equal to $(0.6, 0.8, 0.9, 1)$ times δ). The number of categories of the ordinal variables were $3, 4, 5, 10, 50$ with two different types of decomposition (i.e. two different values of relative effects) described at the end of Section 4.1.

4.3 Formula

It was found that the required sample size depends on the maxi-min sample size computed for the ANOVA F-test and normally distributed variables almost linearly, for the given a, $\frac{\sigma}{\delta}$, the relative effect p, and the number of categories r. Fits of many linear models were explored for estimation of the required sample size. The model below was chosen as the most acceptable (good fit and not too many parameters).

Given the type-I-risk $\alpha = 0.05$, the maxi-min sample size for the Kruskal – Wallis test can be computed as

$$n(\beta) = 3.054 \cdot n_0(\beta) - 47.737 \cdot \frac{\delta}{\sigma} + 51.288 \cdot p^2 + 82.050 \cdot \frac{1}{r} +$$

$$+2.336 \cdot n_0(\beta) \cdot \frac{\delta}{\sigma} - 7.428 \cdot n_0(\beta) \cdot p^2 - 0.535 \cdot n_0(\beta) \cdot \frac{1}{r} +$$

$$+29.708 \cdot \frac{\delta}{\sigma} \cdot p^2 + 56.102 \cdot \frac{\delta}{\sigma} \cdot \frac{1}{r} - 223.770 \cdot p^2 \cdot \frac{1}{r}, \tag{4}$$

where the $n_0(\beta) = n_0(\beta, a, \delta, \sigma)$ is the maxi-min sample size for the F-test.

Formula (4) fits the sample size very well; only 4.8% of the residuals are larger than 20% of the relevant fitted value. Further, 9.0% percent are higher than 15% of the fitted value, 16.6% percent are higher than 10% and 30.8% percent of the residuals are higher than 5% of the fitted value.

Negative residuals are not so dangerous because it follows that the actual type-II-risk would be lower than that required. Using formula (4) 48% of the residuals are negative.

In Table 2, are listed the actual sample sizes and the sample sizes estimated using relation (4) for $\beta = 0.20$ and for some chosen values of the parameters. In

Table 2. Comparison of required sample sizes both simulated and calculated by (4) for $\beta = 0.2$ and some values of the other parameters. The columns record consecutively the number of groups, $\frac{\delta}{\sigma}$, identification of underlying distribution, the relative effect of the distribution of thecategorical variables and their number of categories, maxi-min sample size of the F-test for normal variables and the maxi-min sample sizes for the Kruskal – Wallis test based on the simulation and calculated by formula (4).

Groups	$\frac{\delta}{\sigma}$	Distribution	Rel. effect	Categories	$n_0(\beta)$	$nSIM$	$nFIT$
2	1	1	0.66	3	16.71	31	35
2	1	1	0.78	5	16.71	14	15
2	1.67	1	0.77	3	6.76	11	11
2	1.67	1	0.89	5	6.76	7	7
6	1	1	0.66	3	26.59	47	55
6	1	1	0.78	5	26.59	22	22
6	1.67	1	0.77	3	10.2	16	19
6	1.67	1	0.89	5	10.2	10	10
2	1	3	0.69	3	16.71	28	30
2	1	3	0.77	5	16.71	17	17
2	1.67	3	0.8	3	6.76	11	10
2	1.67	3	0.88	5	6.76	7	7
6	1	3	0.69	3	26.59	44	46
6	1	3	0.77	5	26.59	27	26
6	1.67	3	0.8	3	10.2	16	16
6	1.67	3	0.88	5	10.2	11	10

Figure 1 the properties of residuals in model (4) can be seen. The residuals increase with increasing (simulated and estimated) sample size. The proportion of residuals in the estimated sample size is almost constant with increasing estimated sample size, if the sample size is larger than approximately 25.

4.4 Discussion

The formula for determination of the required sample size, given in the previous paragraph, was derived for some specific cases. The question about legitimacy of its generalization arises. The categorical variables were generated by decomposition of several continuous variables. From a practical point of view this does imply a loss ofgenerality, because usually a continuous property is measured on an ordinal scale and the six continuous distributions used, with different shapes and two decompositions, provide eleven different distributions for categorical variables.

It should be remembered that the formula has been checked for four values of δ/σ between 1 and 1.7. It is hoped that the formula can be interpolated for all values in this interval, which is that usually used in applications. Similarly, it is assumed that the number of categories r can be interpolated for all integer

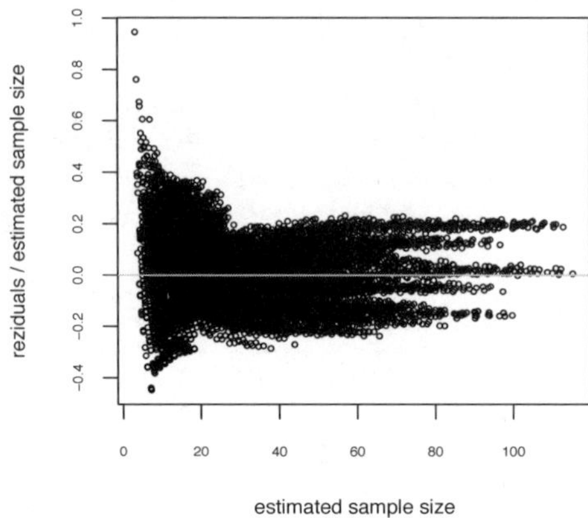

Fig. 1. Relation between the residuals of model (4) and the required sample size estimated from this model. The ratio of the residuals and the estimated sample sizes is plotted on the y-axis, the estimated sample sizes are plotted on the x-axis.

values between 3 and 50. With $r \to \infty$ there is decreasing influence of r on the required size of an experiment (the distribution tends to a continuous one which is reflected by the presence of $\frac{1}{r}$ in the formula). Therefore for r larger than 50 the formula can also be used.

To summarize, the required size of an experiment with categorical variables can, for given type-I-risk $\alpha = 0.05$, type-II-risk β from the interval $[0.05, 0.4]$, and $\frac{\delta}{\sigma}$ in an interesting range of practical values, be calculated by formula (4) for numbers of compared groups ranging between 2 and 10. There are no restrictions on the other parameters.

References

Brunner E, Munzel U (2002) Nichtparametrische Datenanalyse - unverbundene Stichproben. Springer, Berlin

Chakraborti S, Hong B, van de Wiel M (2006) A note on sample size determination for a nonparametric test of location. Technometrics 48:88–94

Herrendörfer G, Rasch D, Schmidt K, Wang M (1997) Determination of the size of an experiment for the F-test in the analysis of variance – mixed model. In: Wegman, E. J. and Azen, P. A.: Computing Science and Statistics, Pasadena, vol 29, 2, pp 547–550

Lehmann EL (1975) Nonparametrics: Statistical Methods Based on Ranks. Holden-Day, INC., San Francisco

R Development Core Team (2005) R: A language and environment for statistical computing. R Foundation for Statistical Computing, Vienna, Austria

Rasch D, Guiard V (2004) The robustness of parametric statistical methods. Psychology Science 46:175–208

Bayesian D_s-Optimal Designs for Generalized Linear Models with Varying Dispersion Parameter

Edmilson Rodrigues Pinto[1] and Antonio Ponce de Leon[2]

[1] Department of Mathematics, Federal University of Uberlândia, Av. João Naves de Ávila 2121, Santa Mônica, Uberlândia, MG - Brazil
`Edmilson@famat.ufu.br`

[2] Department of Epidemiology, Institute of Social Medicine, Rio de Janeiro State University, Rua São Francisco Xavier, 524 / 7013D, Maracanã, Rio de Janeiro, 20550 900 Brazil   `ponce@ims.uerj.br`

Summary. In this article we extend the theory of optimum designs for generalized linear models, addressing the optimality of designs for parameter estimation in a location-dispersion model when either not all p parameters in the mean model or not all q parameters in the dispersion model are of interest. The criterion of Bayesian D_s-optimality is adopted and its properties are derived. The theory is illustrated with an example from the coffee industry.

Key words: D_s-optimum designs, Bayesian designs, extended quasi-likelihood

1 Introduction

Sometimes not all parameters in a statistical model are required to estimate precisely. Rather the focus may be on estimating a given number of parameters very precisely, while the remaining are treated as nuisance parameters. Alternatively one can think of two competing models: the first in which all model parameters are regarded to be important, the full model, whereas under the rival model some of these parameters are set to zero, so that the latter model becomes nested within the former. In order to discriminate between the rival nested models, the researcher seeks a design for which the sets of expected responses under the competing models lie as far apart from each other as possible. This criterion of optimality is known in the literature of optimum designs as T-optimality, however for the case of nested models, it is equivalent to the criterion of D_s-optimality.

The joint modeling of mean and dispersion (JMMD) in generalized linear models (GLMs) consists of regarding a full model specification for the response variance, in parallel to the usual specification, i.e. a link function, a linear predictor, and an error distribution assumption for the response. This is important when the response shows some degree of over or under dispersion. In the complete joint model p parameters are regarded to describe the mean response and q parameters to account for the dispersion, thus the parametric space consists of $p + q$ parameters.

However, in what follows the focus is on estimating s_m and s_d parameters, respectively to the mean and the dispersion models, where either $s_m < p$ or $s_d < q$. To be more specific in this article the problem of finding Bayesian D_s-optimum designs for GLMs with a varying dispersion parameter is discussed and an illustration is provided. This work extends that of Pinto and Ponce de Leon (2004) in which they used extended quasi likelihood in order to build a theory that allows searching for Bayesian D-optimum designs for GLMs with overdispersion. Under that approach, the likelihood is supposed to be unknown, however the specification of the first two moments is known.

2 The experimental matrix and the standardized variance for the JMMD

Let y_i be the ith response of interest, let $\boldsymbol{f}_i(\boldsymbol{x}_i)$ be the $p \times 1$ vector representing the ith setting of the covariates presumed to influence the expected value of the response, and let $\boldsymbol{g}_i(\boldsymbol{z}_i)$ be the $q \times 1$ vector representing the ith setting of the factors presumed to influence the response variance. We allow $\boldsymbol{z}_i$ to contain some or all of the components of $\boldsymbol{x}_i$ as well as others.

Suppose the distribution of the response is unknown, nevertheless suppose $E(Y_i) = \mu_i$ and $Var(Y_i) = \phi_i V(\mu_i)$ and that full expressions can be specified for the mean and variance functions as follows. Let k be the link function for the mean model, i.e. $\eta_i = k(\mu_i) = \boldsymbol{f}_i^t\boldsymbol{\beta}$, where $\boldsymbol{\beta}$ is a $p \times 1$ vector of unknown parameters; and let h be the link function for the dispersion model, i.e. $\tau_i = h(\phi_i) = \boldsymbol{g}_i^t\boldsymbol{\gamma}$, where $\boldsymbol{\gamma}$ is a $q \times 1$ vector of the unknown parameters.

We regard the Extended Quasi Likelihood (EQL), proposed by Nelder and Pregibon (1987) as a criterion to fit the joint model, since the distribution of the response vector is unknown and ϕ is allowed to vary.

The notation used in the models is the following. For $i = 1, \dots, n$,
$$\boldsymbol{X}^t = (\boldsymbol{f}_1(\boldsymbol{x}_1), \dots, \boldsymbol{f}_n(\boldsymbol{x}_n)) \text{ and } \boldsymbol{Z}^t = (\boldsymbol{g}_1(\boldsymbol{z}_1), \dots, \boldsymbol{g}_n(\boldsymbol{z}_n)), \quad \text{with}$$
$$\boldsymbol{f}_i^{\ t}(\boldsymbol{x}_i) = (f_{i1}(\boldsymbol{x}_i), \dots, f_{ip}(\boldsymbol{x}_i)) \text{ and } \boldsymbol{g}_i^{\ t}(\boldsymbol{z}_i) = (g_{i1}(\boldsymbol{z}_i), \dots, g_{iq}(\boldsymbol{z}_i))$$
where $f_{i1}(\boldsymbol{x}_i) = g_{i1}(\boldsymbol{z}_i) = 1 \quad \forall i$.

Using EQL as the estimation criterion and assuming a gamma distribution for the dispersion (see McCullagh and Nelder (1989)), the pseudo Fisher information matrix is as follows.

$$I(\boldsymbol{X}, \boldsymbol{Z}|\boldsymbol{\beta}, \boldsymbol{\gamma}) = \begin{bmatrix} \boldsymbol{X}^t\boldsymbol{W}\boldsymbol{X} & 0 \\ 0 & \boldsymbol{Z}^t\boldsymbol{V}\boldsymbol{Z} \end{bmatrix} \tag{1}$$

where $\boldsymbol{W} = diag(w_i)$, $\boldsymbol{V} = diag(v_i)$, $w_i = \left(\frac{\partial \mu_i}{\partial \eta_i}\right)^2 \frac{1}{\phi_i V(\mu_i)}$, and $v_i = \left(\frac{\partial \phi_i}{\partial \tau_i}\right)^2 \frac{1}{2\phi_i^2}$. Here $diag(.)$ represents the diagonal matrix.

The experimental matrix $\boldsymbol{M}(\boldsymbol{\theta}|\xi)$ is the expectation w.r.t. the design measure ξ on $\mathcal{X}$ of the information per observation,

$$\boldsymbol{M}(\boldsymbol{\theta}|\xi) = \int_{\mathcal{X}} \boldsymbol{I}(\boldsymbol{u}|\boldsymbol{\theta})d\xi(\boldsymbol{u}) \tag{2}$$

Regard the pseudo Fisher information, given in (3), and the following discrete design.

$$\xi = \left\{ \begin{array}{ccc} \boldsymbol{x}_1 & \ldots & \boldsymbol{x}_n \\ p_1 & \ldots & p_n \end{array} \right\} \tag{3}$$

where $(\boldsymbol{x}_1, \ldots, \boldsymbol{x}_n)$ are the design points and $(p_1, \ldots, p_n)$ are the associated design weights. Then the experimental matrix is as follows.

$$\boldsymbol{M}(\boldsymbol{\theta}|\xi) = \begin{bmatrix} \boldsymbol{X}^t \boldsymbol{W} \boldsymbol{P} \boldsymbol{X} & 0 \\ 0 & \boldsymbol{Z}^t \boldsymbol{V} \boldsymbol{P} \boldsymbol{Z} \end{bmatrix}, \tag{4}$$

where $\boldsymbol{P} = diag(p_i)$ for $i = 1, \ldots, n$.

Regarding the joint model and a discrete design like (3) the standardized variance related to the ith design point, corresponding to the ith row of matrices $\boldsymbol{X}$ and $\boldsymbol{Z}$ is given as follows.

$$d_i = w_i \boldsymbol{f}_i^t(\boldsymbol{x}_i)(\boldsymbol{X}^t \boldsymbol{W} \boldsymbol{P} \boldsymbol{X})^{-1} \boldsymbol{f}_i(\boldsymbol{x}_i) + v_i \boldsymbol{g}_i^t(\boldsymbol{z}_i)(\boldsymbol{Z}^t \boldsymbol{V} \boldsymbol{P} \boldsymbol{Z})^{-1} \boldsymbol{g}_i(\boldsymbol{z}_i) . \tag{5}$$

For more details about the construction of the experimental matrix and the standardized variance for the JMMD, see Pinto and Ponce de Leon (2004).

3 D_s-optimality

D_s-optimality concerns maximizing precision of parameter estimates for only a reduced number of the complete set of unknown model parameters. As stated previously an equivalent criterion is that of T-optimality to discriminate between two nested models. For more about the D_s-optimality criterion, see Atkinson and Donev (1992).

The aim of this paper is to develop the required theory so as Bayesian D_s-optimum designs for GLMs with a varying dispersion parameter can be obtained.

In the JMMD two GLMs are proposed, the first for the mean and the second for the dispersion. Let the linear predictor for the mean model be $\eta(\boldsymbol{\beta}, \boldsymbol{x}) = \boldsymbol{f}^t(\boldsymbol{x})\boldsymbol{\beta}$, let $\tau(\boldsymbol{\gamma}, \boldsymbol{z}) = \boldsymbol{g}^t(\boldsymbol{z})\boldsymbol{\gamma}$ be the linear predictor for the dispersion model, where $\boldsymbol{\beta}$ and $\boldsymbol{\gamma}$ are vectors of unknown parameters. The vectors

$f^t(x)$ and $g^t(z)$ are of dimension $1 \times p$ and $1 \times q$, respectively. Thus, the experimental matrix for the JMMD, denoted by M_C is represented as follows.

$$M_C = \begin{bmatrix} M & 0 \\ 0 & D \end{bmatrix} = \begin{bmatrix} M_{11} & M_{12} & 0 & 0 \\ M_{21} & M_{22} & 0 & 0 \\ 0 & 0 & D_{11} & D_{12} \\ 0 & 0 & D_{21} & D_{22} \end{bmatrix} \tag{6}$$

where $M_{p \times p}$ is the experimental matrix for the mean and $D_{q \times q}$ is the experimental matrix for the dispersion, whereas the partitions are explained below.

Provided that the main interest lies in estimating s_m parameters related to the mean and s_d parameters related to the dispersion, and that $1 \leq s_m < p$ and $1 \leq s_d < q$, the dimensions of M_{11}, M_{12}, M_{21} and M_{22} are, respectively, $s_m \times s_m$, $s_m \times (p-s_m)$, $(p-s_m) \times s_m$ and $(p-s_m) \times (p-s_m)$; and the dimensions of D_{11}, D_{12}, D_{21} and those of D_{22} are, respectively, $s_d \times s_d$, $s_d \times (q - s_d)$, $(q - s_d) \times s_d$ and $(q - s_d) \times (q - s_d)$.

In addition to the above other possible configurations for the number of parameters are: (i)$s_m = p$ and $s_d = q$, (ii)$s_m = p$ and $s_d < q$, and (iii)$s_m < p$ and $s_d = q$. The first is the case of D-optimality, as for the second and third cases, either M_{12}, M_{21}, and M_{22} or D_{12}, D_{21}, and D_{22} do not exist, thus in the following expressions some adjustments must be made. Thus, bearing in mind the restrictions imposed by the matrices dimensions, matrix M_C^{-1} can be written as follows.

$$M_C^{-1} = \begin{bmatrix} M^{11} & M^{12} & 0 & 0 \\ M^{21} & M^{22} & 0 & 0 \\ 0 & 0 & D^{11} & D^{12} \\ 0 & 0 & D^{21} & D^{22} \end{bmatrix} \tag{7}$$

where, for $i = j = 1, 2$, M^{ij} and D^{ij} have the same dimensions as M_{ij} and D_{ij}, respectively. Since we are interested in s_m parameters in the model for the mean and in s_d parameters in the dispersion model, we consider the matrix A^t, whose dimensions are $(s_m + s_d) \times (p + q)$, i.e.

$$A^t = \begin{bmatrix} I_{s_m} & 0 & 0 & 0 \\ 0 & 0 & I_{s_d} & 0 \end{bmatrix} \tag{8}$$

With a bit of algebra, we find that $A^t M_C^{-1} A$, with dimensions $(s_m + s_d) \times (s_m + s_d)$, is expressed as follows.

$$A^t M_C^{-1} A = \begin{bmatrix} M^{11} & 0 \\ 0 & D^{11} \end{bmatrix} \tag{9}$$

Thus, based on well-known results from linear algebra, $|A^t M_C^{-1} A| = |M^{11}||D^{11}| = |(M_{11} - M_{12} M_{22}^{-1} M_{21})^{-1}||(D_{11} - D_{12} D_{22}^{-1} D_{21})^{-1}| = |M_{11} -$

$M_{12}M_{22}^{-1}M_{21}|^{-1}|D_{11} - D_{12}D_{22}^{-1}\,D_{21}|^{-1}$. Therefore, the criterion function has the form:

$$\varphi_s(M_C) = \ln\left\{\frac{|M|}{|M_{22}|}\right\} + \ln\left\{\frac{|D|}{|D_{22}|}\right\} \qquad (10)$$

In order to compute the criterion function, besides M and D, we only need to obtain M_{22} and D_{22}, i.e. the matrices related to the covariates which we do not have interest.

Let the function $\varphi_s : \mathcal{M} \longrightarrow \mathbb{R}$, where $\mathcal{M} = \{M_C(\theta|\xi) : \theta \in \Theta; \xi \in \Xi\}$, Θ is the parametric space, and Ξ is the class of all probability measures on the experimental region $\mathcal{X}$. Under JMMD, the Fréchet derivative of φ_s at M_{C1} in the direction of M_{C2}, where $M_{C1} = M_C(\theta|\xi_1)$, $M_{C2} = M_C(\theta|\xi_2)$ and $\xi_1, \xi_2 \in \Xi$, is as follows.

$$\begin{aligned}
F_{\varphi s}(M_{C1}, M_{C2}) = Tr(M_2 M_1^{-1}) &- Tr(M_{2,22} M_{1,22}^{-1}) \\
&+ Tr(D_2 D_1^{-1}) - Tr(D_{2,22} D_{1,22}^{-1}) - s
\end{aligned} \qquad (11)$$

with $s = s_m + s_d$.

The information matrix for the ith observation is: $I_i(u_i|\theta) = \sum_{j=1}^{2} h_{ij}h_{ij}^{t}$, with $h_{i1}^{t} = (f_i^t(x_i)(w_i)^{1/2}, 0^t)$, $h_{i2}^{t} = (0^t, (v_i)^{1/2}g_i^t(z_i))$, $u_i^{t} = (f_i^t(x_i), g_i^t(z_i))$ and $\theta^t = (\beta^t, \gamma^t)$. The Fréchet derivative for the ith observation, taking $M_2 = f_i(x_i)w_i f_i^t(x_i)$, $M_{2,22} = f_{2i}(x_i)w_i f_{2i}^t(x_i)$, $D_2 = g_i(z_i)v_i g_i^t(z_i)$, $D_{2,22} = g_{2i}(z_i)v_i g_{2i}^t(z_i)$, $M_1^{-1} = M^{-1}$, $M_{1,22}^{-1} = M_{22}^{-1}$, $D_1^{-1} = D^{-1}$ and $D_{1,22}^{-1} = D_{22}^{-1}$ is given by:

$$\begin{aligned}
F_{\varphi s_i}(M_{C1}, M_{C2}) = Tr(f_i(x_i)w_i f_i^t(x_i)M^{-1}) &+ Tr(g_i(z_i)v_i g_i^t(z_i)D^{-1}) \\
- Tr(f_{2i}(x_i)w_i f_{2i}^t(x_i)M_{22}^{-1}) &- Tr(g_{2i}(z_i)v_i g_{2i}^t(z_i)D_{22}^{-1}) - s = \\
= w_i f_i^t(x_i)M^{-1}f_i(x_i) &+ v_i g_i^t(z_i)D^{-1}g_i(z_i) \\
- w_i f_{2i}^t(x_i)M_{22}^{-1}f_{2i}(x_i) &- v_i g_{2i}^t(z_i)D_{22}^{-1}g_{2i}(z_i) - s.
\end{aligned} \qquad (12)$$

Thus, the standardized variance for the ith observation is the following.

$$\begin{aligned}
d_{s_i}(x_i, \xi) = w_i[f_i^t(x_i)M^{-1}f_i(x_i) &- f_{2i}^t(x_i)M_{22}^{-1}f_{2i}(x_i)] \\
+ v_i[g_i^t(z_i)D^{-1}g_i(z_i) &- g_{2i}^t(z_i)D_{22}^{-1}g_{2i}(z_i)].
\end{aligned} \qquad (13)$$

For the D_s-optimum design ξ^*, with $s = s_m + s_d$, we must have $d_s(x, \xi^*) \leq s$, $\forall x \in \mathcal{X}$, where equality occurs at the optimal design support points. In order to compute the standardized variance for the D_S-optimality criterion we need only find the matrices M_{22} and D_{22} of the partitions of M and D, where M_{22} and D_{22} are generated by the columns corresponding to the parameters of no interest. This result eases the computation burden of the problem.

4 Example: application to the coffee industry

This example was taken from Pinto (2005), when applying JMMD to data rising from a experiment based on a complete 2^3 factorial, for which the objective was to identify factors affecting the response and factors affecting the dispersion, in a problem related with the coffee industry. The response variable is the amount of trigoneline found in the coffee and the factors were: temperature of drying (x_1), temperature of toasting (x_2), and air speed in the drying of the coffee (x_3). The levels considered for x_1 were 300°C (high level) and 100°C (low level); for x_2 were 600°C (high level) and 300°C (low level); and for x_3 were 1850 rpm (high level) and 1300 rpm (low level). The data are shown in Table 1.

Table 1. Factorial 2^3 for trigoneline response

x_1	x_2	x_3	Y
1	1	1	0.38 0.45 0.40
1	1	-1	0.63 0.59 0.65
1	-1	1	0.73 0.68 0.66
1	-1	-1	0.69 0.68 0.70
-1	1	1	0.39 0.37 0.40
-1	1	-1	0.65 0.65 0.64
-1	-1	1	0.70 0.71 0.75
-1	-1	-1	0.67 0.68 0.79

Pinto (2005) tackles the problem of selecting the covariates that supposedly affect the response and those affecting the dispersion, using JMMD. The identity link function together with the variance function $V(\mu) = 1$ were considered for the mean. As for the dispersion a GLM from the gamma family of distributions and a logarithmic link function were considered.

The final models estimated for the mean and dispersion were as follows.

$$\hat{\mu} = 2.374 - 0.374x_2 - 0.167x_3 - 0.261x_2x_3 \tag{14}$$

$$\hat{\phi} = \hat{\sigma}^2 = \exp\{-6.240 - 1.106x_2x_3\}. \tag{15}$$

Now suppose that the central interest lies in testing whether the covariates x_2, x_3 and their interaction term x_2x_3 are indeed significant for the response and whether the interaction x_2x_3 is significant for the dispersion.

The systematic components for the mean and dispersion models are $\eta(\boldsymbol{x}, \boldsymbol{\beta}) = \beta_0 + \beta_1x_2 + \beta_2x_3 + \beta_3x_2x_3$ and $\tau(\boldsymbol{x}, \boldsymbol{\gamma}) = \gamma_0 + \gamma_1x_2x_3$. However we are only interested in the parameters β_1, β_2 and β_3 in the mean model $(s_m = 3)$ and γ_1 in the dispersion model $(s_d = 1)$, hence the matrix $\boldsymbol{M}_{22}$ is formed by the parameters of no interest in the mean model, namely, only

β_0 whereas $\boldsymbol{D}_{22}$ is formed by the parameters of no interest in the dispersion model, namely, only γ_0. Further, there are four parameters in the mean model and two parameters in the dispersion model, hence $\boldsymbol{M}$ has dimension 4×4 and $\boldsymbol{D}$, 2×2.

A prior distribution for the parameters related to the mean and dispersion models was constructed based on the estimates found in the JMMD, as shown in equations (14) and (15). We use a uniform discrete prior distribution for both models. The process of construction of a prior distribution was the same as in Pinto and Ponce de Leon (2004) in which α is deducted and added to the (estimated) parameter values related to the mean, and δ to the (estimated) parameter values related to the dispersion. The values for α and δ were 0.25. The Bayesian D_S-optimal design was:

$$\xi_S^* = \left\{ \begin{matrix} (1.0,\,1.0) & (1.0,\,0.7) & (-1.0,\,-1.0) & (-1.0,\,1.0) & (-1.0,\,0.7) & (1.0,\,-1.0) & (-0.7,\,-1.0) & (0.7,\,1.0) \\ 0.15 & 0.08 & 0.15 & 0.19 & 0.08 & 0.19 & 0.08 & 0.08 \end{matrix} \right\}$$

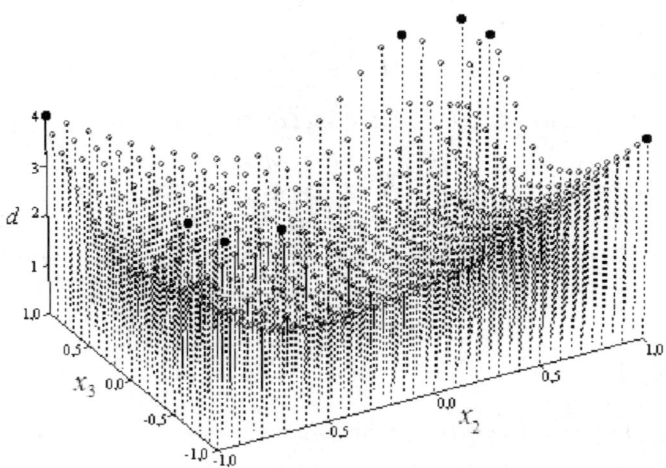

Fig. 1. Graph of standardized variance

Figure 1 shows the the standardized variance as a function of the covariates x_2 and x_3. Notice that at the optimum design points (full points in Figure 1) the standardized variance attains 4.0, the number of parameters that we are interested in estimating in the models for the mean and dispersion, and that nowhere in the design region is the standardized variance greater than 4.0.

5 Final considerations

We discussed and implemented the criterion of Bayesian D_S-optimality when there are two interlinked GLMs, one for the response and one for the dispersion. The main interest lies in estimating only s_m parameters in the model for the mean and s_d parameters in the dispersion model. In case all parameters of the mean and dispersion models are of interest, then Bayesian D-optimality could be applied, as shown in Pinto and Ponce de Leon (2004). From the results in this paper, local D_S-optimal designs are obtained in a straightforward manner.

Appendix: Fréchet derivative for the D_s-optimality

In order to prove result (11), we must calculate the Fréchet derivative for the D_S-optimal criterion applied to the JMMD. Using φ_S as the criterion function, its Fréchet derivative is obtained in the following.

$$F_{\varphi_S}(\boldsymbol{M}_{C1}, \boldsymbol{M}_{C2}) = \lim_{\epsilon \to 0^+} \frac{1}{\epsilon} \{\varphi_S \{(1 - \epsilon)\boldsymbol{M}_{C1} + \epsilon\boldsymbol{M}_{C2}\} - \varphi_S(\boldsymbol{M}_{C1})\} =$$

$$\lim_{\epsilon \to 0^+} \frac{1}{\epsilon}\{\psi_p\{(1 - \epsilon)\boldsymbol{M}_1 + \epsilon\boldsymbol{M}_2\} - \psi_{p-s_m}\{(1 - \epsilon)\boldsymbol{M}_{1,22} + \epsilon\boldsymbol{M}_{2,22}\} + \psi_q\{(1 - \epsilon)\boldsymbol{D}_1 + \epsilon\boldsymbol{D}_2\} - \psi_{q-s_d}\{(1 - \epsilon)\boldsymbol{D}_{1,22} + \epsilon\boldsymbol{D}_{2,22}\} - [\psi_p(\boldsymbol{M}_1) - \psi_{p-s_m}(\boldsymbol{M}_{1,22}) + \psi_q(\boldsymbol{D}_1) - \psi_{q-s_d}(\boldsymbol{D}_{1,22})]\} =$$

$$\lim_{\epsilon \to 0^+} \frac{1}{\epsilon}\{\psi_p\{(1-\epsilon)\boldsymbol{M}_1 + \epsilon\boldsymbol{M}_2\} - \psi_p(\boldsymbol{M}_1)\} - \lim_{\epsilon \to 0^+} \frac{1}{\epsilon}\{\psi_{p-s_m}\{(1-\epsilon)\boldsymbol{M}_{1,22} + \epsilon\boldsymbol{M}_{2,22}\} - \psi_{p-s_m}(\boldsymbol{M}_{1,22})\} + \lim_{\epsilon \to 0^+} \frac{1}{\epsilon}\{\psi_q\{(1 - \epsilon)\boldsymbol{D}_1 + \epsilon\boldsymbol{D}_2\} - \psi_q(\boldsymbol{D}_1)\} - \lim_{\epsilon \to 0^+} \frac{1}{\epsilon}\{\psi_{q-s_d}\{(1 - \epsilon)\boldsymbol{D}_{1,22} + \epsilon\boldsymbol{D}_{2,22}\} - \psi_{q-s_d}(\boldsymbol{D}_{1,22})\} = F_{\psi_p}(\boldsymbol{M}_1, \boldsymbol{M}_2) - F_{\psi_{p-s_m}}(\boldsymbol{M}_{1,22}, \boldsymbol{M}_{2,22}) + F_{\psi_q}(\boldsymbol{D}_1, \boldsymbol{D}_2) - F_{\psi_{q-s_d}}(\boldsymbol{D}_{1,22}, \boldsymbol{D}_{2,22}) = Tr(\boldsymbol{M}_2\boldsymbol{M}_1^{-1}) - p + Tr(\boldsymbol{D}_2\boldsymbol{D}_1^{-1}) - q - [Tr(\boldsymbol{M}_{2,22}\boldsymbol{M}_{1,22}^{-1}) - (p - s_m)] - [Tr(\boldsymbol{D}_{2,22}\boldsymbol{D}_{1,22}^{-1}) - (q - s_d)] = Tr(\boldsymbol{M}_2\boldsymbol{M}_1^{-1}) + Tr(\boldsymbol{D}_2\boldsymbol{D}_1^{-1}) - Tr(\boldsymbol{M}_{2,22}\boldsymbol{M}_{1,22}^{-1}) - Tr(\boldsymbol{D}_{2,22}\boldsymbol{D}_{1,22}^{-1}) - (s_m + s_d) = Tr(\boldsymbol{M}_2\boldsymbol{M}_1^{-1}) + Tr(\boldsymbol{D}_2\boldsymbol{D}_1^{-1}) - Tr(\boldsymbol{M}_{2,22}\boldsymbol{M}_{1,22}^{-1}) - Tr(\boldsymbol{D}_{2,22}\boldsymbol{D}_{1,22}^{-1}) - s.$$

References

Atkinson AC, Donev AN (1992) Optimum Experimental Designs. Clarendon Press, Oxford

McCullagh P, Nelder JA (1989) Generalized Linear Models, 2nd edition. Chapman and Hall, London

Nelder JA, Pregibon D (1987) An extended quasi-likelihood function. Biometrika 74:221–231

Pinto ER (2005) Optimum experimental designs for generalized linear models with varying dispersion parameter (in portuguese). PhD thesis, Federal University of Rio de Janeiro, Rio de Janeiro, Brazil

Pinto ER, Ponce de Leon A (2004) Model Oriented Data Analysis, mODa7, Physica Verlag, Eindhoven, chap Bayesian D-optimum designs for generalized linear models with varying dispersion parameter

Some Curiosities in Optimal Designs for Random Slopes

Thomas Schmelter[1], Norbert Benda[2], and Rainer Schwabe[3]

[1] Institute for Mathematical Stochastics, Otto-von-Guericke-University, PF 4120, 39 016 Magdeburg, Germany / Bayer Schering Pharma AG, Clinical Statistics Europe, 13 342 Berlin, Germany `Thomas.Schmelter@schering.de`
[2] Novartis Pharma AG, Statistical Methodology, Lichtstrasse 35, 4056 Basel, Switzerland `norbert.benda@novartis.com`
[3] Institute for Mathematical Stochastics, Otto-von-Guericke-University, PF 4120, D–39 016 Magdeburg, Germany
`rainer.schwabe@mathematik.uni-magdeburg.de`

Summary. The purpose of this note is to show by a simple example that some of the favourite results in optimal design theory do not necessarily carry over if random effects are involved. In particular, the usage of the popular D-criterion appears to be doubtful.

Key words: optimal design, mixed linear model, random coefficient regression

1 Introduction

Mixed models have attracted growing interest in the biosciences, when replicated measurements are available from different individuals. While the corresponding statistical analysis is well-developed, only a few results are available on optimal designs for such experiments. For a recent survey on the particular setting of random coefficient regression see Entholzner et al (2005).

The most popular criterion in applications is the D-criterion in analogous fixed-effects models, where it has some nice properties, but which do not necessarily carry over to mixed models. Fedorov and Hackl (1997) p. 75, provide an equivalence theorem for this situation, which has been extended by Schmelter (2006). While results are quite obvious for random intercept models (Schwabe and Schmelter (2006)), the optimisation may lead to apparently counter-intuitive solutions, if there is randomness in the treatment effects (see e. g. Fedorov and Leonov (2004)).

In the present note we will indicate how various standard criteria are influenced by the presence of random individual effects.

2 The model

To keep notation as simple as possible we discuss a straight line regression model on the unit interval, in which only the slopes are affected by random effects. More specifically, we consider n individuals with m observations each, and the jth observation Y_{ij} of individual i is described by

$$Y_{ij} = \mu + b_i\, x_{ij} + \varepsilon_{ij}\ ,$$

where x_{ij} is the corresponding experimental setting, $0 \leq x_{ij} \leq 1$, $i = 1,\ldots,n, j = 1,\ldots,m$. The individual random slopes b_i are assumed to be *iid* with unknown population mean β and known variance σ_β^2. A typical example for a bunch of the conditional individual mean response lines $\mu + b_i x$ is given in the spaghetti plot of Figure 1. Our interest will be only in the population parameters μ and β or, equivalently, the mean response $\mu + \beta x$ across the individuals, rather than in prediction.

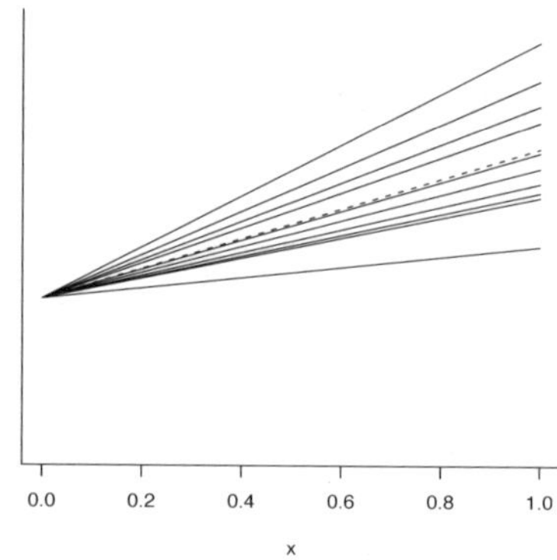

Fig. 1. Population (dashed line) and individual mean response curves (solid lines)

Furthermore the observational errors ε_{ij} are assumed to be homoscedastic (*iid*) with zero mean and known variance σ^2 and to be independent of the random slope parameters b_i. We define the dispersion factor $d = \sigma_\beta^2/\sigma^2$ as the variance ratio of the slope compared to the observational error.

Denote by $\mathbf{Y}_i = (Y_{i1},\ldots,Y_{im})^\top$ and $\mathbf{x}_i = (x_{i1},\ldots,x_{im})^\top$ the vector of observations and corresponding experimental settings, respectively, and by $\mathbf{F}_i = (\mathbf{1}_m\,|\,\mathbf{x}_i)$ the associated design matrix for individual i. Here $\mathbf{1}_m$ is a vector of length m with all entries equal to one. The covariance matrix $\mathbf{V}_i$ of the observations $\mathbf{Y}_i$ is given by $\mathbf{V}_i = \sigma^2(\mathbf{I}_m + \mathbf{F}_i \left(\begin{smallmatrix} 0 & 0 \\ 0 & d \end{smallmatrix}\right) \mathbf{F}_i^\top) = \sigma^2(\mathbf{I}_m + d\,\mathbf{x}_i\mathbf{x}_i^\top)$, where $\mathbf{I}_m$ is the m-dimensional identity matrix. For simplicity we will assume from now on, without loss of generality, that σ^2 is equal to 1.

Then the weighted least squares estimator, which is the best linear unbiased estimator for $(\mu, \beta)^\top$, is given by $(\sum_{i=1}^n \mathbf{F}_i^\top \mathbf{V}_i^{-1} \mathbf{F}_i)^{-1} \sum_{i=1}^n \mathbf{F}_i^\top \mathbf{V}_i^{-1} \mathbf{Y}_i$.

The corresponding covariance matrix is equal to the inverse of the information matrix $\mathbf{M}_d = \sum_{i=1}^{n} \mathbf{F}_i^{\top} \mathbf{V}_i^{-1} \mathbf{F}_i$, which depends on the variance ratio d through the inverse $\mathbf{V}_i^{-1} = \mathbf{I}_m - \frac{d}{1+d\mathbf{x}_i^{\top}\mathbf{x}_i}\,\mathbf{x}_i\,\mathbf{x}_i^{\top}$ of the observational covariance matrix as well as on the experimental settings $\mathbf{x}_1, \ldots, \mathbf{x}_n$. Note that the individual information $\mathbf{F}_i^{\top} \mathbf{V}_i^{-1} \mathbf{F}_i$ is equal to $\frac{m}{1+md\,\nu_{i2}} \begin{pmatrix} 1 + m\,d(\nu_{i2} - \nu_{i1}^2)\,\nu_{i1} \\ \nu_{i1} \qquad\qquad \nu_{i2} \end{pmatrix}$, where $\nu_{ik} = \frac{1}{m}\sum_{j=1}^{m} x_{ij}^k$ denotes the kth moment of the experimental setting $\mathbf{x}_i$.

For estimating the mean response $\mu + \beta x$ over the design region $(0 \le x \le 1)$ the variance function is given by $v_d(x) = \mathrm{var}\,(\hat{\mu} + \hat{\beta}x) = (1, x)\,\mathbf{M}_d^{-1}\binom{1}{x}$.

3 Optimal design

Design optimality aims at finding the best experimental settings $\mathbf{x}_i$ to maximise the information $\mathbf{M}_d = \mathbf{M}_d(\mathbf{x}_1, \ldots, \mathbf{x}_n)$ or, equivalently, to minimise the covariance matrix $\mathbf{M}_d^{-1}$. As uniform matrix optimisation is not possible, there are various competing optimality criteria, which are real-valued functionals of the information $\mathbf{M}_d$. One of the most popular is the D-criterion, which aims at maximising the determinant of $\mathbf{M}_d$. For fixed-effects models without individual slopes (i. e. $d = 0$) D-optimality is equivalent to optimisation with respect to the G-criterion, which aims at minimising the maximum $\max_{0 \le x \le 1} v_d(x)$ of the variance function over the design region, according to the Kiefer-Wolfowitz equivalence theorem (Kiefer and Wolfowitz (1960)) within the setup of approximate designs. To avoid discretisations we will deal with such a generalised setup throughout this section. According to Schmelter (2006) optimal designs can be found among those, which are uniform across the individuals, i. e. $\mathbf{x}_i = \mathbf{x}$ and, hence, $\mathbf{F}_i = \mathbf{F}$ for all i. Then the covariance matrix simplifies to $\mathbf{M}_d^{-1} = \frac{1}{n}\left((\mathbf{F}^{\top}\mathbf{F})^{-1} + \binom{0\ 0}{0\ d}\right)$ (see Entholzner et al (2005)). Due to majorisation (see e. g. Pukelsheim (1993) p. 101) we can confine the search for optimal designs to those with observations at the extreme settings $x = 0$ and $x = 1$. Candidates for an optimal design are, thus, characterised by the number m_1 or, equivalently, by the proportion $w = m_1/m$ of observations at the experimental setting $x = 1$, while $m_0 = (1-w)m$ observations are made at the baseline, $x = 0$, for each individual. The corresponding covariance matrix can be calculated as

$$\mathbf{M}_d^{-1} = \frac{1}{nm}\,\frac{1}{w(1-w)}\begin{pmatrix} w & -w \\ -w & 1 + mdw(1-w) \end{pmatrix}.$$

For the optimisation we also allow generalised proportions w, which are not necessarily multiples of $1/m$.

Theorem 1. *The D-optimal proportion w_D^* at $x = 1$ equals $(1 + \sqrt{md+1})^{-1}$.*

Proof. The determinant of $\mathbf{M}_d^{-1}$ is proportional to $(1 + mdw)/(w(1-w))$, which is minimised by w_D^*. $\qquad\qquad\square$

The optimal proportion w_D^* varies continuously with the variance ratio d, and w_D^* tends to zero as d tends to infinity (see Figure 2).

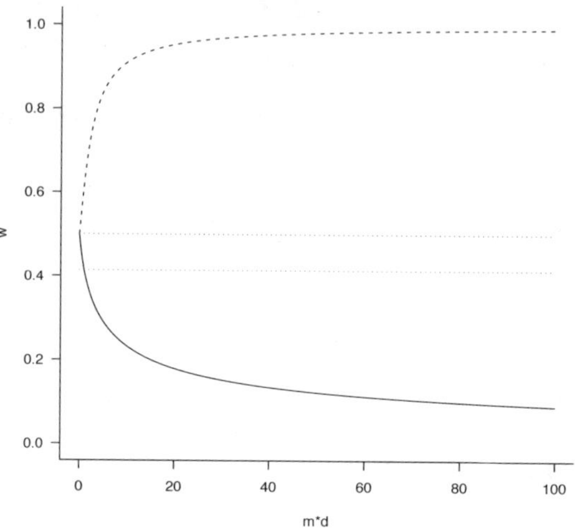

Fig. 2. Optimal proportions of observations in $x = 1$: D-optimal (solid line), G-optimal (dashed line), β- and $IMSE$-optimal (upper horizontal dotted line), and A-optimal (lower horizontal dotted line)

It has to be noted that all optimal designs depend on the number m of replications and the variance ratio d only through their product $m \cdot d$ as does the standardised information matrix $\frac{1}{nm}\mathbf{M}_d$.

Theorem 2. *The G-optimal proportion w_G^* at $x = 1$ equals $\frac{1}{2}(1 - 2(md)^{-1} + \sqrt{1 + 4(md)^{-2}})$ if $d > 0$, and $w_G^* = 1/2$ for $d = 0$.*

Proof. As the variance function $v_d(x) = \frac{1}{nm} \frac{1}{w(1-w)} (w - 2wx + (1 + m\,dw(1 - w))x^2)$ is a polynomial of degree 2 with positive leading term $(0 < w < 1)$, its maximum is attained either at $x = 0$ or $x = 1$, or both, i. e. $\max_{0 \leq x \leq 1} v_d(x) = \max(v_d(0), v_d(1))$. Now, the standardised variance $nmv_d(0) = (1 - w)^{-1}$ is strictly increasing in w while $nmv_d(1) = w^{-1} + md$ is strictly decreasing in w. Thus, $\min_{0 < w < 1} \max(v_d(0), v_d(1))$ is attained when $v_d(0) = v_d(1)$, i. e. $(1 - w)^{-1} = w^{-1} + md$, which is solved by w_G^*. $\square$

The optimal proportion w_G^* varies continuously in d, but, in contrast to the D-optimal proportion, it tends to 1 as d tends to infinity (see Figure 2). Thus D- and G-optimal proportions are very sensitive to the variance ratio d and differ essentially if d is large.

Linear criteria, however, which are of the form $\mathrm{tr}(\mathbf{A}\,\mathbf{M}_d^{-1})$ for some fixed positive semidefinite matrix $\mathbf{A}$ are not affected by the variance ratio d, because $n^{-1}\mathrm{tr}(\mathbf{A}\,\mathbf{M}_d^{-1}) = \mathrm{tr}(\mathbf{A}(\mathbf{F}^\top \mathbf{F})^{-1}) + da_{22}$ decomposes into the corresponding criterion $\mathrm{tr}(\mathbf{A}(\mathbf{F}^\top\mathbf{F})^{-1})$ of the fixed-effects model without individual slopes and a design independent constant da_{22}, where a_{22} is the lower right entry in $\mathbf{A}$. Hence, for such criteria the optimal design is independent of d. Typical examples are the c-criterion for the slope β, $c^\top \mathbf{M}_d^{-1} c$, where $c = (0,1)^\top$, the integrated mean squared error (IMSE) criterion, $\int_0^1 v_d(x)\, dx$ or the A-criterion, $\mathrm{tr}(\mathbf{M}_d^{-1})$.

Theorem 3. *The β- and IMSE-optimal proportions $w_\beta^* = w_{IMSE}^*$ are equal to $1/2$. The A-optimal proportion w_A^* is equal to $\sqrt{2}-1$.*

To judge the impact of design optimisation one is tempted to calculate the efficiency of the proportion $w_0 = 1/2$, which is simultaneously D- and G-optimal for the fixed-effects model without individual slopes (i. e. $d = 0$), when the variance ratio increases. The D-efficiency $(\det \mathbf{M}_d(w_0)/\det \mathbf{M}_d\,(w_D^*))^{1/2} = (1+\sqrt{md+1})/\sqrt{4+2md}$ decreases slowly to $1/\sqrt{2} > 0.70$ if the variance ratio d becomes large. But the G-efficiency, which is equal to $(w_G^{*-1}+md)/(2+md)$, shows a strange behaviour. If d increases the G-efficiency drops very quickly to about 0.86 and, then, increases again and tends ultimately to 1 for d tending to infinity. This strange limiting behaviour may be explained by the fact that the G-efficiency for w_0 is bounded by $(1 + md)/(2 + md)$ from below. The efficiencies are plotted in Figure 3.

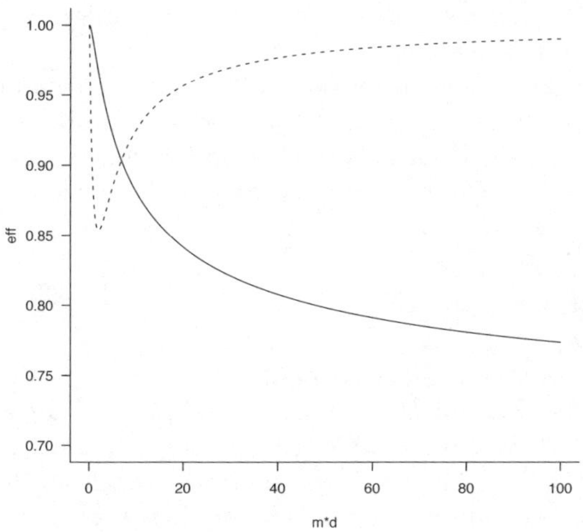

Fig. 3. Efficiency of the equireplicated design $(w = 0.5)$: D-criterion (solid line) and G-criterion (dashed line)

4 Discretisation

For applications it must be required that the proportion w is a multiple of $1/m$, i.e. that the number $m_1 = w \cdot m$ of observations at $x = 1$ is an integer. In general, optimal numbers m_1^* can be found by rounding w^*m to the next smaller or larger integer $[w^*m]$ or $[w^*m] + 1$, respectively, where w^* is the optimal generalised proportion obtained in the previous section and $[\cdot]$ denotes the integer part.

For example, in the case $m = 2$ the optimal value m_1^* for the number of observations at $x = 1$ is equal to 1 independently of d for every reasonable criterion considered in section 3 due to estimability requirements. In the case $m = 4$ the D-optimal number m_1^* is equal to 2 for small variance ratios $d \leq 1/2$ and has to be chosen as 1 for larger values, $d \geq 1/2$. Note that in this case the generalised solution $w_D^* \, m$ yields the integer value 1 for $d = 2$.

In the situation of non-integer w^*m it may additionally turn out that it is more favourable to apply non-uniform designs, in which experimental settings may differ from individual to individual. It seems reasonable that for a certain proportion α of individuals m_1 is chosen to be equal to $[w^*m] + 1$ while for the remaining $(1 - \alpha) \cdot n$ individuals m_1 equals $[w^*m]$ in order to improve the performance of the design. For example, in the case $m = 4$ it is D-optimal to have a proportion of $1 - d$ individuals with $m_1 = 2$ observations at 1 and a proportion of d individuals with $m_1 = 1$ as long as the variance ratio d is smaller than 1. For larger d, $d \geq 1$, the uniform design becomes D-optimal with $m_1 = 1$ for all individuals.

For $m = 2$ it can be shown by the multivariate version of the equivalence theorem (see Fedorov (1972) p. 212) that the uniform design with $m_1 = 1$ is simultaneously D-optimal for all values of the variance ratio d.

Analogous findings can be obtained for the G-criterion.

5 Discussion

In the present simple model of straight line regression with random slopes neither the commonly used D-criterion nor its pretended counterpart the G-criterion seem to show a reasonable behaviour, particularly, if the variability is large. While the D-criterion yields solutions, which are lightweight in the sense that most observations are made where it is 'easy', i.e. where the variance is small, the G-criterion overemphasises 'difficult' observations, where the variation is large and cannot be substantially reduced by increasing the number of intra-individual replications. Moreover, the G-criterion exhibits a strange non-monotonic efficiency behaviour. In fact, it can be shown that for every regular design its G-efficiency tends to 1 if the variance ratio tends to infinity. This indicates that, with respect to the G-criterion, all designs are equally good - or equally bad - if d is large.

Although this last statement also applies to linear criteria like the IMSE-criterion they seem to be a reasonable compromise and, in particular, have the advantage of resulting in optimal designs, which are independent of the magnitude of the variance ratio.

References

Entholzner M, Benda N, Schmelter T, Schwabe R (2005) A note on designs for estimating population parameters. Listy Biometryczne - Biom. Letters 42:25–41

Fedorov VV (1972) Theory of Optimal Experiments. Academic Press, New York

Fedorov VV, Hackl P (1997) Model-Oriented Design of Experiments, Lecture Notes in Statistics, vol 125. Springer, New York

Fedorov VV, Leonov S (2004) Optimal designs for regression models with forced measurements at baseline. In: DiBucchianico A, Läuter H, Wynn HP (eds) mODa 7 - Advances in Model-Oriented Design and Analy., Physica, Heidelberg, 61–69

Kiefer J, Wolfowitz J (1960) The equivalence of two extremum problems. Canadian Journal of Mathematics 12:363–366

Pukelsheim F (1993) Optimal Design of Experiments. Wiley, New York

Schmelter T (2006) The optimality of single-group designs for certain mixed models. Metrika (to appear)

Schwabe R, Schmelter T (2006) On optimal designs in random intercept models. Tatra Mountains Mathematical Publications (to appear)

The Within-B-Swap (BS) Design is A- and D-optimal for Estimating the Linear Contrast for the Treatment Effect in 3-Factorial cDNA Microarray Experiments

Sven Stanzel and Ralf-Dieter Hilgers

Institute of Medical Statistics, Aachen University of Technology, Pauwelsstraße 30, 52074 Aachen, Germany
`sstanzel@ukaachen.de`   `rhilgers@ukaachen.de`

Summary. cDNA microarrays are a powerful tool in gene expression analysis Speed (2003). Landgrebe et al (2006) proposed a special 3-factor model to estimate various effects on the log ratios of measured fluorescence intensities. We demonstrate in this paper that the Within-B-Swap (BS) design introduced by Landgrebe et al (2006) is A- and D-optimal for estimating the linear contrast for the treatment effect in the general case of l treatments and k cell lines.

Key words: A-optimality, BS design, cDNA microarray experiment, D- optimality, equivalence theorem, fixed effects linear model, treatment effect

1 Introduction

In a 3-factorial cDNA microarray experiment, two measurements of fluorescence intensity are carried out using two different fluorescent dyes (green and red) in order to determine the response (i.e the log ratio of the two fluorescence intensity measurements) of a particular gene on a specific array. Moreover, the two dyes (factor C) can be chosen independently from the treatments (factor A) and cell lines (factor B) investigated. It is well-known that, using the same combination of treatment and cell line, the two different dyes can be associated with quite different fluorescence intensity values. However, the primary interest is in an efficient estimation of the treatment-, cell line- as well as treatment$\times$cell line interaction effects, while the two different dyes serve as a block effect. So, keeping the levels of treatment and cell line factor fixed for both dyes used on a particular array would be a useless experimental setup.

Thus, the nature of the experiment is such that on a particular array we either can compare two different treatments (keeping the cell line fixed) or two different cell lines (using the same treatment), or two different treatments in combination with two different cell lines. To further extract the variation due to dyes, we have to carry out another experiment on a different array, switching the dyes used for the same two combinations of treatments and cell lines.

We discuss optimal designs for estimating the parameters of interest in the fixed effects gene-specific linear model for log ratios of measured fluorescence intensities introduced by Landgrebe et al (2006). To estimate special linear contrasts for subsets of these parameters in an efficient manner, we will search for A- and D-optimal designs (Pukelsheim, 1993, p. 135ff). In this paper only linear contrasts for estimating all pairwise treatment differences are of interest. We will make use of the *equivalence theorem for matrix means* proposed by Pukelsheim (1993) to prove optimality.

In section 2, we will introduce some notation and a special type of design called Within-B-Swap (BS) design. A- and D-optimality of the BS design will be shown in section 3. The results will be summarized and discussed in section 4.

2 Notations and examples

As introduced by Landgrebe et al (2006), due to the connection of dye labelling and treatment allocation, specified in the form $x_{i.} = 1$ (treatment i, green dye labelling) or $x_{i.} = -1$ (treatment i, red dye labelling), and the restriction that one has to change either treatment or cell line or both when switching from the green to the red dye channel (or vice versa) on a specific array, the most relevant information (i.e. the most relevant parameters) of the experimental setup is obtained, if the Landgrebe model is written in usual matrix form,

$$Z = X\theta + \epsilon. \tag{1}$$

In model (1), Z denotes the vector of observed log ratios for a single gene on N arrays; ϵ is the corresponding vector of i.i.d. error terms with $E(\epsilon) = 0$ and $Var(\epsilon) = \sigma^2 I_N$. The ordered set of all relevant parameters in the experiment is given by the parameter vector $\theta = [\delta_g, \delta_r, \tau_{11}, ..., \tau_{lk}]^T$ of dimension $(lk+2)$, where δ_g and δ_r are the fixed dye effects respectively of the green the red fluorescent dye and τ_{ij} is the fixed combination effect of treatment i $(i = 1, ..., l)$ and cell line j $(j = 1, ..., k)$. The $N \times (lk + 2)$ design matrix X of a concrete experiment using N arrays contains all relevant design information about the Landgrebe model. Corresponding to the parameter vector θ, the first two columns in X describe the dye effects of the two fluorescent dyes, the remaining columns characterize the lk combination effects of treatments and cell lines (see example 2.1). Landgrebe et al (2006)

Because of the nature of 3-factorial microarray experiments only *balanced incomplete block designs (BIBD)* can be realized for estimating the parameters specified in model (1). A special type of these designs was introduced by Landgrebe et al (2006) and is called *Within-B-Swap (BS) design*. In this design, all pairwise comparisons between the levels (treatments) of factor A are conducted twice within the same level (cell line) of factor B; dyes are always swapped in the second of these two repetitions of the same experimental condition.

Due to the restrictions explained in section 1, only $m = 2\binom{kl}{2}$ different discrete design points x are possible for this design. Thus, the discrete design space is given as $\chi = \{x_1, ..., x_m\}$. Each of the m design points x is of the form
$x = [g, r, x_{11}, ..., x_{lk}]^T$, where g and r correspond to the green and red fluorescent dye while x_{ij} indicates whether the combination of treatment i and cell line j on a particular array is labelled with the green dye ($x_{ij} = +1$), the red dye ($x_{ij} = -1$) or not at all ($x_{ij} = 0$). By convention, g is always set to '+1' and r to '-1'; *dye swaps*, i.e. switching the dyes used for the two combinations compared, are obtained by switching the signs of the two combinations (see example 2.1).

The $2k\binom{l}{2}$ support points of the BS design comprise all $\binom{l}{2}$ different pairwise comparisons of two treatments used in combination with each of the k cell lines in each of the two dye constellations. Using model (1), the dye swap character of the BS design, i.e. that the support points are such that each treatment comparison is repeated with the two dyes switched, guarantees the *estimability* of the BS design with respect to the linear contrast for the treatment effect Landgrebe et al (2006).

Table 1. SP of BS design for $l = 3, k = 2$ [left] and $l = 2, k = 3$ [right]

SP	a1	a2	a3	b1	b2	b3
1	1	-1	0	0	0	0
2	1	0	-1	0	0	0
3	0	1	-1	0	0	0
4	-1	1	0	0	0	0
5	-1	0	1	0	0	0
6	0	-1	1	0	0	0
7	0	0	0	1	-1	0
8	0	0	0	1	0	-1
9	0	0	0	0	1	-1
10	0	0	0	-1	1	0
11	0	0	0	-1	0	1
12	0	0	0	0	-1	1

SP	a1	a2	b1	b2	c1	c2
1	1	-1	0	0	0	0
2	-1	1	0	0	0	0
3	0	0	1	-1	0	0
4	0	0	-1	1	0	0
5	0	0	0	0	1	-1
6	0	0	0	0	-1	1

Example 2.1

The respective support points (SP) of the BS designs for, $l = 3$ treatments (1,2,3) and $k = 2$ cell lines (a,b), and for, $l = 2$ treatments (1,2) and $k = 3$ cell lines (a,b,c), are displayed in table 1. The fixed dye codings ($g = 1$, $r = -1$) are non-informative and thus ignored.

Since all pairwise treatment comparisons are considered equally important in estimating the linear contrast for the treatment effect, the support points of the BS design are uniformly weighted ($p_t = 1/2k\binom{l}{2}$). This yields the symmetric moment matrix

$$M = \frac{1}{2k\binom{l}{2}} \begin{bmatrix} 2k\binom{l}{2} & -2k\binom{l}{2} & 0_{kl}^T \\ -2k\binom{l}{2} & 2k\binom{l}{2} & 0_{kl}^T \\ 0_{kl} & 0_{kl} & I_k \otimes (A_l^T A_l) \end{bmatrix} \tag{2}$$

with corresponding g-inverse

$$G = 2k\binom{l}{2} \begin{bmatrix} \frac{1}{8k\binom{l}{2}} & -\frac{1}{8k\binom{l}{2}} & 0_{kl}^T \\ -\frac{1}{8k\binom{l}{2}} & \frac{1}{8k\binom{l}{2}} & 0_{kl}^T \\ 0_{kl} & 0_{kl} & I_k \otimes (\frac{1}{4l^2} A_l^T A_l) \end{bmatrix} \tag{3}$$

of dimension $(lk + 2)$ under the BS design applied to the general situation of l treatments and k cell lines. From this point on, the symbol $\otimes$ denotes the Kronecker product of two matrices (Searle, 1982, p. 265) and I_k the k-dimensional identity matrix, $I_k = diag\{1,...,1\} \in \mathcal{R}^k$. Moreover, the matrix A_l can be partitioned into

$$A_l = \begin{bmatrix} \tilde{A}_l \\ -\tilde{A}_l \end{bmatrix}, \tag{4}$$

where the $\binom{l}{2} \times l$ matrix $\tilde{A}_l$ specifies all $\binom{l}{2}$ possible pairwise comparisons between two out of the l treatments. By convention, in each row of the matrix $\tilde{A}_l$ the first of the two treatments compared is indicated by '+1', the other treatment accordingly by '-1'; the remaining entries of the corresponding row are zero. Thus, the matrix $\tilde{A}_l$ has the following general structure:

$$\tilde{A}_l = \begin{bmatrix} 1_{l-1} & & -I_{l-1} & & \\ 0_{l-2} & 1_{l-2} & & -I_{l-2} & \\ 0_{l-3} & 0_{l-3} & 1_{l-3} & -I_{l-3} & \\ \vdots & \ddots & \ddots & & \ddots \\ 0_2 & \cdots & 0_2 & 1_1 & -I_1 \\ 0_1 & & \cdots & & 0_1 & 1_1 & -I_1 \end{bmatrix}. \tag{5}$$

The treatment effect will be tested by means of the linear contrast $K^T\theta$, where θ is the vector of the $(lk+2)$ interesting parameters specified in model (1) and

$$K = \left[0_{\binom{l}{2}} \ 0_{\binom{l}{2}} \ 1_k^T \otimes \tilde{A}_l \right]^T \tag{6}$$

is the $(lk+2) \times \binom{l}{2}$ contrast matrix for estimating all possible pairwise differences between the l treatments of interest. Thus, it can be seen that even in the definition of the matrix K the matrix $\tilde{A}_l$ given in (5) has a role to play.

Example 2.2

For $l = 2$ treatments and $k = 3$ cell lines the contrast matrix of the linear contrast for the treatment effect is given as

$$K_{23} = [0, 0, 1, -1, 1, -1, 1, -1]^T,$$

for $l = 3$ treatments and $k = 2$ cell lines it is

$$K_{32} = \begin{bmatrix} 0 & 0 & 1 & -1 & 0 & 1 & -1 & 0 \\ 0 & 0 & 1 & 0 & -1 & 1 & 0 & -1 \\ 0 & 0 & 0 & 1 & -1 & 0 & 1 & -1 \end{bmatrix}^T .$$

3 A- and D-optimality of the BS design

The *equivalence theorem for matrix means* introduced by Pukelsheim (Pukelsheim, 1993, p. 180) can be used to prove A- and D-optimality of the BS design. To avoid problems that might appear due to singular covariance matrices K^TGK, we had to *modify the normality inequality* of this theorem in such a way that it can be used for both, *regular* and *singular*, covariance matrices. This can be achieved by replacing the matrix means ϕ_p by their corresponding *rank deficient matrix means* ϕ_p' and was also demonstrated by Pukelsheim (Pukelsheim, 1993, p. 205).

A-optimality of the BS design with respect to the linear contrast for the treatment effect can be obtained from theorem 1:

Theorem 1 (A-optimality of the BS design).

The Within-B-Swap (BS) design is **A-optimal** *for estimating the linear contrast $K^T\theta$ for the treatment effect in model (1).*

Proof.

To prove A-optimality , we use the equivalence theorem for matrix means with singularity modification ($p = -1$). Thus, we have to show

$$x^T GK[K^TGK]^+[K^TGK]^2[K^TGK]^+K^TG^Tx \leq tr\left\{[K^TGK]^+[K^TGK]^2\right\} \tag{7}$$

for all x, where strict equality holds true for all support points of the BS design. Here K is the $(lk + 2) \times 2$ contrast matrix for estimating all pairwise treatment differences (see (6)), G denotes the g-inverse of the moment matrix M corresponding to the BS design given respectively in (3), (2), and

$$\left[K^T G K\right]^+ = \frac{1}{k^2 l \binom{l}{2}} \tilde{A}_l \tilde{A}_l^T \tag{8}$$

is the Moore-Penrose inverse of the covariance matrix

$$K^T G K = \frac{k^2}{l} \binom{l}{2} \tilde{A}_l \tilde{A}_l^T . \tag{9}$$

With the notation and definitions outlined in section 2 it can be shown by extensive use of some rules and theorems already known from the theory of matrix algebra Searle (1982); Harville (1997) that

$$x^T G K \tilde{A}_l \tilde{A}_l^T K^T G^T x = k^2 \binom{l}{2}^2 \sum_{q=1}^{l} \sum_{r=1}^{k} \left[x_{qr} \sum_{s=1}^{k} (x_{qs} - \bar{x}_{\cdot s}) \right] , \tag{10}$$

where $\bar{x}_{\cdot s} = \frac{1}{l} \sum_{q=1}^{l} x_{qs}$. Using the results given in (8)-(10), it can be derived by straightforward computation that inequality (7) is equivalent to

$$\sum_{q=1}^{l} \sum_{r=1}^{k} \left[x_{qr} \sum_{s=1}^{k} (x_{qs} - \bar{x}_{\cdot s}) \right] \leq 2. \tag{11}$$

To show (11), we have to distinguish three cases with respect to the possible design points x:

Case 1: Support points of BS

All $2k\binom{l}{2}$ design points comparing two treatments $i \neq i'; i, i' \in \{1, ..., l\}$ within the same cell line $j \in \{1, ..., k\}$.

Let, without loss of generality, $x_{ij} := +1$, $x_{i'j} := -1$, $x_{qr} := 0$ otherwise. Then, $\bar{x}_{\cdot j} = 0$ and (11) holds true because

$$\sum_{q=1}^{l} \sum_{r=1}^{k} \left[x_{qr} \sum_{s=1}^{k} (x_{qs} - \bar{x}_{\cdot s}) \right] = x_{ij} (x_{ij}) + x_{i'j} (x_{i'j}) = 2.$$

Case 2:

All non-support points comparing two treatments $i \neq i'; i, i' \in \{1, ..., l\}$ for two different cell lines $j \neq j'; j, j' \in \{1, ..., k\}$.

Let $x_{ij} := +1$, $x_{i'j'} := -1$, $x_{qr} := 0$ otherwise. Then, $\bar{x}_{\cdot j} = 1/l, \bar{x}_{\cdot j'} = -1/l$, and

$$\sum_{q=1}^{l} \sum_{r=1}^{k} \left[x_{qr} \sum_{s=1}^{k} (x_{qs} - \bar{x}_{\cdot s}) \right] = x_{ij} \left[\left(x_{ij} - \tfrac{1}{l}\right) + \left(x_{ij'} + \tfrac{1}{l}\right) \right] +$$

$$x_{i'j'}\left[\left(x_{i'j}-\tfrac{1}{l}\right)+\left(x_{i'j'}+\tfrac{1}{l}\right)\right]=x_{ij}\left(x_{ij}\right)+x_{i'j'}\left(x_{i'j'}\right)=2.$$

Case 3:

All non-support points comparing the same treatment $i \in \{1,...,l\}$ in two different cell lines $j \neq j'; j, j' \in \{1,...,k\}$.

Let $x_{ij} := +1$, $x_{ij'} := -1$, $x_{qr} := 0$ otherwise. Then, $\bar{x}_{.j} = 1/l$, $\bar{x}_{.j'} = -1/l$, and

$$\sum_{q=1}^{l}\sum_{r=1}^{k}\left[x_{qr}\sum_{s=1}^{k}\left(x_{qs}-\bar{x}_{.s}\right)\right]=x_{ij}\left[\left(x_{ij}-\tfrac{1}{l}\right)+\left(x_{ij'}+\tfrac{1}{l}\right)\right]+$$

$$x_{ij'}\left[\left(x_{ij}-\tfrac{1}{l}\right)+\left(x_{ij'}+\tfrac{1}{l}\right)\right]=\left(x_{ij}+x_{ij'}\right)\left(x_{ij}+x_{ij'}\right)=0.$$

$\square$

The next theorem states the *D-optimality* of the BS design with respect to the linear contrast for the treatment effect:

Theorem 2 (D-optimality of the BS design).
The Within-B-Swap (BS) design is **D-optimal** *for estimating the linear contrast $K^T\theta$ for the treatment effect in model (1).*

Proof.
Using $p = 0$ in the equivalence theorem for matrix means (with the singularity modification again), we have to show

$$x^T GK[K^T GK]^+[K^T GK][K^T GK]^+ K^T G^T x \leq tr\left\{[K^T GK]^+[K^T GK]\right\} \qquad (12)$$

for all x, where strict equality holds true for all support points of the BS design. The proof is analogous to the first part of the previous proof in that we find that (12) is equivalent to (11) so that D-optimality holds true according to the second part of that proof.

$\square$

4 Conclusions and discussion

We considered the fixed effects gene-specific linear model for log ratios of fluorescence intensities to show global A- and D-optimality of the Within-B-Swap (BS) design for estimating the treatment effect in 3-factorial cDNA microarray experiments by means of the linear contrast $K^T\theta$, using a modified version of the equivalence theorem for matrix means. Extending these results to different optimality criteria (E-optimality, T-optimality) and linear contrasts (for cell line effects as well as interaction effects) is straightforward.

The independence of the solution from the optimality criterion chosen can be interpreted as a robustness property of the BS design. We further emphasize that in most scenarios this solution does not depend on the number of treatments and cell lines.

It might be of interest to examine whether the solution is also robust against the model selected by taking different fixed and random effects linear models proposed in the microarray literature into account as well.

For practical situations, the results can be transferred into direct recommendations for the choice of an effective concrete cDNA microarray design, depending on the actual interest of the biologist or physician. Consequently, the results obtained might save a large amount of financial resources normally required for these expensive experiments.

References

Harville DA (1997) Matrix algebra from a statistician's perspective. Springer, New York

Landgrebe J, Bretz F, Brunner E (2006) Efficient design and analysis of two colour factorial microarray experiments. Computational Statistics and Data Analysis 50:499–517

Pukelsheim F (1993) Optimal design of experiments. Wiley, New York

Searle SR (1982) Matrix algebra useful for statistics. Wiley, New York

Speed T (2003) Statistical analysis of gene expression microarray data. Chapman and Hall, Boca Raton

D-optimal Designs and Equidistant Designs for Stationary Processes

Milan Stehlík

Department of Applied Statistics, Johannes Kepler University, Freistädter Straße 315, 4040 Linz, Austria   `Milan.Stehlik@jku.at`

Summary. In this paper we discuss the structure of the information matrices of D-optimal experimental designs for the parameters in a stationary process when the parametrized correlation structure satisfies mild conditions. Such conditions are easily fulfilled by many correlation structures, e.g. structures from power exponential family and some members of the Matérn class. We provide a lower bound for information on the mean parameter and prove it to be an increasing function of distances of design points. The design points can collapse under the presence of some covariance structures and a so called nugget effect can be employed in a natural way. We also show that the information of equidistant designs (designs with equally spaced design points) on the covariance parameter is increasing with the number of design points under our conditions on correlations. If only trend parameters are of interest, the designs covering the whole design space non-uniformly are rather efficient.

Key words: correlated errors, regression experiment, power exponential family, Matérn class

1 Introduction

Here we consider the isotropic stationary process

$$Y(x) = \theta + \varepsilon(x)$$

with the design points $x_1, ..., x_N$ are taken from a compact design space $\mathcal{X}$. The mean parameter $E(Y(x)) := \theta \in \Theta$ is unknown, the variance-covariance structure $C(d, r)$ depends on another unknown parameter r and d is the distance between (neighbouring) design points. Let us define $2\gamma(h) = \operatorname{var}(Y(s+h) - Y(s))$. The function $2\gamma(h)$ is called the *variogram* and $\gamma(h)$ is called the *semivariogram* (for more details see Banerjee et al (2004)). In other words, we study a weak stationary process (also called second-order stationary); see (Cressie (1993),p. 53). In such a model we have the Fisher information matrices:

$$M_\theta(n) = 1^T C^{-1}(r)\, 1$$

and (see Pázman (2004) and Xia et al (2006))

$$M_r(n) = \frac{1}{2} tr \left\{ C^{-1}(r) \frac{\partial C(r)}{\partial r} C^{-1}(r) \frac{\partial C(r)}{\partial r^T} \right\}.$$

So for both parameters of interest we have $M(n)(\theta, r) = \begin{pmatrix} M_\theta(n) & 0 \\ 0 & M_r(n) \end{pmatrix}$.

From now on, by 'information' we mean the Fisher information on the parameter of interest of the isotropic random field. In this paper we study the structure of the Fisher information matrices $M_\theta(n)$ and $M_r(n)$ with some regularity assumptions on covariance structures. We assume that

a) $C(d, r) > 0$ for all r and $0 < d < +\infty$,

b) for all r the mapping $d \to C(d, r)$ is continuous and strictly decreasing on $(0, +\infty)$

c) $\lim_{d \to +\infty} C(d, r) = 0$.

These assumptions are fulfilled by many covariance structures. The example of a family satisfying a),b),c) is the power exponential correlation family, with the variogram

$$\gamma(d) = \begin{cases} 0, & \text{for } d = 0, \\ \tau^2 + \sigma^2(1 - \exp(-rd^p)), & \text{otherwise, } 0 < p \le 2, \ r > 0, \end{cases}$$

where τ^2 is the nugget effect and d is distance. This family is by far the most popular family of correlation models in the computer experiments literature (see Santner et al (2003) and Currin et al (1991)). The exponential $\exp(-rd)$ and Gaussian correlation functions $\exp(-rd^2)$ are special cases of the power exponential correlation family. Gaussian correlation is appropriate if the process is smooth (the realizations of Y are infinitely differentiable with probability 1, see Parzen (1967)). Such a situation occurs e.g. when responses are solutions to a system of differential equations and depend smoothly on the rate constants x which form the inputs (see Sacks et al (1989) for more details). For applications with more erratic responses, we would employ different correlation structures; for instance the exponential correlation. The process with exponential correlation can be thought of as a model for functions only required to have one-sided first-order derivatives (see Sacks and Ylvisaker (1966)). Integrating this process yields one that is smoother but less smooth than a Gaussian process and which may be useful for applications in which some differentiability is present but full analyticity may be too strong an assumption.

The second widely used class (within which we can find desirable covariance functions) is the Matérn class of covariance functions

$$C(d, \phi, v) = \frac{1}{2^{v-1}\Gamma(v)} \left(\frac{2\sqrt{v}d}{\phi} \right)^v \mathcal{K}_v \left(\frac{2\sqrt{v}d}{\phi} \right)$$

(see e.g. Handcock and Wallis (1994)). Here ϕ and v are the parameters and $\mathcal{K}_v$ is the modified Bessel function of the third kind and of order v (see Abramowitz and Stegun (1965)).The class is motivated by the smoothness of the spectral density, the wide range of behaviours covered and the interpretability of the parameters. It includes the exponential correlation function as a special case with $v = 0.5$ and the Gaussian correlation function as a limiting case with $v \to \infty$.

Knowledge of information matrix structures could assist in finding a better design. We can find applications of various criteria of design optimality for second-order spatial models in the literature. Here we consider D-optimality, which corresponds to the maximization of the determinant of a standard Fisher information matrix. This method, "plugged" from the widely developed uncorrelated setup, offers considerable potential for automatic implementation, although further development is needed before it can be applied routinely in practice. Theoretical justifications for using the Fisher information for D-optimal designing under a correlation structure can be found in Abt and Welch (1998); Pázman (2004). Zhu and Stein (2005) use simulations (under a Gaussian random field and a Matérn covariance) to study whether the inverse Fisher information matrix is a reasonable approximation to the covariance matrix of maximum likelihood (ML) estimators as well as a reasonable design criterion. They have observed that when the sample size is small, inverse Fisher information matrices underestimate the variance of ML estimators. As sample size increases, the relative error becomes smaller and smaller. They have already observed that the Fisher information matrix does give good estimates of the variance of ML estimators when the sample size is large enough. Although some simulation and theoretical studies shows that the inverse Fisher information matrix is not a good approximation of the covariance matrix of the ML estimates it can still be used as a design criterion if the relationship between these two are monotone, since for the purpose of optimal designing only the correct ordering is important. For instance, Zhu and Stein (2005) observe a monotone relationship between them.

Currently there are two main asymptotical frameworks, increasing domain asymptotics and infill asymptotics, for obtaining limiting distributions of maximum likelihood estimators of covariance parameters in Gaussian spatial models with or without a nugget effect. These limiting distributions differ in some cases. Zhang and Zimmerman (2005) have investigated the quality of these approximations both theoretically and empirically. They have found, that for certain consistently estimable parameters of exponential covariograms, approximations corresponding to these two frameworks perform roughly equally well. For those parameters that cannot be estimated consistently, however, the infill asymptotics is preferable. They have also observed, that the Fisher information appears to be a compromise between the infill asymptotic variance and the increasing domain asymptotic variance. For exponential variograms some infill asymptotic justification can be found in Zhang and Zimmerman (2005).

2 $M_\theta(n)$ structure

2.1 Exponential covariance structure

In this section we are concerned with the exponential semivariogram structure $\gamma(d) = 1 - e^{-rd}$, the special case of Matérn semivariograms with a zero value for the nugget parameter (obtained when $v = 0.5$) and introduce a new range parameter $r = \frac{\sqrt{2}}{\phi}$. For the sake of simplicity we fix $r = 1$, although formulae are still rather complex for this simplification. Without loss of generality we consider a design space $X = [-1, 1]$. All formulae for $M_\theta(n)$ hold for a general design space.

The information matrix $M_\theta(2)$ has the form $\frac{2e^d}{1+e^d}$ and this is an increasing function of d. Thus the optimal design is the maximally distant one (the design space is compact). For more details see Stehlík (2004).

If we consider a three-point-design with distances $d_i = x_{i+1} - x_i, i = 1, 2$ then information $M_\theta(3)$ has form

$$1 + \frac{2 + 2e^{-d_1-2d_2} - 2e^{-d_1} + 2e^{-2d_1-d_2} - 2e^{-2(d_1+d_2)} - 2e^{-d_2}}{e^{-2(d_1+d_2)} - e^{-2d_1} - e^{-2d_2} + 1}.$$

In Stehlík (2004) it is proved, that the design $\{-1, 0, 1\}$ is D-optimal.

Consider now a 4-point design with distances $d_i = x_{i+1} - x_i, i = 1, 2, 3$. Then information has the form

$$M_\theta(4) = 2(-2 + e^{-d_3} + e^{-d_1} + e^{-d_2} + e^{-2d_2-d_1-2d_3} + e^{-2d_1} + e^{-2d_2} + e^{-2d_3}$$

$$-e^{-2d_1-2d_2-2d_3} - e^{-d_1-2d_3} + e^{-2d_1-2d_2-d_3} - e^{-d_3-2d_1} - e^{-d_2-2d_3}$$

$$-e^{-2d_1-d_2} - e^{-2d_2-d_3} + e^{-2d_3-2d_1-d_2} - e^{-2d_2-d_1})/$$

$$(-1 + e^{-2d_3} + e^{-2d_2} - e^{-2d_2-2d_3} + e^{-2d_1} - e^{-2d_1-2d_3} - e^{-2d_1-2d_2} + e^{-2d_1-2d_2-2d_3}).$$

Employing the exchange algorithm we have checked numerically that the D-optimum design is the equidistant one with $d_1 = d_2 = d_3 = 2/3$ and the D-optimum design information is $M = 1.964538$.

Due to the knowledge of the analytical form of the information one can employ also Lipschitz and continuous optimization techniques (see Horst and Tuy (1996)), which can be implemented like a net-searching algorithm. The only problem with such an algorithm is the time required. The Fisher information in the case of a 5-point design has much more complicated form and can be found in Stehlík (2006). Therein we have computationally obtained that the D-optimal design is equidistant with $d_1 = d_2 = d_3 = d_4 = 1/2$ and has information $M = 1.979674635$ (note, that the information is increasing with number of design points).

From these results we are motivated to study an equidistant designs. Let us consider an equidistant design with $d = x_{i+1} - x_i$. We have

$$M_\theta(2) = \frac{2e^d}{1+e^d}, \quad M_\theta(3) = \frac{-1+3e^d}{1+e^d}, \quad M_\theta(4) = \frac{-2+4e^d}{1+e^d}.$$

Denote by $a(k, k-1)$ the ratio $M_\theta(k)/M_\theta(k-1)$. In Kiseľák and Stehlík (2007) the relation $M_\theta(k) = \frac{2-k+ke^d}{1+e^d}$ is proved for arbitrary k. The limit $a(k, k-1)(+\infty) = \frac{k}{k-1}$ is proved in Theorem 1 for arbitrary $k \geq 3$ in the class of covariance functions satisfying a),b) and c).

2.2 A lower bound for $M_\theta(n)$

Consider the lower-bound for $M_\theta(n)$ of the form

$$LB(d) := n \inf_x \frac{x^T C^{-1}(d,r) x}{x^T x}.$$

Then the following theorem holds.

Theorem 1. *Let $C(d,r)$ be a covariance structure satisfying a),b) and c). Then*

1) For any design $\{x, x+d_1, x+d_1+d_2, ..., x+d_1+...+d_{n-1}\}$ given by distances $d_i, i = 1, ..., n-1$ and for any subset of distances $d_{i_j}, j = 1, ..., m$ the lower bound function $(d_{i_1}, ..., d_{i_m}) \to LB(d)$ is increasing in the d's. In particular, for any equidistant design ($\forall i : d_i = d$) the function $d \to LB(d)$ is increasing in d.

2) Denote by $a(n, n-1)$ the ratio $M_\theta(n)/M_\theta(n-1)$. Then $\lim_{\forall i:d_i \to +\infty} a(n, n-1) = \frac{n}{n-1}$.

Proof First, let us recall the Frobenius theorem (see Rao (1973), p.46). An irreducible positive matrix always has a positive characteristic value $\lambda_0(A)$ which is a simple root of the characteristic equation and not smaller than the moduli of other characteristic values. Moreover, if $A \geq B \geq 0$ then $\lambda_0(A) \geq \lambda_0(B)$.

Now let $+\infty > d_1 > d_2 \geq 0$. Then $C_{i,j}(d_1, r) \leq C_{i,j}(d_1, r)$ for all $i, j = 1, .., n$ and thus $C(d_2, r) \geq C(d_1, r) \geq 0$. Employing the Frobenius theorem we have $\lambda_0(C(d_2, r)) \geq \lambda_0(C(d_1, r))$. Our matrix is symmetric and real. Thus we have $\lambda_{min}(C^{-1}(d_2, r)) \leq \lambda_{min}(C^{-1}(d_1, r))$, where $\lambda_{min}(A)$ denotes the minimal eigenvalue of matrix A.

Now,

$$M_\theta(n) = 1^T C^{-1}(d,r) 1 \geq n \inf_x \frac{x^T C^{-1}(d,r) x}{x^T x} = n\lambda_{min}(C^{-1}(d,r))$$

and thus we have proved that for an equidistant design the lower bound function $d \to n \inf_x \frac{x^T C^{-1}(d,r)x}{x^T x}$ is increasing in d. We can prove the rest of 1) in a similar manner.

To prove 2) let us consider the open set U of all covariance matrices $C(d,r)$ with bounded inverse in a Banach space of real matrices $n \times n$. Then the identity $I(n) = \lim_{\forall i:d_i \to +\infty} C(d,r) \in U$ and map $C(n) \to C(n)^{-1}$ is smooth.

This implies[1]

$$a(n, n-1)(+\infty) = \lim_{\forall i: d_i \to +\infty} \frac{1^T C(n)^{-1}(d,r)1}{1^T C(n-1)^{-1}(d,r)1} = \frac{1^T I(n)1}{1^T I(n-1)1} = \frac{n}{n-1}.$$

$\square$

To illustrate the previous result let us consider a power exponential covariance family with zero nugget. For the sake of simplicity let us consider equidistant designs. We have

$$M_\theta(2) = \frac{2e^{rd^p}}{1+e^{rd^p}}, \quad M_\theta(3) = \frac{e^{d^p r}\left(e^{d^p r} - 4e^{2^p d^p r} + 3e^{(1+2^p)d^p r}\right)}{e^{2d^p r} - 2e^{2^p d^p r} + e^{(2+2^p)d^p r}}.$$

Denote by $a(k, k-1)$ the ratio $M_\theta(k)/M_\theta(k-1)$. Then $a(3,2)(+\infty) = \frac{3}{2}$ holds.

3 $M_r(n)$ structure

The Fisher information M_r for the covariance parameter r is much more complex than M_θ. Therefore we consider only the exponential correlation structure.

First let us consider a 2-point design. We have

$$M_r(2) = \frac{d^2 \exp(-2rd)(1 + \exp(-2rd))}{(1 - \exp(-2rd))^2}.$$

The maximal Fisher information is obtained for $d = 0$. In other words, a 2-point design collapses to a 1-point design. The collapsing behaviour of an equidistant design is related to the behaviour of the maximum likelihood estimator (MLE) of the covariance parameter r in a Gaussian field with an exponential covariance structure. These properties are easily verified e.g. from the analytical study of the log-likelihood limit for $rd \to 0$.

To avoid such 'inconvenient' behavior we suggest decreasing the non-diagonal elements through multiplication by a factor α, $0 < \alpha < 1$. We can thereby include a nugget effect (micro-scale variation effect) of the form

$$\gamma(d,r) = \begin{cases} 0, & \text{for } d = 0, \\ 1 - \alpha + \alpha(1 - \exp(-rd)), & \text{otherwise.} \end{cases}$$

Then we obtain $M_{r,1-\alpha}(2) = \frac{\alpha^2 d^2 \exp(-2dr)(\alpha^2 \exp(-2dr)+1)}{(1-\alpha^2 \exp(-2dr))^2}$.

In Stehlík et al (2007) it is proved, that the distance d of the optimal design is an increasing function of the nugget effect, namely $1 - \alpha$.

---

[1] The open question is, whether the supremum of the ratio $M_\theta(n)/M_\theta(n-1)$ is $\frac{n}{n-1}$, i.e. whether $M_\theta(n) \leq \frac{n}{n-1} M_\theta(n-1)$ holds, as it seems to for some particular correlation functions.

Now let us consider a 3-point design. $M_r(3)$ has the form

$$(d_1^2 e^{-2rd_1} - 2e^{-2r(d_1+d_2)}d_1^2 - 2e^{-2r(2d_1+d_2)}d_1^2 + e^{-2r(2d_2+d_1)}d_1^2 - 2e^{-2r(d_1+d_2)}d_2^2$$

$$+ d_2^2 e^{-2rd_2} + d_2^2 e^{-2r(2d_1+d_2)} + d_1^2 e^{-4r(d_1+d_2)} + d_2^2 e^{-4r(d_1+d_2)} + d_2^2 e^{-4rd_2}$$

$$+ d_1^2 e^{-4rd_1} - 2d_2^2 e^{-2r(2d_2+d_1)})/(-1 + e^{-2rd_1} + e^{-2rd_2} - e^{-2r(d_1+d_2)})^2.$$

We have observed many numerical obstacles arising from collapsing of designs based on $M_r(n), n > 3$ without a nugget effect, caused by $0/0$ expressions among other numerical difficulties. For more details see Stehlík (2006).

Let us consider a 5-point equidistant design $d = d_1 = d_2 = d_3 = d_4$ when the correlation structure is exponential with a zero nugget effect. Then we have $M_r(5) = 4d^2 \exp(-2rd)\frac{\exp(-2rd)+1}{\exp(-4rd)-2\exp(-2rd)+1}$ and $\lim_{d\to 0+} M_r(5) = 2/r^2$.

Note that for an equidistant design we have $4M_r(2) = M_r(5)$ and $2M_r(2) = M_r(3)$. In Kiselák and Stehlík (2007) the relation $(n-1)M_r(2) = M_r(n)$ is proved for arbitrary n.

However, if the exponential correlation with a positive nugget effect is considered, there is no such nice relation between $M_{r,1-\alpha}(n)$ and $M_{r,1-\alpha}(2)$, as can be seen by computing $M_{r,1-\alpha}(3)$. The latter has the form

$$\frac{2\alpha^2 d^2 \left(6\alpha^4 - 5\alpha e^{4dr} + e^{4dr}\left(2 + e^{2dr}\right) - \alpha^3 \left(7 + 4e^{2dr}\right) + \alpha^2 \left(2 + 3e^{2dr} + 2e^{4dr}\right)\right)}{\left(2\alpha^3 + e^{4dr} - \alpha^2 \left(1 + 2e^{2dr}\right)\right)^2}$$

One finds that $\lim_{\alpha \to 1} M_{r,1-\alpha}(3)/M_{r,1-\alpha}(2) = 2 = M_r(3)/M_r(2)$.

4 Discussion

In this paper we study the structure of the Fisher information matrices for stationary processes. We show that, under mild conditions on covariance structures, the lower bound for $M_\theta(k)$ is an increasing function of the distances between the design points. If only trend parameters are of interest, the designs uniformly covering the whole design space are very efficient. The nugget effect is also discussed.

Acknowledgement. This work was supported by WTZ project Nr. 04/2006.

References

Abramowitz M, Stegun I (1965) Handbook of Mathematical Functions. Dover, New York

Abt M, Welch W (1998) Fisher information and maximum-likelihood estimation of covariance parameters in Gaussian stochastic processes. The Canadian Journal of Statistics 26:127–137

Banerjee S, Carlin B, Gelfand A (2004) Hierarchical Modeling and Analysis for Spatial Data. Chapman & Hall/CRC, New York

Cressie N (1993) Statistics for Spatial Data. Wiley, New York

Currin C, Mitchell T, Morris M, Ylvisaker D (1991) Bayesian prediction of deterministic functions, with applications to the design and analysis of computer experiments. J Amer StatAssoc 86:953–963

Handcock M, Wallis J (1994) An approach to statistical spatial-temporal modeling of meteorological fields. J Amer StatAssoc 89:368–378

Horst R, Tuy H (1996) Global Optimization - Deterministic Approaches. Springer, Berlin

Kiselák J, Stehlík M (2007) Equidistant and D-optimal designs for parameters of Ornstein-Uhlenbeck process. IFAS research report

Parzen E (1967) Time Series Analysis Papers by Emanuel Parzen, Holden-Day, San Francisco, chap Statistical Inference on Time Series by Hilbert Space Methods I, pp 251–382

Pázman A (2004) Correlated optimum design with parametrized covariance function: Justification of the use of the Fisher information matrix and of the method of virtual noise. Research Report Series of the Department of Statistics and Mathematics, Wirtschaftsuniversität Wien (Nr. 5)

Rao C (1973) Linear Statistical Inference and Its Applications, 2nd edn. Wiley & Sons, New York

Sacks J, Ylvisaker D (1966) Designs for regression problems with correlated errors. The Annals of Mathematical Statistics 37:66–89

Sacks J, Schiller S, Welch W (1989) Designs for computer experiments. Technometrics 31(1):41–47

Santner T, Williams B, Notz W (2003) The Design and Analysis of Computer Experiments. Springer, New York

Stehlík M (2004) Some properties of D-optimal designs for random fields with different variograms. Research Report Series of the Department of Statistics and Mathematics, Wirtschaftsuniversität Wien (Nr. 4)

Stehlík M (2006) Some properties of exchange design algorithms under correlation. WU Wien: Research Report Series/ Department of Statistics and Mathematics (Nr. 28)

Stehlík M, Rodríguez-Díaz JM, Müller WG, López-Fidalgo J (2007) Optimal allocation of bioasseys in the case of parametrized covariance functions: an application in lung's retention of radioactive particles. TEST (in press)

Xia G, Miranda M, Gelfand A (2006) Approximately optimal spatial design approaches for environmental health data. Environmetrics 17:363–385

Zhang H, Zimmerman D (2005) Towards reconciling two asymptotic frameworks in spatial statistics. Biometrika 92(4):921–936

Zhu Z, Stein M (2005) Spatial sampling design for parameter estimation of the covariance function. Journal of Statistical Planning and Inference 134(2):583–603

Optimal Designs for Discriminating among Several Non-Normal Models

Chiara Tommasi

Department of Economics, Business and Statistics – University of Milano
via Conservatorio 7, 20122 Milano, Italy   chiara.tommasi@unimi.it

Summary. Typically T-optimality is used to discriminate among several models with Normal errors. In order to discriminate between two non-Normal models, a criterion based on the Kullback-Liebler distance has been proposed, the so called KL-criterion . In this paper, a generalization of the KL-criterion is proposed to deal with discrimination among several non-Normal models. An example where three logistic regression models are compared is provided.

Key words: Kullback-Leibler distance, T-optimality, KL-optimality

1 Introduction

Many results on optimal experimental designs are derived under the assumption that the statistical model is known at the design stage. Thus, the purpose of the experiment is to estimate a specific aspect of that model. However, rarely is the researcher confident that a particular model underlies the data. More often than not he may be confident that one of several models will be adequate, but does not know which. Thus, in the present paper several rival models are assumed to be available. The purpose of the experiment is to determine which of the models is the more adequate. The rival models may be linear or non-linear and its underlying error distribution is not limited to the Normal.

In order to check the adequacy of a linear regression model, Atkinson (1972) proposes to embed the model in a more general model and to design to estimate the additional parameters in the best way. Atkinson and Cox (1974) generalize this approach to the case of comparing several linear models. Another method for discriminating between two or more regression models (linear or not) is the T-criterion proposed by Atkinson and Fedorov (1975a,b). However, this criterion is based on the assumption that the random errors of the model are Gaussian and homoscedastic. A generalization of the T-optimality criterion for heteroscedastic models is provided by Uciński and Bogacka (2004)

but again there is the assumption of Normal random errors. A generalization
of the T-criterion for discriminating between two generalized linear models is
provided by Ponce de Leon and Atkinson (1992). This new criterion is called
generalized T–optimality criterion and consists of maximizing the deviance
arising from the fit of model 2 when data are generated by model 1. In order
to deal with any distribution for the random errors, López-Fidalgo et al (2005,
2007) propose a new criterion based on the Kullback-Liebler distance. This
new criterion is called the KL-criterion and a design which maximizes it is
called a KL-optimal design.

There are many measures of distance between probability distributions.
Here, the Kullback-Liebler distance is considered because it is the logical basis
for model selection as defined by Akaike, i.e. it is the basis for the derivation
of the well known Akaike's information criterion (AIC). For more details see,
for instance, Burnham and Anderson (1998). Furthermore, the KL-criterion
is a very general criterion, which includes as special cases both the T-criterion
(in the homoscedastic case) and the generalization provided by Uciński and
Bogacka (2004) (in the heteroscedastic case), whenever the error distribution
is Normal (the details are given in López-Fidalgo et al (2007)). In addition,
López-Fidalgo et al (2007) prove that when the discrimination is between two
binary response models then the KL–optimality criterion coincides with the
generalized T–optimality criterion proposed by Ponce de Leon and Atkinson
(1992).

The main aim of this paper is to generalize the KL–criterion to the case
of more than two rival models. In Section 2 the generalized KL-criterion is
defined. In Section 3 an illustrative example is given where three logistic regression models are considered.

2 The generalized KL–criterion

In a more general context than regression models, from now on a statistical
model is a family of probability density functions. Let the i-th statistical
model, $i = 1, \ldots, k$, be denoted by $f_i(y, x, \theta_i)$, where y is the response variable,
x is a vector of experimental conditions and $\theta_i \in \Omega_i \subset \mathbb{R}^{m_i}$ is the unknown
parameter vector.

Following the same idea of Atkinson and Cox (1974), in order to compare these k rival models an extended model which includes them is considered. In other words the k models are embedded in a more general model,
$f_{k+1}(y, x, \theta_{k+1})$. To detect departures from the i-th model in the direction of
the other models the KL-criterion for discriminating between the i-th model
and $f_{k+1}(y, x, \theta_{k+1})$ is used. In this paper the parameters of the extended
model are assumed to be known. Thus local optimal designs are computed.
Other interesting possibilities are Bayesian optimal designs and maximin optimal designs that could be derived from the ideas proposed in this paper.

The i-th KL–optimality criterion function is

$$I_{i,k+1}(\xi) = \min_{\theta_i \in \Omega_i} \int_{\mathcal{X}} \mathcal{I}\left[f_{k+1}(y,x,\theta_{k+1}), f_i(y,x,\theta_i)\right] \xi(dx), \qquad (1)$$

where

$$\mathcal{I}\left[f_{k+1}(y,x,\theta_{k+1}), f_i(y,x,\theta_i)\right] = \int f_{k+1}(y,x,\theta_{k+1}) \log\left[\frac{f_{k+1}(y,x,\theta_{k+1})}{f_i(y,x,\theta_i)}\right] dy$$

is the Kullback–Leibler distance between the true model $f_{k+1}(y,x,\theta_{k+1})$ and the alternative model $f_i(y,x,\theta_i)$.

If ξ is any design, a measure of the efficiency of ξ for detecting departures from the i-th model is the ratio of the criterion function (1) at ξ to its maximum value, i.e.

$$\text{Eff}_{i,k+1}(\xi) = \frac{I_{i,k+1}(\xi)}{I_{i,k+1}(\xi_i^*)}, \quad i = 1,\ldots,k$$

where

$$\xi_i^* = \arg\max_{\xi} I_{i,k+1}(\xi)$$

is the KL–optimum design for discriminating model i from the general model. Let the following linear combination of efficiencies

$$I_\alpha(\xi) = \sum_{i=1}^{k} \alpha_i \cdot \text{Eff}_{i,k+1}(\xi) \qquad (2)$$

be the generalized KL–criterion function which is useful for comparing several models. The subscript α is the $k \times 1$ vector whose items are the coefficients α_i, which are such that $0 \le \alpha_i \le 1$ for $i = 1,\ldots,k$ and $\sum_{i=1}^{k} \alpha_i = 1$. These coefficients measure the relative importance of departures from the k models. If no information about this relative importance is available then the KL–criterion function (2) becomes the arithmetic mean of the k efficiencies. Another possibility could be to impose the constraint of equal efficiencies and then to maximize this common value. But a design which is equally efficient for all the KL–criterion functions (1) may not exist or equality may occur where the efficiency is low. For this reason the KL–criterion function (2) with $\alpha_i = 1/k$ may be preferred when there is no information about the relative importance of the k models.

Instead of using the linear combination (2) the geometric mean of efficiencies could be used, defining the following criterion function,

$$I_\alpha^{GM}(\xi) = \prod_{i=1}^{k} \left[\text{Eff}_{i,k+1}(\xi)\right]^{\alpha_i}.$$

The corresponding optimum design, $\xi_{GM}^* = \arg\max_{\xi} I_\alpha^{GM}(\xi)$, could be found by maximizing

$$\log I_\alpha^{GM}(\xi) = \sum_{i=1}^{k} \alpha_i \cdot \log[\mathrm{Eff}_{i,k+1}(\xi)] = \sum_{i=1}^{k} \alpha_i \cdot \log[I_{i,k+1}(\xi)] - \sum_{i=1}^{k} \alpha_i \cdot \log[I_{i,k+1}(\xi_i^*)].$$

Thus, ξ_{GM}^* actually maximizes the geometric mean of the criterion functions $I_{i,k+1}(\xi)$, $i = 1, \ldots, k$, without regard for their possibly different magnitudes. For this reason a linear combination of efficiencies is preferred here.

Another interesting criterion function would be

$$I_m(\xi) = \min_{i=1,\ldots,k} \mathrm{Eff}_{i,k+1}(\xi).$$

It seems very appropriate for discriminating purposes since it naturally gives equal efficiencies to the models between which it is most difficult to discriminate. Its properties will be studied in depth in a future paper.

Let ξ_α^* be the generalized KL–optimum design for discriminating among the k models, i.e.

$$\xi_\alpha^* = \arg\max_\xi I_\alpha(\xi).$$

Theoretical results are given for known values of α_i in (2). In the numerical example both the cases of constant α_i and equal efficiencies will be considered.

A design for which the sets

$$\Omega_i(\xi) = \left\{ \hat{\theta}_i : \hat{\theta}_i(\xi) = \arg\min_{\theta_i \in \Omega_i} \int_{\mathcal{X}} \mathcal{I}\left[f_{k+1}(y, x, \theta_{k+1}), f_i(y, x, \theta_i)\right] \xi(dx) \right\},$$
$$i = 1, \ldots, k \tag{3}$$

are singletons is called a regular design, otherwise it is called singular design. From now on the generalized KL–optimum design ξ_α^* is assumed to be regular.

Let ξ and $\bar{\xi}$ be any two designs, then the directional derivatives of $I_{i,k+1}(\xi)$ at ξ in the direction $\bar{\xi} - \xi$ is

$$\partial I_{i,k+1}(\xi, \bar{\xi}) = \lim_{\beta \to 0^+} \frac{I_{i,k+1}[(1-\beta)\xi + \beta\bar{\xi}] - I_{i,k+1}(\xi)}{\beta}.$$

Let ξ_x be a design which puts the whole mass at point x. Assuming that ξ is regular, then

$$\partial I_{i,k+1}(\xi, \bar{\xi}) = \int_{\mathcal{X}} \psi_{i,k+1}(x, \xi)\, \bar{\xi}(dx)$$

where the function

$$\psi_{i,k+1}(x, \xi) = \mathcal{I}\left[f_{k+1}(y, x, \theta_{k+1}), f_i(y, x, \hat{\theta}_i)\right]$$
$$- \int_{\mathcal{X}} \mathcal{I}\left[f_{k+1}(y, x, \theta_{k+1}), f_i(y, x, \hat{\theta}_i)\right] \xi(dx)$$

is the directional derivative of $I_{i,k+1}(\xi)$ at ξ in the direction $\xi_x - \xi$ and $\hat{\theta}_i$ is the unique element of $\Omega_i(\xi)$ defined by (3). With this notation the directional derivative of $I_\alpha(\xi)$ at ξ in the direction $\bar{\xi} - \xi$ is

$$\partial I_\alpha(\xi, \bar{\xi}) = \sum_{i=1}^{k} \alpha_i \frac{\partial I_{i,k+1}(\xi, \bar{\xi})}{I_{i,k+1}(\xi_i^*)} = \int_\mathcal{X} \psi_\alpha(x, \xi)\, \bar{\xi}(dx),$$

where

$$\psi_\alpha(x, \xi) = \sum_{i=1}^{k} \alpha_i \frac{\psi_{i,k+1}(x, \xi)}{I_{i,k+1}(\xi_i^*)}$$

is the directional derivative of $I_\alpha(\xi)$ at ξ in the direction $\xi_x - \xi$.

Theorem 1. *Let ξ_α^* be a regular design for discriminating among k models. Then a necessary and sufficient condition for the design ξ_α^* to be a generalized KL–optimum is $\psi_\alpha(x, \xi_\alpha^*) \leq 0, \quad x \in \mathcal{X}$.*

The proof is a straightforward generalization of the proof of Theorem 1 in López-Fidalgo et al (2007).

The analytical construction of generalized KL–optimal designs is intractable. For this reason numerical procedures must be adopted in practice. In this paper the classical steepest ascent algorithm described by Wynn (1970) and Fedorov (1972) is used.

Remark 1. In order to prove Theorem 1, the concavity of the criterion function (2) is required. Let ξ_1 and ξ_2 be any two designs and $0 < \lambda < 1$ be a constant. From the definition of the KL-criterion function it follows that

$$\begin{aligned}
I_{i,k+1}[\lambda\xi_1 + (1-\lambda)\xi_2] = \min_{\theta_i \in \Omega_i} &\left\{ \lambda \int_\mathcal{X} \mathcal{I}\left[f_{k+1}(y, x, \theta_{k+1}), f_i(y, x, \theta_i)\right] \xi_1(dx) \right. \\
&\left. + (1-\lambda) \int_\mathcal{X} \mathcal{I}\left[f_{k+1}(y, x, \theta_{k+1}), f_i(y, x, \theta_i)\right] \xi_2(dx) \right\} \\
\geq\ & \lambda\, I_{i,k+1}(\xi_1) + (1-\lambda)\, I_{i,k+1}(\xi_2),
\end{aligned}$$

where the last inequality follows immediately by replacing each term

$$\int_\mathcal{X} \mathcal{I}\left[f_{k+1}(y, x, \theta_{k+1}), f_i(y, x, \theta_i)\right] \xi_j(dx), \quad j = 1, 2$$

with its minimum $I_{i,k+1}(\xi_j)$, $j = 1, 2$. Criterion function (2) is a linear combination of KL-criterion functions thus, it is also concave.

3 An example

In this section a classical example given by Atkinson and Cox (1974) in the context of linear regression models is generalized in order to apply the above theoretical results. More specifically, Atkinson and Cox (1974) consider the following rival models for the expected response: $\eta_1 = \beta_1 x$, $\eta_2 = \beta_0 + \beta_1 x$, $\eta_3 = \beta_1 x + \beta_2 x^2$ and the combined model $\eta_4 = \beta_0 + \beta_1 x + \beta_2 x^2$.

In this paper, the experimental conditions are assumed to vary in the interval $[0, 1]$ and logistic regression models are considered instead of linear regression models. In other words, y is a binary response variable such that

$$P(y = 1, x, \theta_i) = F(\eta_i) = \frac{e^{\eta_i}}{1 + e^{\eta_i}}.$$

For the parameters of the combined model the following nominal values have been used, $\beta_0 = \beta_1 = \beta_2 = 1$.

For this problem the criterion function (2) becomes

$$I_\alpha(\xi) = \alpha_1 \frac{I_{1,4}(\xi)}{0.1098} + \alpha_2 \frac{I_{2,4}(\xi)}{0.0026} + (1 - \alpha_1 - \alpha_2) \frac{I_{3,4}(\xi)}{0.1102}.$$

Generalized KL–optimal designs for different values of the coefficients α_i are listed in Table 1. The efficiencies of each design for detecting departures from the i-th model are given in the last three columns.

Table 1. Generalized KL–optimal designs and efficiencies

α_1	α_2	Generalized KL–optimal design	$\mathrm{Eff}_{1,4}(\xi^*_\alpha)$	$\mathrm{Eff}_{2,4}(\xi^*_\alpha)$	$\mathrm{Eff}_{3,4}(\xi^*_\alpha)$
1	0	$\left\{\begin{array}{ccc} 0 & 0.5 & 1 \\ 0.9993 & 0.0003 & 0.0003 \end{array}\right\}$	0.99934	0.00040	0.99933
0	1	$\left\{\begin{array}{ccc} 0 & 0.4160 & 1 \\ 0.2201 & 0.4655 & 0.3144 \end{array}\right\}$	0.23853	1	0.22009
0	0	$\left\{\begin{array}{ccc} 0 & 0.5 & 1 \\ 0.9993 & 0.0003 & 0.0003 \end{array}\right\}$	0.99934	0.00040	0.99933
1/3	1/3	$\left\{\begin{array}{ccc} 0 & 0.4 & 1 \\ 0.9360 & 0.0400 & 0.0240 \end{array}\right\}$	0.93835	0.10710	0.93662
0.3	0.4	$\left\{\begin{array}{ccc} 0 & 0.368 & 1 \\ 0.6162 & 0.2391 & 0.1446 \end{array}\right\}$	0.63075	0.62131	0.61665

The first three rows of Table 1 list the generalized KL–optimal designs for discriminating between each one of the three competing models and the combined model. Thus these generalized KL-optimal designs in reality are KL-optimal designs. For both the first and the third models the KL–optimal design, ξ^*_i $(i = 1, 3)$, is the singular design with measure concentrated at the point zero. Let us denote this singular design by ξ_0, then $\xi^*_i = \xi_0$ with $i = 1, 3$. Since, with a singular design, no unique parameter estimation is possible, a regularization procedure is used; see for instance Fedorov and Hackl (1997, §3.2). In more detail, instead of using the KL–optimality criterion (1), the following criterion function

$$I^{\gamma}_{i,k+1}(\xi) = I_{i,k+1}[(1-\gamma)\xi + \gamma\tilde{\xi}]$$

is used, where $\tilde{\xi}$ is a regular design and $0 < \gamma < 1$. Let $\xi^*_{i\gamma}$ be the optimal design for $I^{\gamma}_{i,k+1}(\xi)$. Then, from the concavity of the KL-criterion function (1),

$$0 \le I_{i,k+1}(\xi^*_i) - I_{i,k+1}[(1-\gamma)\xi^*_{i\gamma} + \gamma\tilde{\xi}] \le \gamma\,[I_{i,k+1}(\xi^*_i) - I_{i,k+1}(\tilde{\xi})]. \qquad (4)$$

From inequality (4) it follows that $\xi^*_{in} = (1-\gamma)\xi^*_{i\gamma} + \gamma\tilde{\xi}$ is a "nearly" KL-optimal design and in particular the smaller γ the better this design.

For both the first and the third models, $\xi^*_{i\gamma} = \xi^*_i = \xi_0$. Thus, using the same regular design

$$\tilde{\xi} = \left\{ \begin{matrix} 0 & 0.5 & 1 \\ 1/3 & 1/3 & 1/3 \end{matrix} \right\}$$

and $\gamma = 0.001$, the regularization procedure leads to the following nearly KL–optimal design,

$$\xi^*_{in} = \left\{ \begin{matrix} 0 & 0.5 & 1 \\ 0.9993 & 0.0003 & 0.0003 \end{matrix} \right\}, \quad i = 1, 3.$$

This nearly KL-optimal design puts almost the whole mass at zero since both models do not have intercepts while the combined model has. Thus, zero is the point where there is the best discrimination. The other two support points make unique estimation of the parameters possible, whichever the chosen model is.

In the forth row of Table 1 there is the optimal design corresponding to constant weights $\alpha_i = 1/3$, $i = 1, 2, 3$. In this example the generalized KL–optimal design corresponding to the arithmetic mean of the efficiencies is very efficient for detecting departures from the first and the third models but absolutely inefficient for detecting departures from the second model. On the other hand, an almost equally efficient design for all the KL–optimal criterion functions is given in the fifth row and the common efficiency is about 60%. Let this almost equally efficient optimal design be denoted by $\xi^*_{(0.3,0.4)}$.

Table 2 lists the KL-optimal designs for discriminating between pairs of models. The last column of Table 2 provides the efficiency of $\xi^*_{(0.3,0.4)}$ with respect to these KL-optimal designs.

Comparing $\xi^*_{(0.3,0.4)}$ with the optimal designs given in Table 2, it seems that the support point zero essentially allows us to discriminate between the first (and the third) model and the second one, while the other two support points allow us to discriminate between the first and the third models, admittedly with a low efficiency in the last case.

Acknowledgement. The author is very grateful to both the anonymous referees for the useful comments and suggestions which have improved an earlier version of this paper.

Table 2. KL–optimal designs for discriminating between two models

Comparisons	KL–optimal design	$\mathrm{Eff}\left[\xi^*_{(0.3,0.4)}\right]$
η_1 *versus* η_2	ξ_0	0.6695
η_3 *versus* η_2	ξ_0	0.6166
η_2 *versus* η_3	$\left\{\begin{array}{ccc} 0 & 0.4538 & 1 \\ 0.2396 & 0.4667 & 0.2937 \end{array}\right\}$	0.6357
η_1 *versus* η_3	$\left\{\begin{array}{cc} 0.3683 & 1 \\ 0.6599 & 0.3401 \end{array}\right\}$	0.3815

References

Atkinson A (1972) Planning experiments to detect inadequate regression models. Biometrika 59:275–293

Atkinson AC, Cox D (1974) Planning experiments for discriminating between models. JR Statist Soc B 36:321–348

Atkinson AC, Fedorov V (1975a) The designs of experiments for discriminating between two rival models. Biometrika 62:57–70

Atkinson AC, Fedorov V (1975b) Optimal design: experiments for discriminating between several models. Biometrika 62:289–303

Burnham K, Anderson D (1998) Model selection and inference: a practical information-theoretic approach. Springer-Verlag, New York

Fedorov V (1972) Theory of optimal experiments. Academic Press, New York

Fedorov V, Hackl P (1997) Model-oriented design of experiments. Springer-Verlag, New York

Ponce de Leon A, Atkinson A (1992) The design of experiments to discriminate between two rival generalized linear models. In: Lecture Notes in Statistics - Advances in GLM and Statistical Modelling, Springer-Verlag, New York, pp 159–164

López-Fidalgo J, Tommasi C, Trandafir P (2005) Optimal designs for discriminating between heteroscedastic models. In: Proceedings of the 5th St.Petersburg Workshop on Simulation, NII Chemestry Saint Petersburg University Publishers, Saint Petersburg, pp 429–436

López-Fidalgo J, Tommasi C, Trandafir P (2007) An optimal experimental design criterion for discriminating between non-normal models. JRStatistSoc B 69(2):1–12

Uciński D, Bogacka B (2004) T-optimum designs for multiresponse heteroscedastic models. In: mODa 7 - Advances in Model-Oriented Design and Analysis, Physica-Verlag, Heidelberg, New York, pp 191–199)

Wynn H (1970) The sequential generation of d-optimal experimental designs. Ann Math Statist 41:1655–1664

Optimal Orthogonal Three-Level Factorial Designs for Factor Screening and Response Surface Exploration

Kenny Q. Ye[1], Ko-Jen Tsai[2], and William Li[3]

[1] Department of Epidemiology and Population Health, Albert Einstein College of Medicine, 1300 Morris Park Ave, Bronx, NY 10461, U.S.A.
`kye@aecom.yu.edu`
[2] Yes Solutions, 66 Exeter St, Williston Park, NY 11596, U.S.A.
`ko-jen.tsai@yessolutions.com`
[3] Operations and Management Science Dept., University of Minnesota, 321 19th Ave. S, Room 3-225, Minneapolis, MN 55455, U.S.A.
`wli@csom.umn.edu`

Summary. Three-level factorial designs can be used to perform factor screening and subsequently response surface exploration on its projections in a single stage experiment. Here we select optimal designs for this approach from 18-run and 27-run orthogonal designs. Our choices are based on two types of design criteria. Besides commonly used model estimation criteria, we also consider model discrimination criteria.

Key words: model estimation, model discrimination, orthogonal arrays, geometric isomorphism, average optimal scale

1 Background

Traditionally, response surface methodology takes two stages of experiments, a factor screening experiment to screen out unimportant factors, usually using an orthogonal two-level factorial design, followed by an experiment to fit a second order model on a smaller set of factors, with the central composite design as a common choice. Cheng and Wu (2001) proposed a new approach for response surface studies with a single-stage three-level factorial experiment that facilitates the two-stage data analysis. The first stage of the data analysis uses established methods for factor screening. At the second stage, a second-order model

$$y = \beta_0 + \sum_{i=1}^{k} \beta_i x_i + \sum_{i=1}^{k} \beta_{ii} x_i^2 + \sum_{1 \le i < j \le k} \beta_{ij} x_i x_j + \epsilon \tag{1}$$

is fitted on a projection of the three-level factorial design.

Projection Properties and Geometric Isomorphism

The key of this new approach is the projection properties of three-level factorial designs. As pointed out by Box and Wilson (1951), in general, regular 3^{k-p} type of factorial designs do not facilitate the estimation of second-order models on their projections well. However, Cheng and Wu (2001) showed that nonregular three-level factorial designs usually have good projection properties to facilitate the estimation of second-order models. They also noted that level permutations of an orthogonal array may result in changes of estimability and estimation efficiencies of the second-order model on its projections. This observation led to the work by Cheng and Ye (2004), which defined the geometric isomorphism of three-level factorial designs. They argued that two geometrically isomorphic designs have the same geometric structures, and therefore share the same design properties. They also developed an indicator function representation of factorial designs and showed that the coefficients of the indicator functions of two geometrically non-isomorphic designs have different patterns. From the indicator functions, they derived a measure called β-Word Length Pattern(β-WLP) as a simple way of classifying factorial designs. Two designs with different β-WLPs must be geometrically non-isomorphic, but the reverse is not necessarily true.

Data Analysis for Factor Screening

At the first stage of a data analysis of a response surface study, one can simply perform a main-effect analysis for factor screening. However, ignoring interaction effects may lead to the false exclusion of some important factors. A more comprehensive Bayesian approach was given by Box and Meyer (1993). Their main idea is to evaluate the posterior probability of 2^p models, each of which corresponds to a subset of the p candidate factors and includes all main effects and interaction effects of these factors. That is,

$$p(M_i|\mathbf{y}) = \frac{p(M_i)f(\mathbf{y}|M_i)}{\displaystyle\sum_{M_j \in \mathcal{M}} p(M_j)f(\mathbf{y}|M_j)} \tag{2}$$

where $f(\mathbf{y}|M_i)$ is the likelihood of observed response $\mathbf{y}$ under model M_i and $\mathcal{M}$ is the collection of 2^p models. The importance of the factors are evaluated by their marginal posterior probability, which is the sum of posterior probabilities of all models with the jth factor being active. This idea was originally introduced to analyze two-level designs, but can be easily modified to analyze three-level designs as proposed by Ye et al (2006). They use a model space of $\binom{p}{k}$ models, each of which is a second-order model of k factors. They also suggest that k should be chosen so that models are estimable.

Selection of optimal three-level designs

Selection of the optimal three-level factorial designs for response surface studies has been discussed by several papers. Cheng and Wu (2001) considered the estimability of the second-order model on all projections of a design and used average D-efficiency as the criterion. They studied regular 27-run orthogonal designs, projections of a non-regular $OA(27, 3^{13})$, an $OA(18, 3^7)$ and an $OA(36, 3^{12})$. Xu et al (2004) slightly revised the selection procedure and considered projections of three combinatorially non-isomorphic $OA(18, 3^7)$s and three combinatorially non-isomorphic $OA(27, 3^{13})$s. They first screened out poor OAs using the *generalized minimum aberration* criterion, proposed by Xu and Wu (2001), and a new *projection aberration* criterion, then considered designs obtained by level permutations from the remaining OAs using the same projection efficiency criterion used by Cheng and Wu (2001). A similar study by Tsai et al (2000) used average A-efficiency of second-order models to select optimal 18-run orthogonal designs. They also used a different strategy to generate the candidate designs. Instead of taking projections of known OAs, they generated designs by augmenting columns to non-isomorphic designs with fewer factors. Moreover, Cheng and Ye (2004) used β-WLP to rank the designs that are projections of an $OA(18, 2^1 3^7)$, including the mixed-level designs. The work we present here differs from previous contributions mainly in that we consider the model discrimination properties, in addition to the projection estimability and estimation efficiency.

The remainder of the paper is organized as following. Section 2 presents the criteria used to compare the designs. The procedures for selecting optimal designs and the results are presented in Section 3. Section 4 gives some concluding remarks.

2 Design criteria

We consider two sets of design criteria, model estimation criteria and model discrimination criteria. For each 18-run design, we consider all second-order models of 2, 3 and 4 factors. For each 27-run design, we consider all second-order models of 2, 3, 4 and 5 factors. For models of a given number of factors k, we consider the following two model estimation criteria, *Estimation Capacity* (EC) and *Information Capacity* (IC). The first is the proportion of estimable models, and the second is the average D-efficiency of all models. That is, $\sum_i |X_i^T X_i| / \binom{p}{k}$, where X_i is the model matrix of the second-order model of the ith subset of k factors. *Information Capacity* was proposed by Sun (1993), and was then modified by Li and Nachtsheim (2000) to construct efficient two-level designs. The definition of IC given here is the same as the one used by the latter paper. We should note here that for mixed-level 18-run designs, we average the models with and without the two-level factor separately as the quantities $\sum_i |X_i^T X_i|$ of the two groups are not directly comparable as their model sizes differ. The second-order model with a two-level factor has one less parameter.

Although model discrimination, as a property, has been largely ignored in the recent literature of optimal designs for factor screening and model selection, this topic received considerable attention in 1960s and 1970s. See Hill (1978) for a comprehensive review on these early works. Two very recent papers, Bingham and Chipman (2007) and Jones et al (2007), revisited this topic when investigating optimal factorial designs for model selection. Both focused on the prediction difference between two models. The former developed a Bayesian criterion and the latter proposed six non-Bayesian criteria. Here we use two criteria proposed in the latter, which are *Average Expected Prediction Differences*

$$AEPD = \frac{1}{\binom{n}{2}} \sum_{1 \leq i < j \leq n} E(\|\hat{\mathbf{y}}_i - \hat{\mathbf{y}}_j\| \, | \, \|\mathbf{y} = 1\|) \tag{3}$$

and *Minimum Maximum Prediction Differences*

$$MMPD = \min_{1 \leq i < j \leq n} \max_{\|\mathbf{y}=1\|} \|\hat{\mathbf{y}}_i - \hat{\mathbf{y}}_j\| \tag{4}$$

where n is the number of candidate models, $\mathbf{y}$ is the response vector, and $\hat{\mathbf{y}}_i$ is the fitted value of the ith model. The expectation in (3) is over $\mathbf{y}$ uniformly distributed on a unit ball. Both criteria can be very efficiently computed compared to the Bayesian criterion proposed by Bingham and Chipman (2007). Details about these two criteria can be found in Jones et al (2007). For each design, we evaluate these criteria for each number of factors k separately, while n equals the number of estimable k-factor second-order models.

Since different design criteria do not always agree with each other, here we propose an overall measure of design optimality, *Average Optimal Scale* (AOS), defined as

$$AOS = (\prod_{i=1}^{q} \mathcal{C}^i / \mathcal{C}^i_{\max})^{1/q}, \tag{5}$$

where q is the total number of criteria to be considered, and $\mathcal{C}^i_{\max}$ is the maximum value of the ith criterion over all candidate designs. To select optimal designs for factor screening, we use AOS which combines the IC and the AEPD criteria values of different model sizes.

3 Optimal 18-run and 27-run orthogonal designs

3.1 Select 18-run orthogonal arrays

A complete catalog of geometrically non-isomorphic 18-run orthogonal designs is constructed by Tsai (2005). Therefore, we only need to examine all designs in this catalog. The number of designs and the number of designs with full EC (*i.e.*, EC = 100%) are listed in Table 1. Note that there is no OA(18,3^7) and no OA(18,$2^1 3^7$) design with full estimation capacity. Therefore, we do not recommend them for factor screening in response surface studies. If a seventh three-level factor has to be included in the study, one might want to consider non-orthogonal designs.

Table 1. Number of OA18s with full estimation capacity

s	OA(18, 3^s)					OA(18, $2^1 3^s$)				
	3	4	5	6	7	3	4	5	6	7
Distinct Designs	13	133	332	478	284	119	1836	1332	1617	762
$EC_4 = 1$	NA	98	132	67	0	109	979	369	67	0
$EC_3 = 1$	11	122	276	224	0	116	1253	1008	649	0

Table 2. Some optimal 18-run orthogonal designs

3^3			3^4		3^5			3^6				$2^1 3^3$		$2^1 3^4$			$2^1 3^5$				$2^1 3^6$				
$v1$	$v2$	$v3$	$v3$	$v4$	$v3$	$v4$	$v5$	$v3$	$v4$	$v5$	$v6$	$v0$	$v3$	$v0$	$v3$	$v4$	$v0$	$v3$	$v4$	$v5$	$v0$	$v3$	$v4$	$v5$	$v6$
0	0	0	0	0	0	0	0	0	0	0	0	1	0	1	0	0	1	0	0	0	1	0	0	0	0
0	0	1	1	1	1	1	1	1	1	2	2	-1	1	-1	1	1	-1	0	1	2	-1	0	1	2	2
0	1	1	0	2	0	1	2	0	2	1	2	-1	0	-1	0	2	-1	1	0	1	-1	1	0	1	1
0	1	2	2	2	2	2	2	2	0	2	1	1	2	1	2	2	-1	2	2	2	1	2	2	0	2
0	2	0	1	1	1	2	0	1	2	0	1	1	1	1	1	1	1	1	2	0	-1	1	2	2	0
0	2	2	2	0	2	0	1	2	1	1	0	-1	2	-1	2	0	1	2	1	1	1	2	1	1	1
1	0	2	1	2	0	2	2	0	2	2	1	-1	0	1	1	2	1	1	2	1	1	1	1	0	1
1	0	2	2	0	2	1	1	2	0	1	2	1	2	1	2	0	1	2	0	2	-1	2	2	2	1
1	1	0	0	1	1	2	1	1	1	1	1	1	1	1	0	1	-1	0	2	0	-1	0	1	1	0
1	1	0	2	1	2	0	0	2	2	0	0	-1	2	-1	2	1	1	2	1	0	1	1	0	2	2
1	2	1	0	0	0	1	0	0	1	0	2	1	0	-1	0	0	-1	0	1	1	1	0	2	1	2
1	2	1	1	2	1	0	2	1	0	2	0	-1	1	-1	1	2	-1	1	0	2	-1	2	0	0	0
2	0	0	0	1	1	0	2	1	2	1	0	1	1	-1	0	1	-1	1	1	0	1	1	2	1	0
2	0	1	2	2	2	2	0	2	1	0	1	-1	2	-1	2	2	-1	2	2	1	-1	2	0	1	2
2	1	1	1	0	0	0	1	0	1	2	0	1	0	1	1	0	1	0	0	1	-1	0	2	0	1
2	1	2	1	0	1	1	0	1	0	0	2	-1	1	-1	1	0	1	1	1	2	1	2	1	2	0
2	2	0	0	2	0	2	1	0	0	1	1	-1	0	1	0	2	1	0	2	2	1	0	0	2	1
2	2	2	2	1	2	1	2	2	2	2	2	1	2	1	2	1	-1	2	0	0	-1	1	1	0	2

Design criteria IC, AEPD and MMPD are evaluated only on those designs with full estimation capacity. The designs with the best AOS values are listed in Table 2 and their properties are given in Table 3. Note in Table 3 the IC and MMPD values for models having two factors are not given because they are both equal to 1 for all candidate designs. The IC-criterion values of the mixed-level designs are listed separately for models with and without the two-level factor (by the first and the second numbers in the parenthesis respectively). For example, the best OA(18,$2^1 3^3$) design gives the average D-efficiency of three 3-factor models with 1 two-level and 2 three-level factors at the value 1 and the D-efficiency of the 3-factor models with all 3 three-level factors at the value 0.63. In Table 2, columns $v1$ and $v2$ are shared by all listed designs, and hence are listed only once.

Table 3. Properties of optimal $OA(18, 3^s)$ designs

	IC		AEPD			MMPD		AOS
	3f	4f	2f	3f	4f	3f	4f	
$OA(18, 3^3)$	0.82	NA	0.32	NA	NA	NA	NA	1.000
$OA(18, 3^4)$	0.66	0.13	0.35	0.38	NA	1	NA	0.964
$OA(18, 3^5)$	.63	0.058	0.36	0.49	0.22	.99	1	0.945
$OA(18, 3^6)$	.63	0.058	0.38	0.43	0.26	.99	1	0.990
$OA(18, 2^1 3^3)$	(1,.63)	0.44	0.33	0.38	NA	1	NA	0.942
$OA(18, 2^1 3^4)$	(.78,.66)	(.16,.13)	0.35	0.41	0.30	0.95	0.98	0.851
$OA(18, 2^1 3^5)$	(.68,.60)	(.087,.054)	0.36	0.42	0.29	.88	.86	0.884
$OA(18, 2^1 3^6)$	(.63,.60)	(.079,.044)	0.37	0.44	0.29	.82	.76	0.894

3.2 Select optimal 27-run orthogonal designs

To find the optimal OA27 designs, we examine the projections of three $OA(27, 3^{13})$s considered by Xu et al (2004), and listed in Table 4. Design III is the regular design. For each of the three $OA(27, 3^{13})$s, we first consider all its projections and apply all possible level permutations to each projection, and obtain a set of designs with distinct β-WLP. We then evaluate design properties of these designs, which are geometrically non-isomorphic. Note that we may miss many geometrically non-isomorphic designs since some of them share the same β-WLPs.

Table 4. Non-regular $OA(27, 3^{13})$ chosen from Xu et al (2004)

I													II													III (Regular)												
0	0	0	0	0	0	0	0	0	0	0	0	0	0	0	0	0	0	1	2	0	0	0	0	0	0	0	0	0	0	0	0	0	0	0	0	0	0
0	1	1	2	0	1	1	2	0	1	1	2	0	0	1	1	2	2	2	2	2	2	2	1	0	0	0	0	1	0	1	1	1	2	2	2	0	1	2
0	2	2	1	0	2	2	1	0	2	2	1	0	0	2	2	1	2	1	0	0	0	0	1	2	1	0	0	2	0	2	2	2	1	1	1	0	2	1
1	0	1	1	1	0	1	1	1	0	1	1	0	0	0	1	1	2	1	0	1	1	0	1	1	1	0	1	0	1	0	1	1	0	1	1	2	2	2
1	1	2	0	1	1	2	0	1	1	2	0	0	0	1	2	2	0	1	1	1	1	1	2	0	2	0	1	1	1	1	2	2	2	0	0	2	0	1
1	2	0	2	1	2	0	2	1	2	0	2	0	0	2	0	0	2	2	0	2	1	1	0	2	2	0	1	2	1	2	0	0	1	2	2	2	1	0
2	0	2	2	2	0	2	2	2	0	2	2	0	0	0	0	1	0	0	1	2	2	2	2	2	1	0	2	0	2	0	2	2	0	2	2	1	1	1
2	1	0	1	2	1	0	1	2	1	0	1	0	0	1	2	1	1	0	0	0	2	1	0	1	0	0	2	1	2	1	0	0	2	1	1	1	2	0
2	2	1	0	2	2	1	0	2	2	1	0	0	0	2	1	0	1	0	2	1	0	2	2	1	2	0	2	2	2	2	1	1	1	0	0	1	0	2
0	0	2	1	1	1	0	2	2	2	1	0	1	1	0	1	2	1	1	0	2	0	1	2	2	0	1	0	0	1	1	0	1	1	0	1	1	1	1
0	1	1	0	1	2	2	1	2	0	0	2	1	1	1	0	0	1	2	0	1	2	0	2	0	1	1	0	1	1	2	1	2	0	2	0	1	2	0
0	2	0	2	1	0	1	0	2	1	2	1	1	1	2	1	2	0	0	0	0	1	2	0	0	1	1	0	2	1	0	2	0	2	1	2	1	0	2
1	0	1	2	2	1	2	0	0	2	0	1	1	1	0	2	2	2	0	2	1	2	0	0	2	2	1	1	0	2	1	1	2	1	1	2	0	0	0
1	1	0	1	2	2	1	2	0	0	2	0	1	1	1	0	1	1	1	2	0	1	2	1	2	2	1	1	1	2	2	2	0	0	0	1	0	1	2
1	2	2	0	2	0	0	1	0	1	1	2	1	1	2	0	1	2	0	1	1	0	1	1	0	0	1	1	2	2	0	0	1	2	2	0	0	2	1
2	0	0	0	0	1	1	1	1	2	2	2	1	1	0	1	0	0	2	1	0	2	1	1	1	2	1	2	0	0	1	2	0	1	2	0	2	2	2
2	1	2	2	0	2	0	0	1	0	1	1	1	1	1	2	0	2	1	1	2	0	2	0	1	1	1	2	1	0	2	0	1	0	1	2	2	0	1
2	2	1	1	0	0	2	2	1	1	0	0	1	1	2	2	1	0	2	2	2	1	0	2	1	0	1	2	2	0	0	1	2	2	0	1	2	1	0
0	0	0	0	2	2	2	2	1	1	1	1	2	2	0	0	2	1	2	1	1	1	2	0	1	0	2	0	0	2	2	0	2	2	0	2	2	2	2
0	1	1	2	2	0	0	1	1	2	2	0	2	2	1	1	0	2	0	1	0	1	0	2	2	0	2	0	1	2	0	1	0	1	2	1	2	0	1
0	2	2	1	2	1	1	0	1	0	0	2	2	2	2	2	0	0	1	0	1	2	2	1	2	0	2	0	2	2	1	2	1	0	1	0	2	1	0
1	0	1	1	0	2	0	0	2	1	2	2	2	2	0	2	1	2	2	0	0	0	2	2	0	2	2	1	0	0	2	1	0	2	1	0	1	1	1
1	1	2	0	0	0	1	2	2	2	0	1	2	2	1	1	1	0	2	2	1	0	1	0	2	1	2	1	1	0	0	2	1	1	0	2	1	2	0
1	2	0	2	0	1	2	1	2	0	1	0	2	2	2	0	2	2	1	2	0	2	1	2	1	1	2	1	2	0	1	0	2	0	2	1	1	0	2
2	0	2	2	1	2	1	1	0	1	0	0	2	2	0	2	0	1	0	2	2	1	1	1	0	1	2	2	0	1	2	2	1	2	2	1	0	0	0
2	1	0	1	1	0	2	0	0	2	1	2	2	2	1	0	2	0	0	0	2	0	0	1	1	2	2	2	1	1	0	0	2	1	1	0	0	1	2
2	2	1	0	1	1	0	2	0	0	2	1	2	2	2	1	1	1	1	1	2	2	0	0	0	2	2	2	2	1	1	1	0	0	0	2	0	2	1

Table 5 lists the number of designs with distinct β-WLP and the number of designs with full EC. Note that the regular designs have EC=0 when the number of factors exceeds four, explained by Cheng and Wu (2001) as *curse of three-letter words*. Since evaluating AEPD and MMPD is much more time consuming than evaluating estimation efficiency, we only evaluate the model discrimination criteria of OA$(27,3^p)$s with top 10% IC-criterion values when $p \geq 8$. The best designs and their properties are listed in Table 6. The designs are presented as projections of OA$(27,3^{13})$s with level permutations applied to some of its factors. The superscripts 1, 2, and 3 in Table 6 correspond to no permutation, (0 1 2) $\rightarrow$ (1 2 0), and (0 1 2) $\rightarrow$ (2 0 1), respectively.

Table 5. Number of designs with different β-WLP and number of geometrically non-isomorphic designs with full estimation capacity of 5-factor second order models

	OA27 I	OA27 II	OA27 III	Overall
3 Factor	12/14	11/11	1/3	17/19
4 Factor	102/106	310/310	2/6	373/377
5 Factor	367/602	3858/3894	0/9	4206/4475
6 Factor	974/2884	15695/16065	0/21	16669/18953
7 Factor	1262/9659	46028/48181	0/41	47290/57852
8 Factor	437/23083	92972/108265	0/59	93409/131407
9 Factor	0/39809	135401/180442	0/91	135401/220343
10 Factor	0/47910	132456/216548	0/102	132456/264560
11 Factor	0/37642	80740/176984	0/73	80740/214699
12 Factor	0/17645	26922/88275	0/50	26922/105970
13 Factor	0/3442	3520/19829	0/25	3520/23296

Table 6. Optimal orthogonal 27-run designs

factors	Design	IC			AEPD			MMPD			AOS
		3f	4f	5f	3f	4f	5f	3f	4f	5f	
4	$\mathrm{II}(1^2 2^2 7^3 8^1)$	1	.82	NA	.29	NA	NA	1	NA	NA	0.995
5	$\mathrm{II}(1^3 2^1 3^2 4^1 9^3)$	.83	.47	.065	.33	.31	NA	.99	.99	NA	0.853
6	$\mathrm{II}(1^3 2^1 3^2 5^2 9^3 11^2)$	.82	.34	.021	.36	.38	.28	.97	.97	.91	0.804
7	$\mathrm{II}(1^1 2^1 3^1 4^2 7^2 8^2 11^2)$	.77	.25	.0098	.38	.40	.30	.91	.94	.90	0.846
8	$\mathrm{II}(1^1 2^1 3^2 4^3 5^3 7^2 8^2 9^1)$	.75	.23	.0063	.39	.42	.30	.86	.86	.89	0.840
9	$\mathrm{II}(1^1 2^2 3^2 4^2 5^1 6^1 7^2 8^2 12^1)$	.74	.22	.0048	.40	.43	.31	.91	.88	0.75	0.914
10	$\mathrm{II}(1^1 2^2 3^1 4^1 5^2 6^2 7^1 8^1 11^1 13^3)$	.75	.20	.0034	.41	.44	.32	.92	.81	.74	0.988
11	$\mathrm{II}(1^1 2^2 3^3 4^1 5^2 6^2 7^1 8^3 9^1 10^3 12^2)$	.74	.19	.0032	.42	.45	.33	.92	.80	.73	0.990
12	$\mathrm{II}(1^1 2^1 3^1 4^1 5^2 6^2 7^1 8^1 9^2 10^3 11^2 12^3)$	.74	.19	.0030	.43	.46	.33	.92	.80	.71	0.977
13	$\mathrm{II}(1^1 2^1 3^1 4^1 5^1 6^2 7^2 8^3 9^1 10^2 11^1 12^2 13^2)$	.73	.19	.0029	.43	.47	.34	.92	.80	.73	1.000

4 Concluding remarks

We are the first to consider both model estimation and model discrimination criteria for selecting optimal three-level factorial designs for factor screening and response surface studies. Although the discussion in this paper is based on the two-stage analysis strategy, the consideration of both types of criteria applies to any perceivable methods for model selection and factor screening. For 18-run designs, the recently constructed catalog allows us to examine all possible orthogonal designs. Since none of the $OA(18,3^7)$ designs have full EC, non-orthogonal designs might be considered to accommodate 7 three-level factors. For 27-run designs, we examined projections of three $OA(27,3^{13})$s, which only represent a small fraction of a total of 68 combinatoric non-isomorphic $OA(27,3^{13})$s, as given by Lam and Tonchev (1996). More elaborative computation can be used to select better OA27 designs.

References

Bingham D, Chipman H (2007) Optimal designs for model selection. Technometrics to appear

Box GEP, Meyer RD (1993) Finding the active factors in fractionated screening experiments. Journal of Quality Technology 25:94–105

Box GEP, Wilson KB (1951) On the experimental attainment of optimum conditions (with disscussion). J Roy Statist Soc Ser B 13:1–45

Cheng SW, Wu CFJ (2001) Factor screening and response surface exploration (with discussion). Statistica Sinica 11:553–604

Cheng SW, Ye KQ (2004) Geometric isomorphism and minimum aberration for factorial designs with quantitative factors. Annals of Statistics 32:2168–2185

Hill PDH (1978) Experimental design procedures for regression model discrimination. Technometrics 20:15–21

Jones BA, Li W, Nachtsheim CJ, Ye KQ (2007) Model discrimination - another perspective on model-robust design. Journal of Statistical Planning and Inference in press

Lam C, Tonchev VD (1996) Classification of affine resolvable 2-(27,9,4) designs. Journal of Statistical Planning and Inference 56:187–202

Li W, Nachtsheim CJ (2000) Model-robust factorial designs. Technometrics 42:345–352

Sun DX (1993) Estimation capacity and related topics in experimental designs. PhD thesis, University of Waterloo

Tsai KJ (2005) Construction of optimal three-level factorial design for response surface model selection. PhD thesis, State University of New York at Stony Brook

Tsai PW, Gilmour SG, Mead R (2000) Projective three-level main effects designs robust to model uncertainty. Biometrika 87:467–475

Xu H, Wu CFJ (2001) Generalized minimum aberration for asymmetrical fractional factorial designs. Annals of Statistics 29:1066–1077

Xu H, Cheng SW, Wu CFJ (2004) Optimal projective three-level designs for factor screening and interaction detection. Technometrics 46:280–292

Ye KQ, Cheng SW, Kim N (2006) A simple two-stage analysis for designed experiment with complex aliasing, submitted

List of Contributors

Vladimir V. Anisimov
GlaxoSmithKline
New Frontiers Science Park (South)
Third Avenue, Harlow, Essex
CM19 5AW, U.K.
Vladimir.V.Anisimov@gsk.com

Rosa Arboretti Giancristofaro
Department of Mathematics
University of Ferrara
Via Macchiavelli 35
44100 Ferrara, Italy
rosa.arboretti@unife.it

Dario Basso
Department of Statistics
University of Padova
Via C. Battisti 241
35121 Padova, Italy
dario@stat.unipd.it

Norbert Benda
Novartis Pharma AG
Statistical Methodology
Lichtstrasse 35
4056 Basel, Switzerland
norbert.benda@novartis.com

Atanu Biswas
Indian Statistical Institute
Applied Statistics Unit
203 B.T. Road
Kolkata – 700 108, India
atanu@isical.ac.in

Barbara Bogacka
School of Mathematical Sciences
Queen Mary
University of London
Mile End Road
London E1 4NS, U.K.
B.Bogacka@qmul.ac.uk

Stefano Bonnini
Center for Modelling
Computing and Statistics
University of Ferrara
Via Macchiavelli 35
44100 Ferrara, Italy
bnnsfn@unife.it

Roberto Dorta-Guerra
Departamento de Estadística
Investigación Operativa y
Computación
Universidad de La Laguna
c/Astrofísico Francisco Sánchez
38271 La Laguna, Tenerife, Spain
rodorta@ull.es

Darryl Downing
GlaxoSmithKline

1250 So Collegeville Rd
PO Box 5089
Collegeville, PA 19426-0989, U.S.A.
darryl.j.downing@gsk.com

Younis Fathy
Fakultaet V
Institut fuer Mathematik
Carl von Ossietzky University
Oldenburg
Postfach 2503
26111 Oldenburg, Germany
younis.fathy@
mail.uni-oldenburg.de

Valerii V. Fedorov
GlaxoSmithKline
1250 So Collegeville Rd
PO Box 5089
Collegeville, PA 19426-0989, U.S.A.
Valeri.V.Fedorov@gsk.com

Nancy Flournoy
Department of Statistics
University of Missouri - Columbia
146 Middlebush Hall, Columbia
MO 65211-6100, U.S.A.
flournoyn@missouri.edu

Richard D. Gill
Mathematical Institute
Leiden University
P.O. Box 9512
2300 RA Leiden, Netherlands
gill@math.leidenuniv.nl

Josep Ginebra
Departament d'Estadística
i Investigació Operativa
Universitat Politècnica
de Catalunya
Avgda. Diagonal 647
08028 Barcelona, Spain
josep.ginebra@upc.edu

Enrique González-Dávila
Departamento de Estadística
Investigación Operativa y
Computación
Universidad de La Laguna
c/Astrofísico Francisco Sánchez
38271 La Laguna, Tenerife, Spain
egonzale@ull.es

Ulrike Graßhoff
Institute for Mathematical
Stochastics
Otto-von-Guericke-University
PF 4120, 39 016 Magdeburg
Germany
ulrike.grasshoff@
mathematik.uni-magdeburg.de

Heiko Großmann
School of Mathematical Sciences
Queen Mary
University of London
Mile End Road
London E1 4NS, U.K.
h.grossmann@qmul.ac.uk

Linda M. Haines
Department of Statistical Sciences
University of Cape Town
Rondebosch 7700, South Africa
lhaines@stats.uct.ac.za

Ralf-Dieter Hilgers
Institute of Medical Statistics
Aachen University of Technology
Pauwelsstr. 30, 52074 Aachen,
Germany
rhilgers@ukaachen.de

Heinz Holling
Psychologisches Institut IV,
Westfälische Wilhelms-
Universität Münster
Fliednerstr. 21,
48149 Münster, Germany
holling@psy.uni-muenster.de

Gaëtan Kabera
School of Statistics
and Actuarial Sciences
University of KwaZulu-Natal
Pietermaritzburg 3200, South Africa
201291190@ukzn.ac.za

Patrick J. Laycock
School of Mathematics
University of Manchester
Sackville St
Manchester M601QD, U.K.
pjlaycock@manchester.ac.uk

Sergei L. Leonov
GlaxoSmithKline
1250 So Collegeville Rd
PO Box 5089
Collegeville, PA 19426-0989, U.S.A.
Sergei.2.Leonov@gsk.com

William Li
Operations and Management
Science Departament
University of Minnesota
321 19th Ave. S, Room 3-225
Minneapolis, MN 55455, U.S.A.
wli@csom.umn.edu

J. López-Fidalgo
Department of Mathematics
University of Castilla-La Mancha
Avda. Camilo José Cela 3
13071–Ciudad Real, Spain
jesus.lopezfidalgo@uclm.es

Saumen Mandal
Department of Statistics
University of Manitoba
338 Machray Hall
Winnipeg, MB R3T 2N2, Canada
saumen_mandal@umanitoba.ca

Ignacio Martínez
Dpto. de Estadística
y Matemática Aplicada
Universidad de Almería
Edificio CITE-III
Cra. Sacramento s/n
La Cañada de San Urbano
04120 Almería, Spain
ijmartin@ual.es

Hugo Maruri-Aguilar
Department of Statistics
The University of Warwick
Gibbet Hill Road
Warwick CV4 7AL, U.K.
H.Maruri-Aguilar@warwick.ac.uk

Viatcheslav B. Melas
Faculty of Mathematics and
Mechanics
St.Petersburg State University
University avenue 28, Petrodvoretz
198504 St. Petersburg, Russia
v.melas@pobox.spbu.ru

José A. Moler
Departamento de Estadística
e Investigación Operativa
Universidad Pública de Navarra
Campus de Arrosadia s/n
31006-Pamplona, SPAIN
jmoler@unavarra.es

Christine Müller
Department of Mathematics
and Computer Science
University of Kassel
Heinrich-Plett-Str. 40
34132 Kassel, Germany
cmueller@
mathematik.uni-kassel.de

Werner G. Müller
Department of Applied Stat. (IFAS)
Johannes-Kepler-University
Freistädter Straße 315
4040 Linz, Austria
werner.mueller@jku.at

Principal Ndlovu
School of Statistics
and Actuarial Science
University of KwaZulu-Natal
Pietermaritzburg 3200, South Africa
ndlovup@ukzn.ac.za

The Nguyen
Department of Statistics
University of Glasgow
15 University Gardens
Glasgow G12 8QW, U.K.
the@stats.gla.ac.uk

Timothy E. O'Brien
Department of Mathematics
and Statistics
Loyola University Chicago
6525 N. Sheridan Road
Chicago, Illinois 60626, U.S.A.
teobrien@gmail.com

Isabel Ortiz
Dpto. de Estadística
y Matemática Aplicada
Universidad de Almería
Edificio CITE-III
Cra. Sacramento s/n
La Cañada de San Urbano
04120 Almería, Spain
iortiz@ual.es

Maciej Patan
Institute of Control and
Computation Engineering
University of Zielona Góra
ul. Podgorna 50
65-246 Zielona Gora
Poland
M.Patan@issi.uz.zgora.pl

Andrej Pázman
Department of Applied Mathematics
and Statistics
Faculty of Mathematics

Physics and Informatics
Comenius University
84248 Bratislava, Slovakia
pazman@center.fmph.uniba.sk

Andrey Pepelyshev
Faculty of Mathematics and
Mechanics
St.Petersburg State University
University avenue 28, Petrodvoretz
198504 St. Petersburg, Russia
andrey@ap7236.spbu.ru

Fortunato Pesarin
Department of Statistical Sciences
University of Padova
Via C. Battisti, 241-243
35121 Padova, Italy
fortunato.pesarin@unipd.it

Philipp Pluch
Dept. of Statistics
University of Klagenfurt
University Street 65-67
9020 Klagenfurt
Austria
philipp.pluch@uni-klu.ac.at

Antonio Ponce de Leon
Department of Epidemiology
Institute of Social Medicine
Rio de Janeiro State University
Rua São Francisco Xavier
524/7013D, Maracanã
Rio de Janeiro
20550 900 Brazil
ponce@ims.uerj.br

Luc Pronzato
Laboratoire I3S
Les Algorithmes, Bâtiment Euclide
2000 route des lucioles, BP 121
06903 Sophia Antopolis cedex
France
pronzato@i3s.unice.fr

Dieter Rasch
University of Natural Resources
and Applied Life Sciences, Vienna
Department of Landscape, Spatial
and Infrastructure Sciences, IASC
Peter Jordanstraße 82
1190 Wien, Austria
dieter.rasch@boku.ac.at

Eva Riccomagno
Department of Mathematics
Polytechnic of Turin
Corso Duca degli Abruzzi, 24
10129, Torino, Italy
Eva.Riccomagno@polito.it

Edmilson Rodrigues Pinto
Department of Mathematics
Federal University of Uberlândia
Av. João Naves de Ávila, 2121
Santa Mônica, Uberlândia
MG - Brazil
Edmilson@famat.ufu.br

Carmelo Rodríguez
Dpto. de Estadística
y Matemática Aplicada
Universidad de Almería
Edificio CITE-III
Cra. Sacramento s/n
La Cañada de San Urbano
04120 Almería, Spain
crt@ual.es

Luigi Salmaso
Department of Management
and Engineering
University of Padova
Stradella S. Nicola 3
36100 Vicenza - Italy
salmaso@gest.unipd.it

Thomas Schmelter
Clinical Statistics Europe
Bayer Schering Pharma AG
Mllerstr. 178
13 353 Berlin, Germany
Thomas.Schmelter@schering.de

Rainer Schwabe
Institute for Mathematical
Stochastics
Otto-von-Guericke-University
PF 4120, 39 016 Magdeburg
Germany
rainer.schwabe@
mathematik.uni-magdeburg.de

Marie Šimečková
Institute of Animal Science
Pratelstvi 815
104 00 Prague Uhrineves
Czechia
simeckova.marie@vuzv.cz

Sven Stanzel
Institute of Medical Statistics
Aachen University of Technology
Pauwelsstraße 30
52074 Aachen, Germany
sstanzel@ukaachen.de

Milan Stehlík
Department of Applied Statistics
Johannes Kepler University
Freistädter Straße 315
4040 Linz, Austria
Milan.Stehlik@jku.at

Chiara Tommasi
Department of Economics
Business and Statistics
University of Milano
via Conservatorio 7
20122 Milano, Italy
chiara.tommasi@unimi.it

Ben Torsney
Department of Statistics
University of Glasgow
15 University Gardens
Glasgow G12 8QW, U.K.
bent@stats.gla.ac.uk

Ko-Jen Tsai
Yes Solutions
66 Exeter St
Williston Park, NY 11596, U.S.A.
ko-jen.tsai@yessolutions.com

Yuehui Wu
Research Statistical Unit
SQS, R&D, GlaxoSmithKline
Collegeville, P.O.Box 5089
Collegeville, PA 19426, U.S.A.
Yuehui.2.Wu@gsk.com

Kenny Q. Ye
Department of Epidemiology
and Population Health
Albert Einstein College of Medicine
1300 Morris Park Ave
Bronx, NY 10461, U.S.A.
kye@aecom.yu.edu

List of Referees

Anthony C. Atkinson
Department of Statistics
Columbia House
London School of Economics
Houghton Street
London WC2A 2AE, U.K.
A.C.Atkinson@lse.ac.uk

Barbara Bogacka
School of Mathematical Sciences
Queen Mary
University of London
Mile End Road
London E1 4NS, U.K.
B.Bogacka@qmul.ac.uk

Ching-Shui Cheng
University of California
Berkeley
CA-5106427892, U.S.A.
cheng@stat.berkeley.edu

Alessandro Di Bucchianico
Eindhoven University of Technology
Department of Mathematics
HG 10.17, P.O. Box 513
5600 MB Eindhoven
The Netherlands
A.d.Bucchianico@tue.nl

Darryl Downing
GlaxoSmithKline
1250 So Collegeville Rd
PO Box 5089
Collegeville, PA 19426-0989, U.S.A.
darryl_j_downing@sbphrd.com

Valerii V. Fedorov
GlaxoSmithKline
1250 So Collegeville Rd
PO Box 5089
Collegeville, PA 19426-0989, U.S.A.
Valeri.V.Fedorov@gsk.com

Nancy Flournoy
Departament of Statistics
University of Missouri - Columbia
146 Middlebush Hall, Columbia
MO 65211-6100, U.S.A.
flournoyn@missouri.edu

Alessandra Giovagnoli
Dipartimento di Scienze Statistiche
Universita' di Bologna
via Belle Arti 41
40126 Bologna, Italy
alessandra.giovagnoli@unibo.it

Linda M. Haines
Department of Statistical Sciences
University of Cape Town
Rondebosch 7700, South Africa
lhaines@stats.uct.ac.za

Ralf-Dieter Hilgers
Institute of Medical Statistics
Aachen University of Technology
Pauwelsstr. 30, 52074 Aachen,
Germany
rhilgers@ukaachen.de

Christos P. Kitsos
Department of Mathematics
Technological Education Institute
of Athens
Ag. Spyridonos and Palikaridi St
Egaleo 122 10, Athens, Greece
xkitsos@teiath.gr

Peter W. Lane
GlaxoSmithKline
NFSP South, Third Avenue
Harlow CM19 5AW, U.K.
peter.w.lane@gsk.com

Henning Läuter
Institut für Mathematik
Universität Potsdam
Am Neuen Palais 10
144 69 Potsdam, Germany
laeuter@rz.uni-potsdam.de

Sergei L. Leonov
GlaxoSmithKline
1250 So Collegeville Rd
PO Box 5089
Collegeville, PA 19426-0989, U.S.A.
Sergei.2.Leonov@gsk.com

J. López-Fidalgo
Department of Mathematics
University of Castilla-La Mancha
Avda. Camilo José Cela 3
13071–Ciudad Real, Spain
jesus.lopezfidalgo@uclm.es

Viatcheslav B. Melas
Faculty of Mathematics and
Mechanics
St.Petersburg State University
University avenue 28, Petrodvoretz
198504 St. Petersburg, Russia
v.melas@pobox.spbu.ru

Christine Müller
Department of Mathematics
and Computer Science
University of Kassel
Heinrich-Plett-Str. 40
34132 Kassel, Germany
cmueller@
mathematik.uni-1kassel.de

Werner G. Müller
Department of Applied Stat. (IFAS)
Johannes-Kepler-University
Freistädter Straße 315
4040 Linz, Austria
werner.mueller@jku.at

Andrej Pázman
Department of Applied Mathematics
and Statistics
Faculty of Mathematics
Physics and Informatics
Comenius University
84248 Bratislava, Slovakia
pazman@center.fmph.uniba.sk

Fortunato Pesarin
Department of Statistical Sciences
University of Padova
Via C. Battisti, 241-243
35121 Padova, Italy
fortunato.pesarin@unipd.it

A. Ponce de Leon
Department of Epidemiology
Institute of Social Medicine
Rio de Janeiro State University
Rua São Francisco Xavier
524/7013D, Maracanã
Rio de Janeiro
20550 900 Brazil
ponce@ims.uerj.br

Luc Pronzato
Laboratoire I3S
Les Algorithmes, Bâtiment Euclide
2000 route des lucioles, BP 121
06903 Sophia Antopolis cedex
France
pronzato@i3s.unice.fr

Friedrich Pukelsheim
Institut für Mathematik
Universität Augsburg
86135 Augsburg, Germany
Pukelsheim@Math.Uni-Augsburg.De

Eva Riccomagno
Department of Mathematics
Polytechnic of Turin
Corso Duca degli Abruzzi 24
10129, Torino, Italy
Eva.Riccomagno@polito.it

Juan M. Rodríguez Díaz
Department of Statistics
University of Salamanca
Pl. de los Caídos s/n
37008 Salamanca, Spain
juanmrod@usal.es

Rainer Schwabe
Institute for Mathematical
Stochastics
Otto-von-Guericke-University
PF 4120, 39016 Magdeburg
Germany
rainer.schwabe@
mathematik.uni-magdeburg.de

Milan Stehlík
Department of Applied Statistics
Johannes Kepler University
Freistädter Straße 315
4040 Linz, Austria
Milan.Stehlik@jku.at

Chiara Tommasi
Department of Economics
Business and Statistics
University of Milano
via Conservatorio 7
20122 Milano, Italy
chiara.tommasi@unimi.it

Ben Torsney
Department of Statistics
University of Glasgow
15 University Gardens
Glasgow G12 8QW, U.K.
bent@stats.gla.ac.uk

Martina Vandebroek
Research Center for Operations
Research and Business Statistics
K.U.Leuven
Naamsestraat 69
3000 Leuven, Belgium
martina.vandebroek@
econ.kuleuven.be

Ivan Vuchkov
Euroquality Centre
University of Chemical Technology
and Metallurgy
boul. "Kl. Ohridski" 8
1756 Sofia, Bulgaria.
qstat@dir.bg

H.P. Wynn
London School of Economics
Houghton Street
London, WC2A 2AE, U.K.
h.wynn@lse.ac.uk

A.A. Zhigljavsky
School of Mathematics
Cardiff University
Senghennydd Road
Cardiff, CF24 4YH, U.K.
ZhigljavskyAA@cardiff.ac.uk

Index

Printing: Krips bv, Meppel
Binding: Stürtz, Würzburg